Dynamics and Biogenesis of Membranes

NATO ASI Series

Advanced Science Institutes Series

A series presenting the results of activities sponsored by the NATO Science Committee, which aims at the dissemination of advanced scientific and technological knowledge, with a view to strengthening links between scientific communities.

The Series is published by an international board of publishers in conjunction with the NATO Scientific Affairs Division

A Life Sciences B Physics	Plenum Publishing Corporation London and New York
C Mathematical and Physical Sciences D Behavioural and Social Sciences E Applied Sciences	Kluwer Academic Publishers Dordrecht, Boston and London
F Computer and Systems Sciences G Ecological Sciences H Cell Biology	Springer-Verlag Berlin Heidelberg New York London Paris Tokyo Hong Kong

Series H: Cell Biology Vol. 40

Dynamics and Biogenesis of Membranes

Edited by

J. A. F. Op den Kamp

Centre for Biomembranes and Lipid Enzymology
Padualaan 8, 3584 CH Utrecht, The Netherlands

Springer-Verlag Berlin Heidelberg New York
London Paris Tokyo Hong Kong
Published in cooperation with NATO Scientific Affairs Division

Proceedings of the NATO Advanced Study Institute on Dynamics and Biogenesis of Membranes held in Cargèse, Corsica, June 12–23, 1989

ISBN-13:978-3-642-74196-8 e-ISBN-13:978-3-642-74194-4
DOI: 10.1007/978-3-642-74194-4

Library of Congress Cataloging-in-Publication Data. NATO Advanced Study Institute on Dynamics and Biogenesis of Membranes (1989 : Cargèse, France) Dynamics and biogenesis of membranes / edited by J. A. F. Op den Kamp. p. cm.—(NATO ASI series. Series H, Cell biology ; vol. 40) "Proceedings of the NATO Advanced Study Institute on Dynamics and Biogenesis of Membranes held in Cargèse, Corsica, June 12–23, 1989"—T.p verso. "Published in cooperation with NATO Scientific Affairs Division."
ISBN-13:978-3-642-74196-8 (U.S. : alk. paper)
1. Cell membranes—Formation—Congresses. 2. Glycolipids—Synthesis—Congresses. 3. Membrane proteins—Synthesis—Congresses. I. Kamp, Jos A. F. Op den (Jos Arnoldus Franciscus), 1939– . II. North Atlantic Treaty Organization. Scientific Affairs Division. III. Title. IV. Series. [DNLM: 1. Biogenesis—congresses. 2. Cell Membrane—physiology—congresses. 3. Glycolipids—biosynthesis—congresses. 4. Membrane Proteins—biosynthesis—congresses. QH 601 N279d] QH601.N373 1989 574.87'5—dc20 DNLM/DLC for Library of Congress 89-26335

Softcover reprint of the hardcover 1st edition 1990

2131/3140-543210 – Printed on acid-free-paper

PREFACE

The NATO Advanced Study Institute on "Dynamics and Biogenesis of Membranes" intended to provide an overview on both the structural and dynamic aspects of individual membrane constituents as well as on the processes and mechanisms involved in the assembly of these compounds into well organised biological membranes. Emphasis was laid in particular on the mutual interactions between the proteins, lipids and their glycosylated derivatives. The progressive increase in our knowledge of membrane biogenesis is due in particular to the application of new techniques and approaches in biophysics, genetics, biochemistry and molecular biology. In order to illustrate this and to achieve an extensive overview and a broad perspective of this complex field, lecturers originating from various disciplines contributed to the ASI. The cross fertilization of these disciplines and the combination and integration of the new developments is reflected in the present proceedings.

Much progress has been made in recent years in the elucidation of the structure, biosynthesis and functional properties of fatty acylated proteins and of complex, glycosylated lipids such as the glycosphingolipids and the glycosylated phosphatidylinositols. Furthermore, the biosynthetic routes of glycosylated proteins, in particular the topological aspects, have been studied in great detail. A substantial portion of this information originated from studies of mutants and detailed overviews on these and other important aspects of glycolipid and glycoprotein biogenesis are presented.

Mutual interactions between lipids and between lipids and proteins are the major stabilizing forces responsible for the structure of biological membranes. The various physico-chemical aspects of these interactions, including the possible occurrence of non-bilayer configurations of membrane lipids, are presented and the implications of the data for both membrane protein function as well as for protein insertion and protein translocation are discussed in detail.

Another aspect of membrane biogenesis which has drawn much attention in the past decade concerns the routes and mechanisms of intracellular traffic of membrane(fractions) and individual membrane constituents. Continuous movement of membranous material is essential for membrane repair, membrane multiplication during cell division, endo- and exo-cytosis and various other intracellular events. In particular the mechanisms by which proteins move through the cell and its membranes have been studied in great depth. A fair number of reports and overviews, dealing with protein signalling, targeting, export, translocation and insertion in a variety of biological systems, gives an impression of the current state of affairs in this fast evolving area of research.

Utrecht, October 1989

Jos A.F. Op den Kamp

CONTENTS

GLYCOLIPIDS - INTRACELLULAR MOVEMENT AND STORAGE DISEASES

K. Sandhoff and P. Leinekugel
Institut für Organische Chemie
und Biochemie der Universität Bonn
Gerhard-Domagk-Str. 1
5300 Bonn 1
FRGermany

Glycosphingolipids (GSL) are components of plasmamembranes of animal cells. They are anchored in the cellular membrane by their hydrophobic ceramide (N-acylsphingosine) moiety, while the hydrophilic mono- or oligosaccharide part faces the extracellular space. Together with glycoproteins and glycosaminoglycanes glycosphingolipids form the glycocalix of cell surfaces. The GSL-patterns are characteristic for individual cell types, stages of differentiation and oncogenic transformation (Hakomori 1980; van Echten and Sandhoff, 1989). Though some of the sialic acid containing GSL have been identified as binding sites for toxins and viruses (Yamakawa and Nagai, 1978; Markwell et al., 1981), their physiological functions remain obscure. The structure of about 100 GSL could be elucidated thus far. Sialic acid-containing GSL, called gangliosides, are typical lipids of neuronal surfaces and are predominant in the grey matter (Lowden and Wolfe, 1964; Derry and Wolfe, 1967). Sulfatide and galactosylceramide as main components of myelin prevail on oligodendrocytes while glycolipids of the globo-series predominate on fibroblasts. Any disorder in metabolism of GSL would mainly affect those tissues in which the correspondent GSL is concentrated.

In order to get information about localization, intracellular movement and metabolism of GSL in cultured cells, different derivatives such as spin-, radio-, fluorescent- and biotin-labelled GSL-species, especially of gangliosides, were synthesized (Schwarzmann and Sandhoff, 1987). Studies with exogenous spin-labelled gangliosides in cell culture showed that they are

NATO ASI Series, Vol. H 40
Dynamics and Biogenesis of Membranes
Edited by J.A.F. Op den Kamp

slowly incorporated into the plasma membrane, where they obtain a position which is similar to that of endogenous gangliosides (Schwarzmann et al., 1983, 1984).

Feeding and metabolic experiments with different derivatives of gangliosides show that they participate in intracellular membrane flow after insertion into the plasma membrane. They may enter the lysosomal compartment to get degraded by the action of hydrolases, or, to a smaller extent, reach the Golgi, where they can be used for the biosynthesis of more complex GSL. (Sonderfeld et al., 1985; Sandhoff et al., 1987). Experiments with double labelled gangliosides indicate that those GSL reaching the Golgi-complex are transported directly to this compartment, without passing through lysosomes (Klein et al., 1988). Direct glycosylation in the Golgi was observed with gangliosides G_{M2}, G_{M1} and G_{D1a} as well as with the synthetic amides of G_{M2} and G_{M1}. Electron microscopy studies with biotinylated ganglioside G_{M1} in fibroblasts support this view (Schwarzmann et al., 1986). In contrast, when Farber disease cells were fed with ganglioside G_{M3} or normal cells with G_{M3}-amide respectively, no synthesis of more complex gangliosides could be observed. This indicates that these GSL do not enter the biosynthetic pathway even when reaching the Golgi complex (Klein et al., 1987). However, due to the low activity of N-acetylgalactosaminyl-transferase in Golgi vesicles from rat liver (Pohlentz et al., 1988; Iber et al., 1989) it cannot be ruled out that reglycosylation of these GSL takes place very slowly, below the detection limits of the procedures used. The current understanding of intracellular routing of GSL is summarized in figure 1.

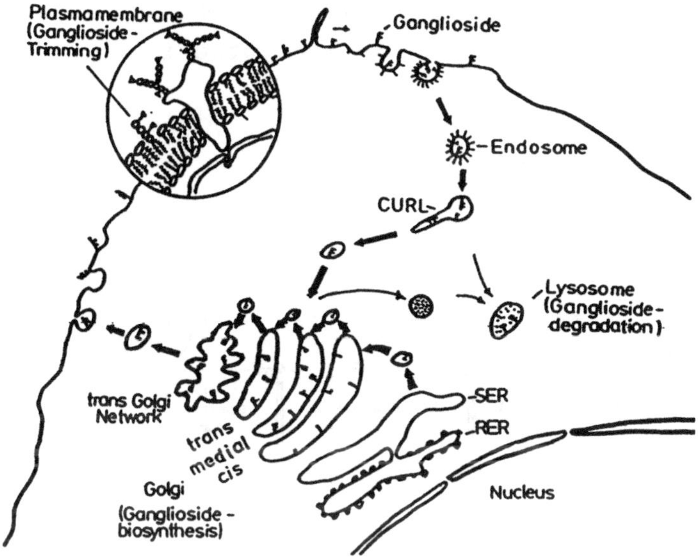

Fig. 1. Model of intracellular transport of exogenous gangliosides in cultured cells (Sandhoff et al., 1988).

Biosynthesis of GSL starts with ceramide formation in the endoplasmatic reticulum. Wether the synthesis of either glucosylceramide or galactosylceramide also takes place in the endoplasmatic reticulum (Suzuki et al., 1984) or at the cytosolic side of the Golgi apparatus (Coste et al. 1985, 1986) is a question which yet has to be clarified. All the other glycosyl- and sialyltransferases involved in GSL-biosynthesis are located in Golgi-cisternae. These membrane-bound enzymes act by sequential addition of galactose, N-acetylgalactosamine and sialic acid to the growing oligosaccharide chain, using the corresponding sugar nucleotides. Inhibitors of sugar nucleotide transport across the Golgi membrane such as tunicamycin, block ganglioside biosynthesis (Yusuf et al., 1983). Competition experiments using lactosylceramide, ganglioside G_{M3} and ganglioside G_{D3} as substrates suggested that N-acetylgalactosamine transfer in rat liver Golgi, leading to gangliosides G_{A2}, G_{M2} and G_{D2}, respectively, is catalyzed by a single enzyme. The same could be shown for the reaction leading to gangliosides G_{M1b}, G_{D1a} and G_{T1b} (Pohlentz et al., 1988). Very recent studies proved G_{A1}-, G_{M1}- and G_{D1b}-synthases to be identical enzymes in rat liver Golgi (Iber et al., 1989), so that a new model for ganglioside biosynthesis, combining those results, can be proposed (fig. 2).

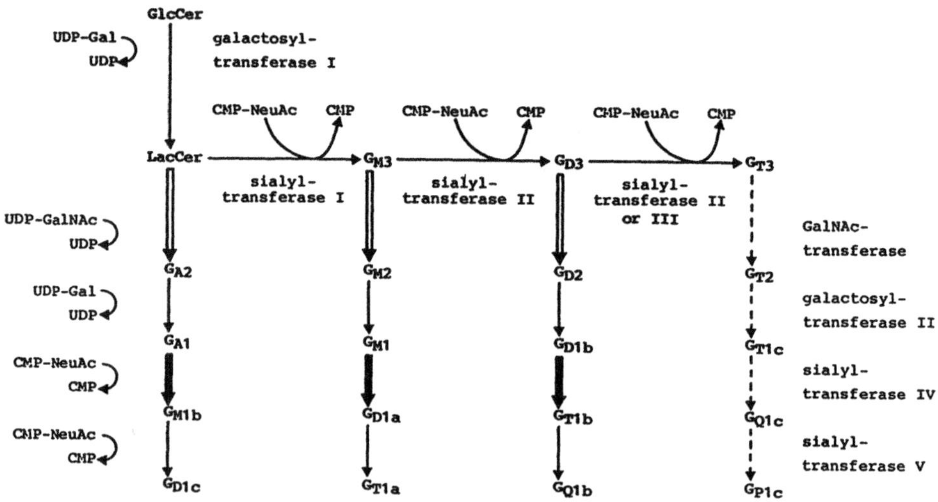

Fig. 2. Proposed model for ganglioside biosynthesis (Pohlentz et al., 1988; Iber et al., 1989). In this model successive transfer of N-acetylgalactosamine, galactose, and sialic acid to lactosylceramide, G_{M3} and G_{D3} (and possibly G_{T3}) leading to the asialo, a and b (and possibly c series) gangliosides is catalyzed by the same set of glycosyltransferases. ⟶, reactions not yet demonstrated in rat liver Golgi. ⇨, reactions catalyzed by one and the same N-acetylgalactosaminyl-transferase. ➡, reactions catalyzed by one and the same sialyltransferase.

Biosynthesis of glycosphingolipids is accompanied by an intracellular vesicle-bound membrane flow. The growing molecule is transported from the endoplasmatic reticulum through the Golgi cisternae to the plasma membrane, following the principle that individual biosynthetic steps correlate with compartmentation. Drugs affecting intracellular membrane flow or modulators of the cytoskeleton have strong inhibitory effects on ganglioside biosynthesis in murine cerebellar cells (van Echten and Sandhoff, 1989). However, the regulation of ganglioside biosynthesis is poorly understood. There is evidence, that ganglioside biosynthesis is under transcriptional control of the respective glycosyltransferases (Hashimoto et al., 1989 a, b; Nakakuma et al., 1984). On the other hand, in vitro studies allow for an additional feedback control (Yusuf et. al., 1987).

GSL degradation takes place in the lysosome and is catalyzed by exohydrolases, which remove the sugar residues in a stepwise manner from the non-reducing end of the oligosaccharide chain (Sandhoff and Conzelmann, 1984; Sandhoff et al., 1987) (fig. 3).

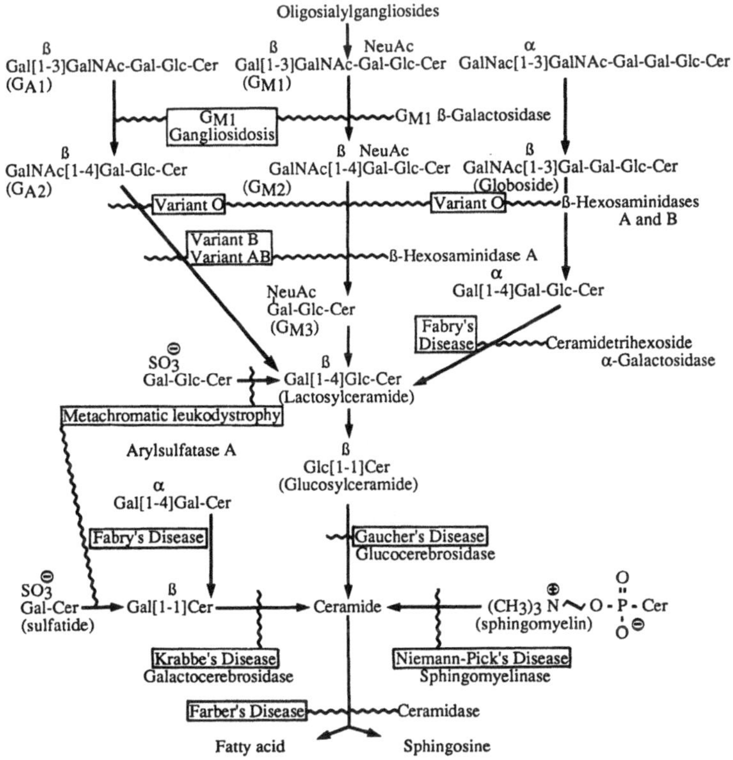

Fig. 3. Degradation scheme of sphingolipids denoting metabolic blocks of known diseases (Sandhoff and Christomanou, 1979).

Any defect of one of these enzymes results in the accumulation of the lipid compounds, which, due to their poor water solubility, precipitate within the lysosome, leading to progressive storage, swelling and finally dysfunction and death of the cells. The clinical symptoms of the resulting diseases depend on the kind of affected tissue or cell type. Though the reason for sphingolipidoses is simply an inherited defect of one catabolic step, there is a great heterogeneity of the respective diseases at the biochemical as well as at the clinical level (for reviews, see, e.g. Conzelmann and Sandhoff, 1984, Sandhoff et al. 1989).

In vitro analysis revealed that many of the lysosomal hydrolases are stimulated by the action of sphingolipid activator proteins (Conzelmann and Sandhoff, 1987 a, b). The function of some activator proteins is to extract the lipid from the membrane, forming a water-soluble complex, which enables the lysosomal enzyme to act upon its substrate. The G_{M2} activator protein is one of the best characterized lysosomal sphingolipid binding proteins; it is a monomeric glycoprotein consisting of one subunit containing 162 amino acids and one N-linked oligosaccharide chain (Fürst et al., 1989). This activator protein forms a stoichiometric complex with ganglioside G_{M2} and its analogues, utilizing a hydrophobic binding site for the ceramide residue and a hydrophilic one for the N-acetylgalactosamine and sialic acid residues of ganglioside G_{M2}. The whole complex is the substrate for the enzyme hexosaminidase A (Hex A) (Sandhoff and Conzelmann, 1984).

Hexosaminidase A is composed of two subunits, α and β, each possessing one catalytic site (Kytzia and Sandhoff, 1985). Only the α subunit is able to degrade activator-bound ganglioside G_{M2}, explaining why patients with variant B_1 of G_{M2}-gangliosidosis accumulate ganglioside G_{M2} in spite of showing Hex A activity against is synthetic substrate 4-methylumbelliferyl-β-D-N-acetylglucosaminide. The Hex A of those patients carries a point mutation in the α active site, the β-chain active site being unaffected (Kytzia et al., 1983, 1985; Tanaka et al., 1988). The model shown in Fig. 4 explains the biochemical heterogeneinity of G_{M2}-gangliosidosis. Mutations in any of the three polypeptides (α-subunit, β-subunit or activator protein) would lead to a defect in G_{M2}-ganglioside catabolism, causing accumulation of this lipid in neuronal tissues of the affected patient.

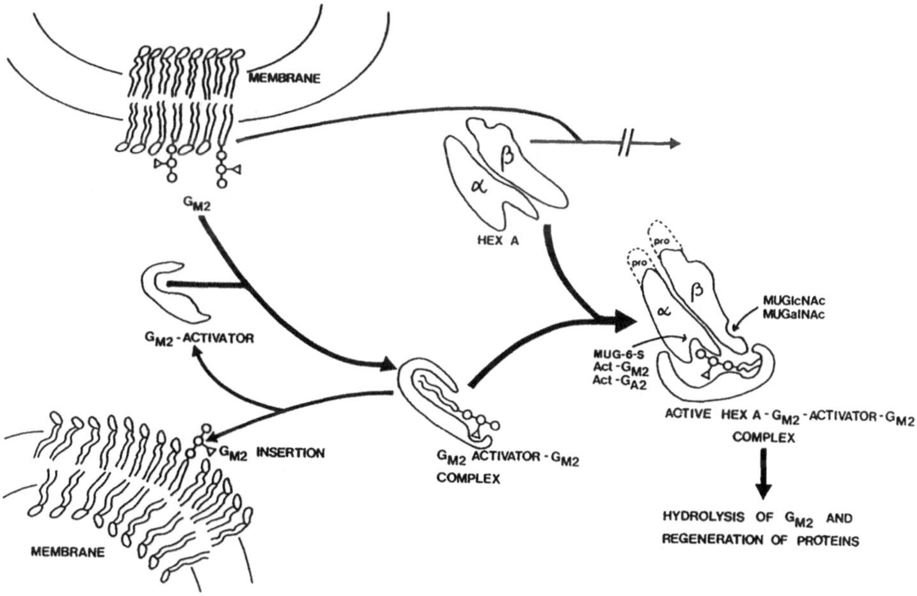

Fig. 4 Model for the lysosomal catabolism of ganglioside G_{M2} (Conzelmann and Sandhoff, 1979; Sandhoff and Conzelmann, 1984; Conzelmann et al., 1982). Hexosaminidase A cannot attack membrane-bound ganglioside G_{M2}. Instead, the ganglioside is extracted from the membrane by the activator protein and the water-soluble activator/lipid complex is the substrate for the enzymatic reaction. Of the two catalytic sites on hexosaminidase A, only the one on the α-subunit cleaves ganglioside G_{M2}. The hexosaminidase precursor ("prohex A") is also fully active on the activator/G_{M2} complex (Hasilik et al., 1982). After the reaction the product, ganglioside G_{M3}, is reinserted into the membrane and the activator protein is available for another round of catalysis.

Other activator proteins and cofactors in GSL-catabolism have been described. (Conzelmann and Sandhoff 1987 b; d'Azzo et al.,

1982). In contrast to the G_{M2} activator protein, the other proteins examined thus far are less specific and their mechanism of function is less clear. For example, the sulfatide-G_{M1} activator protein is able to bind a variety of glyco- and phospholipids (Conzelmann and Sandhoff, 1987 b). It has been proposed that it acts as a kind of physiological detergent, solubilizing the lipids that would otherwise not be accessible to water-soluble hydrolases (Fischer and Jatzkewitz, 1978). The sulfatide-G_{M1} activator protein is formed by proteolytic cleavage of a large precursor protein (Fürst et al., 1988). Recently it has been shown that this precursor protein is processed to four homologous proteins, the sulfatide activator protein, an activator protein for glucosylceramidase (A_1 activator) (Kleinschmidt et al., 1987), and two proteins with unknown function, respectively (Fürst et al., 1988; O'Brien et al., 1988; Nakano et al., 1989).

Allelic mutations affecting proteins involved in sphingolipid degradation may have consequences on synthesis of mRNA, formation of pre-pro-proteins, their intracellular targeting and processing, catalytic activities and substrate specifities of the mature enzymes, their stability against lysosomal proteases, temperature and / or pH-changes (Scriver et al., 1989). The biochemical heterogeneity is increased by the possibility of alternate splicing of hn-mRNA transcripts (Quintern et al., 1989), which was also observed in the case of α- and β-chain mutations of hexosaminidase A (Ohno and Suzuki, 1988; Navon and Proia, 1989; Sandhoff et al., 1989).

However, the molecular analysis of lysosomal storage diseases at the protein or genomic levels so far could not explain the frequently observed heterogeneity of clinical syndromes, such as the occurrence of infantile, juvenile and adult forms.

In G_{M2}-gangliosidosis as in metachromatic leukodystrophy, a disorder in sulfatide catabolism, these clinical forms show small but consistent differences in the residual activities of the affected enzymes, hexosaminidase A and arylsulfatase A, respectively (Conzelmann et al., 1983; Lee-Vaupel and Conzelmann, 1987). In order to understand the significance of variations in the lower range of residual enzyme acitivities for the development of different clinical forms of a disease, a hypothesis was proposed, based on model calculations performed on the steady-state substrate concentration and the degradation rate of the lipid substrate as a function of the residual enzyme activity (Conzelmann and Sandhoff, 1983/84) (Fig. 5).

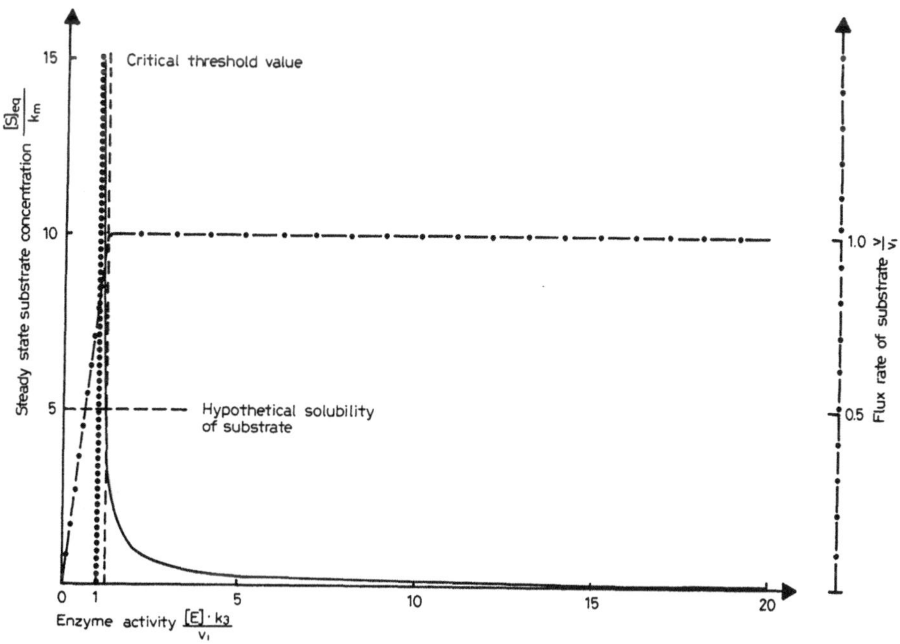

Fig. 5. Steady state substrate concentration as a function of enzyme concentration and activity (Conzelmann and Sandhoff, 1983/84). The model underlying this theoretical calculation assumes influx of the substrate into a compartment at a constant rate (v_i) and its subsequent utilization by the enzyme. ——— = $[S]_{eq}$, steady state concentration; ····· = theoretical threshold of enzyme avtivity; --- = critical threshold value, taking limited solubility of substrate into account; ·-·-· = turnover rate of substrate (flux rate)

Measurements on the degradation rate of lipid substrates and residual enzyme activities in cultured skin fibroblasts from patients with different clinical forms of metachromatic leukodystrophy or G_{M2}-gangliosidosis, variant B, respectively, could confirm this hypothesis (Conzelmann et al., 1988; Leinekugel P., Michel S., Conzelmann, E., Sandhoff K. manuscript in preparation). Supporting the proposed model, the results show that variation of the enzyme activity lead within a wide range only to changes in the steady state substrate concentration but not to accumulation of substrate. However, if the enzyme activity falls below a critical threshold, the decrease of activity can no longer be compensated for by a higher saturation of the enzyme, thus lysosomal storage of the substrate occurs. Below this critical threshold, small differences in residual enzyme activities have significant effects on substrate accumulation. This may explain that patients with small differences in the activities of the affected enzyme show a wide variation in age of onset, progression and severity of the disease.

References:

Conzelmann E and Sandhoff K (1979) Purification and characterization of an activator protein for the degradation of glycolipids G_{M2} and G_{A2} by hexosaminidase A. Hoppe Seyler's Z Physiol Chem 360: 1837-1849

Conzelmann E, Burg J, Stephan G, Sandhoff K (1982) Complexing of glycolipids and their transfer between membranes by the activator protein for lysosomal ganglioside G_{M2} degradation. Eur J Biochem 123: 455-464

Conzelmann E and Sandhoff K (1983/84) Partial enzyme deficiencies: Residual activities and the development of neurological disorders. Dev Neurosci 6: 58-71

Conzelmann E, Sandhoff K (1987a) Activator proteins for lysosomal glycolipid hydrolysis. Meth Enzymol 138, 792-815

Conzelmann E, Sandhoff K (1987b) Glycolipid and glycoprotein degradation. Adv Enzymol 60, 89-216

Conzelmann E, Neumann C, Sandhoff K (1989) Correlation between degradation of sulfatide in cultured skin fibroblasts and residual arylsulfatase A activity. In: Salvayre R, Douste-Blazy L, Gatt S (eds) Lipid storage disorders: Biological and medical aspects. NATO ASI Series, Series A: Life Sciences 150: 267-271

Coste H, Martel MB, Azzar G, Got R (1985) UDP-glucose-ceramide glucosyltransferase from porcine submaxillary glands is associated with the Golgi apparatus. Biochim Biophys Acta 814: 1-7

Coste H, Martel MB, Got R (1986) Topology of glucosylceramide synthesis in Golgi membranes from porcine submaxillary glands. Biochem Biophys Acta 858: 6-12

d'Azzo A, Hoogeveen A, Reuser JJ, Robinson D, Galjaard H, (1982) Molecular defect in combined β-galactosidase and neuraminidase deficiency in man. Proc Natl Sci USA 79: 4535-4539

Derry MD, Wolfe LS (1967) Gangliosides in isolated neurons and glial cells. Science 158: 1450-1452

Fischer G and Jatzkewitz H (1978) The activator protein of cerebroside sulphatase: A model for the activation. Biochem Biophys Acta 528: 69-76

Fürst W, Machleidt W, Sandhoff K (1988) The precursor of sulfatide activator protein is processed to three different proteins. Biol Chem Hoppe-Seyler 369: 317-328

Fürst W, Schubert J, Machleidt W, Sandhoff K (1989) The complete amino acid sequences of G_{M2}-activator protein and sulfatide activator protein. Submitted for publication.

Hakomori SI (1981) Glycosphingolipids in cellular interaction, differentiation and oncogenesis. Ann Rev Biochem 50: 733-764

Hashimoto Y, Otsuka H, Sudo K, Suzuki A, Yamakawa T (1983 a) Genetic regulation of G_{M2} expression in liver of mouse. J Biochem 93: 895-901

Hashimoto Y, Suzuki A, Yamakawa T, Miyashita N, Monikawi K (1983 a) Expression of G_{M1} and G_{D1a} in mouse liver is linked to the H-2 complex as chromosome 17. J Biochem 94: 2043-2048

Hasilik A, von Figura K, Conzelmann E, Nehrkorn H, Sandhoff K (1982) Lysosomal precursors in human fibroblasts. Activation of cathepsin D-precursors in vitro and activity of hexosaminidase A precursor toward ganglioside G_{M2}. Eur J Biochem 125: 317-321

Iber H, Kaufmann R, Pohlentz G, Schwarzmann G, Sandhoff K (1989) Identity of G_{A1}-, G_{M1a}-, and G_{D1b}- synthases in Golgi vesicles from rat liver. FEBS Lett 248: 18-22

Klein D, Pohlentz G, Hinrichs U, Sandhoff K (1987) Metabolism of ganglioside-amides in cultured human fibroblasts. Biol Hoppe-Seyler 368: 1495-1503

Klein D, Leinekugel P, Pohlentz G, Schwarzmann G, Sandhoff K (1988) Metabolism and intracellular transport of gangliosides in cultured fibroblasts. In: Ledeen RW, Hogan EL, Tettamanti G, Yates AJ, Yu EK, (eds) New Trends in Ganglioside Research: Neurochemical and neuroregenerative aspects. Fidia Research Series 14: 247-258

Kleinschmidt T, Christomanou H, Braunitzer G (1987) Complete amino-acid sequence and carbohydrate content of the naturally occuring glucosylceramide activator protein (A_1 activator) absent from a new human Gaucher disease variant. Biol Chem Hoppe-Seyler 368: 1571-1578

Kytzia HJ, Hinrichs U, Maire I, Suzuki K, Sandhoff K (1983) Variant of G_{M2}-gangliosidosis with hexosaminidase A having a severely changed substrate specifity. EMBO J 2: 1201-1205

Kytzia HJ, Sandhoff K (1985) Evidence for two different active sites on human β-hexosaminidase A. Interaction of G_{M2} activator protein with β-hexosaminidase A. J Biol Chem 260: 7568-7572

Lee-Vaupel M and Conzelmann E (1987) Assay for cerebroside sulfate (sulfatide) sulfatase in cultured skin fibroblasts with the natural activator protein. Clin Chim Acta 168: 55-68

Lowden JA, Wolfe LS (1964) Brain gangliosides. II. Evidence for the location of gangliosides specifically in neurons. Can J Biochem 42: 1587-1594

Markwell MAK, Svennerholm L, Paulson JC (1981) Specific gangliosides function as host cell receptors for Sendai virus. Proc Nat Acad Sci USA 78: 5406-5410

Nakakuma H, Sanai Y, Shiroki K, Nagai Y (1984) Gene-regulated expression of glycolipids: Appearance of G_{D3} ganglioside in rat cells on transfection with transforming gene E1 of human adenovirus type 12 DNA and its transcriptional subunits. J Biochem 96: 1471-1480

Nakano T, Sandhoff K, Stümper J, Christomanou H, Suzuki K (1989) Structure of full length cDNA coding for sulfatide activator, a co-β-glucosidase and two other homologous proteins: Two alternate forms of the sulfatide activator. J Biochem 105: 152-154.

Navon R, Proia RL (1989) The mutations in Ashkenazi jews with adult G_{M2} gangliosidosis, the adult form of Tay-Sachs-disease. Science 243: 1471-1474

O'Brien JS, Kretz KA, Dewji NN, Wenger DA, Esch F, Fluharty AL, (1988) Coding of two sphingolipid activator proteins (SAP-1 and SAP-2) by same genetic locus. Science 241: 1098-1101

Ohno K, Suzuki K, (1988) A splicing defect due to an exon-intron junctional mutation results in abnormal β-hexosaminidase α-chain in RNAs in Ashkenazi jewish patients with Tay-Sachs disease. Biochem Biophys Res Comm 153: 463-469

Pohlentz G, Klein D, Schwarzmann G, Schmitz D, Sandhoff K (1988) Both G_{D2}, G_{M2}, and G_{D2} synthases and G_{M1b}, G_{D1a} and G_{T1b} synthases are single enzymes in Golgi vesicles from rat liver. Proc Nat Acad Sci USA 85: 7044-7048

Quintern LE, Schuchmann EH, Levran O, Suchi M, Ferlinz K, Reinke H, Sandhoff K, Desnick RJ (1989) Isolation of cDNA clones encoding human acid sphingomyelinase (occurence of alternatively processed transcripts). Submitted to EMBO J

Sandhoff K and Christomanou H (1979) Biochemistry and genetics of gangliosidoses. Hum Genet 50: 107-143

Sandhoff K and Conzelmann E (1984) The biochemical basis of gangliosidoses. Neuropediatrics 15 (Suppl) 85-92

Sandhoff K, Schwarzmann G, Sarmientos F, Conzelmann E (1987) Fundamentals of ganglioside metabolism. In: Rahmann H (ed) Gangliosides and modulation of neuronal functions. NATO ASI Series H7: 231-250

Sandhoff K, Schwarzmann G, Conzelmann E, Klein D (1988) Gangliosides: Uptake, intracellular transport and metabolism in normal and mutant cells. In: Salvayre R, Douste-Blazy L, Gatt S (eds) Lipid storage disorders: Biological and medical aspects. NATO ASI Series, Series A: Life Sciences 150: 559-562

Sandhoff K, Conzelmann E, Neufeld EF, Kaback MM, Suzuki K (1989) The G_{M2} Gangliosidoses. In: Scriver C, Beaudet AL, Sly WS, Valle D (eds) The metabolic basis of inherited diseases. 6th. edition. McGraw Hill, New York, in press.

Schwarzmann G, Hoffmann-Bleihauer P, Schubert J, Sandhoff K Marsh D (1983) Incorporation of ganglioside analogues into fibroblasts cell membranes. A spin-label study. Biochemistry 22: 5041-5048

Schwarzmann G, Sonderfeld S, Conzelmann E, Marsh D, Sandhoff K (1984) Insertion into cultured cells and metabolism and intracellular transport of exogenous gangliosides. In: Dreyfus H, Massarelli R, Freysz L (eds) Cellular and pathological aspects of glycoconjugate metabolism. INSERM 126: 195-210

Schwarzmann G, Marsh D, Herzog V, Sandhoff K (1986) In vitro incorporation and metabolism of gangliosides. In: Rahmann H (ed) Gangliosides and modulation of neuronal functions. NATO ASI Series H7: 217-229

Schwarzmann G, Sandhoff K (1987) Lysogangliosides: Synthesis and use in preparing labeled gangliosides. In: Ginsburg V (ed) Methods in Enzymology. Academic Press, Orlando/ Florida 138: 319-341

Scriver CR, Beaudet AL, Sly WS, Valle D (eds) The metabolic basis of inherited diseases, 6th edition. McGraw, New York, in press.

Sonderfeld S, Conzelmann E, Schwarzmann G, Burg J, Hinrichs U, Sandhoff K (1985) Incorporation and metabolism of gangliosides in skin fibroblasts from normal and G_{M2} gangliosidosis subjects. Eur J Biochem 149: 247-255

Suzuki Y, Ecker CP, Blough HA (1984) Enzymatic glucosylation of dolichol monophosphate and transfer of glucose from isolated dolichol-D-glucosyl phosphate to ceramides by BHK 21 cell microsomes. Eur J Biochem 143: 447-453

Tanaka A, Ohno K, Suzuki K (1988) G_{M2}-gangliosidosis B_1 variant: A wide geographic and ethnic distribution of the specific β-hexosaminidase α chain mutation originally identified in a Puerto Rican patient. Biochem Biophys Res Comm 156: 1015-1019

van Echten G, Sandhoff K (1989) Modulation of ganglioside biosynthesis in primary cultured neurons. J Neurochem 52: 207-214

Yamakawa T, Nagai Y (1978) Glycolipids at the cell surface and their biological functions. Trds Biochem Sci 3: 128-131

Yusuf HKM, Pohlentz G, Sandhoff K (1983) Tunicamycin inhibits ganglioside biosynthesis in rat liver Golgi apparatus by blocking sugar nucleotide transport across the membrane vesicles. Proc Natl Acad Sci USA 80: 7075-7079

Yusuf HKM, Schwarzmann G, Pohlentz G, Sandhoff K (1987) Oligosialogangliosides inhibit G_{M2}- and G_{M3}-synthesis in isolated Golgi vesicles from rat liver. Biol Chem Hoppe-Seyler 368: 455-462

THE BIOLOGY OF LIPID-CONTAINING MEMBRANE PROTEINS*

Eric N. Olson and Guy James
Department of Biochemistry and Molecular Biology
The University of Texas M.D. Anderson Cancer Center
1515 Holcombe Blvd., Box 117
Houston, TX 77030

A wide range of proteins have been shown to acquire covalent lipids during their maturation. The three most common types of lipid modifications of proteins involve esterification of palmitate to cysteine, attachment of myristate through amide linkage to the amino terminus and attachment of a complex glycosylated phosphatidylinositol anchor to the carboxy terminus. The enzymes responsible for these modifications are localized to different subcellular compartments, as are the proteins that undergo these modifications. Proteins that contain covalent lipids are generally, but not always, associated with membranes and in many cases, the lipid moiety plays an essential role in directing membrane association. Fatty acylation is an especially common modification among proteins that participate in transmembrane regulatory pathways involving multiple interacting components. For these proteins, the protein-associated lipid appears to mediate protein-protein interactions within the bilayer. This review will summarize current concepts regarding the enzymology of protein lipidation and will consider the functional consequences of these unique modifications, particularly for proteins that participate in transmembrane signaling.

INTRODUCTION:

A diverse array of proteins have been found in recent years to be modified by covalent lipids (for reviews, see Ferguson and Williams, 1988; Olson, 1988; Schultz et al., 1988; Towler and Gordon, 1988). The most common lipid moieties found to be associated with proteins are the long chain fatty acids, palmitate and myristate, and a complex glycosylated phosphatidylinositol (GPI) anchor. The structures of these modifications and the linkages through which they are attached to proteins are shown in Figure 1. Each of these lipids are attached to proteins by distinct enzymes that reside in different subcellular compartments. The high degree of specificity of each of these modifications with respect to lipid donors and protein acceptors suggests that protein lipidation is a highly selective type of modification that confers unique properties on proteins. In several cases, protein-bound lipids have been shown to function as hydrophobic recognition signals that mediate the association of otherwise soluble proteins with membranes.

---

*Research in the author's laboratory was supported by grants from The National Institutes of Health (AR 39849) and The American Cancer Society (CD 343). ENO is an Established Investigator of The American Heart Association.

NATO ASI Series, Vol. H 40
Dynamics and Biogenesis of Membranes
Edited by J.A.F. Op den Kamp

A diagramatic representation of the ways in which lipid-containing proteins arethought to interact with membranes is shown in Figure 2. A remarkable number of fatty acylated proteins have been shown to participate in transmembrane signaling pathways that involve dynamic interactions among multiple membrane and cytoplasmic components. For these proteins, the lipid moieties appear to be essential for appropriate membrane association and for mediating protein-protein interactions within the bilayer. In contrast to other mechanisms for association of proteins with membranes, lipidation confers a high degree of mobility on proteins and allows for reversible translocation between the cytosol and membranes thereby facilitating interactions between lipid-containing proteins and other cellular components.

Modification of proteins with covalent fatty acids

At least 50 proteins have been identified in yeast, plants, bacteria, mycoplasma, animal cells and viruses, which are modified covalently with long chain fatty acids, and numerous other proteins which are as yet unidentified, have been visualized by labeling of tissue culture cells with ^{3}H-fatty acids. Palmitate and myristate are the most common fatty acids found to be associated covalently with proteins. Despite the fact that these fatty acids differ by only two saturated carbons, the chemical nature of the linkages through which they are attached to proteins, in addition to the enzymes involved in their attachment are entirely different.

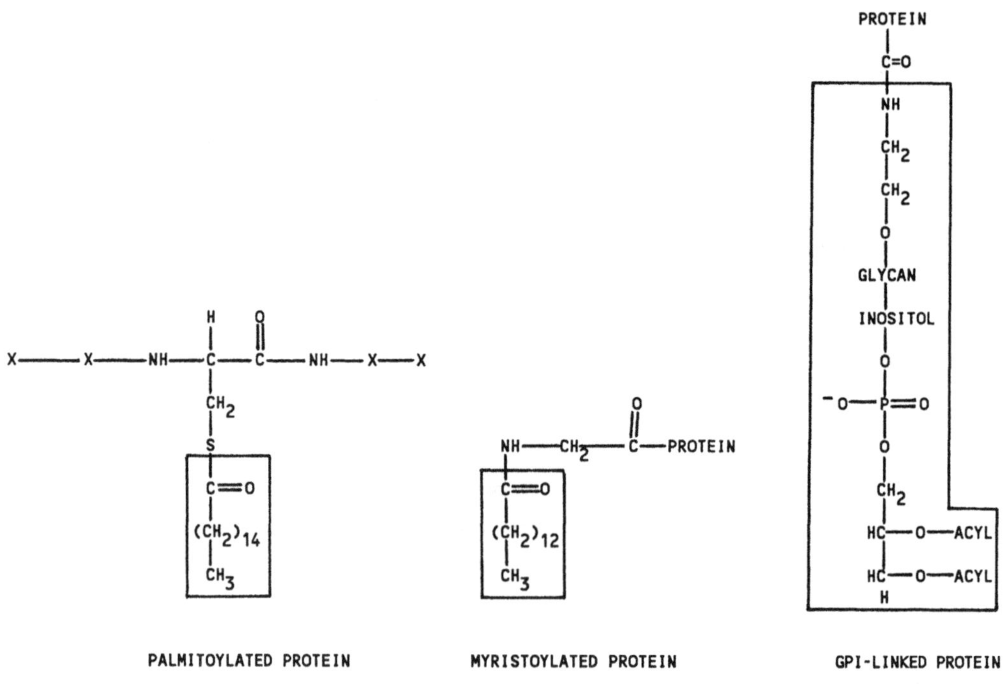

Figure 1. Structures of fatty acyl-protein linkages and glycosyl PI-protein linkages. The lipid moieties are indicated in boxes.

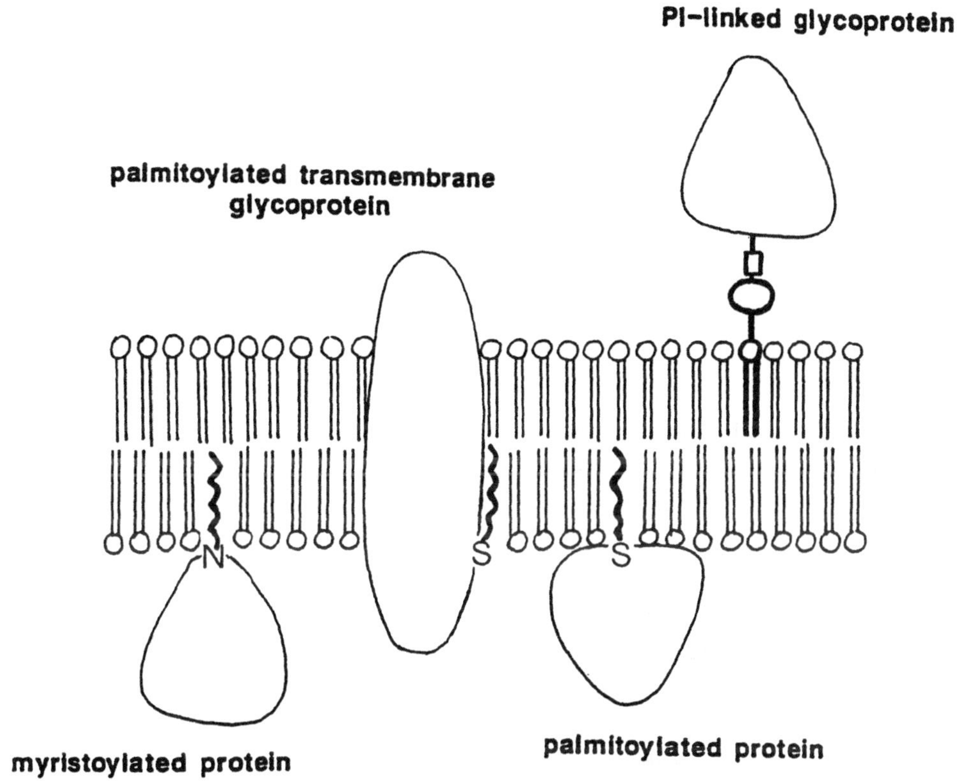

Figure 2. Diagramatic representation of the interactions of lipid-containing proteins with membranes.

Biosynthesis of fatty acylated proteins

Palmitate is generally attached to proteins through highly labile thiol ester linkages, which are readily cleaved by KOH or hydroxylamine at neutral pH (Magee and Courtneidge, 1985; McIlhinney et al., 1985; Olson et al., 1985; Olson and Spizz, 1986). There are rare examples of proteins that contain palmitate linked through O-ester or amide bonds (Olson et al., 1984; Stoffel et al., 1983; Towler and Glaser, 1986); however, the mechanisms involved in modification of these proteins have not been studied in detail. Esterification of palmitate to proteins occurs post-translationally and can occur in at least two distinct subcellular compartments. Whereas palmitate is the preferred substrate for esterification to proteins, the enzymes(s) responsible for this modification will transfer fatty acids of longer and shorter chain length, albeit at reduced efficiency. Several cell surface glycoproteins have have been shown to acquire palmitate as they are transported through the transitional elements of the rough endoplasmic reticulum (RER) or cis-Golgi apparatus, 5 to 15 minutes after synthesis (Schmidt and Schlesinger, 1980). The palmitoyltransferase(s) responsible for modification of glycoproteins is strictly compartmentalized and is unable to acylate proteins that have moved into the medial stacks of the Golgi or beyond. It is intriguing that an N-ethylmaleimide-sensitive factor (NSF) that participates in glycoprotein transport through the Golgi, requires palmitoyl-CoA for activity, suggesting that protein acylation may be involved in one or more steps in vescicular transport (Glick and Rothman, 1987).

The majority of cellular palmitoylated proteins (>90%) are synthesized on free polysomes in the cytosol and do not follow the secretory pathway. Virtually all of the proteins in this class appear to associate with the cytoplasmic surface of the plasmamembrane as a consequence of palmitoylation (see below). Pulse-chase experiments have shown that palmitate can be incorporated into these proteins for several hours after translation (Magee and Courtneidge, 1985; McIlhinney et al., 1985; Olson and Spizz, 1986). The relaxed kinetics for acylation of these proteins relative to cell surface glycoproteins probably reflects the fact that they do not require transport through specific subcellular compartments. Although many palmitoylated proteins in this class are highly charged and do not contain hydrophobic domains that would be expected to interact tightly with membranes, these proteins are resistant to removal from membranes by extraction procedures that normally release peripheral membrane proteins (Olson and Spizz, 1986). These results suggest that the lipid moiety mediates membrane binding and serves to stabilize the tight association of these proteins with membrane lipids or proteins or both.

Considering the differences in kinetics for palmitoylation and the different subcellular distributions of palmitoylated proteins, it is likely that multiple palmitoyltransferases may exist. Although no palmitoyltransferases have been isolated, several in vitro assays for protein palmitoylation have been developed. Using deacylated proteins as substrates for acylation in vitro, palmitoyl-CoA has been shown to serve as acyl donor for this modification (Bizzozero et al., 1987; Grand et al., 1987; Mack et al., 1987; Slomiany et al., 1984; Yoshimura et al., 1987). Whereas palmitoyltransferase(s) exhibit a preference for palmitate in vivo, this acylchain selectively is lost in vitro.

In many cases, the palmitate group on proteins has been shown to turnover, suggesting that an acylation-deacylation cycle may modulate certain properties of acylproteins. The transferrin receptor, for example, has been shown to be palmitoylated late in its lifecycle (Omary and Trowbridge, 1981a). Pulse-chase experiments have revealed that the lipid moiety turns over faster than the receptor itself, suggesting the existence of a dynamic cycle of acylation-deacylation-reacylation (Omary and Trowbridge, 1981b). The exact subcellular locations of the enzymes responsible for acylating and deacylating the transferrin receptor remain to be determined. Ankyrin, as well as other unidentified erythrocyte proteins which are associated with the membrane and cytoskeleton, have also been demonstrated to contain covalent palmitate that undergoes turnover (Staufenbiel, 1987). Particularly intriguing are observations that turnover of covalent palmitate on certain proteins may be accelerated in response to external stimuli (see below). A fatty acyl esterase that removes palmitate from proteins has been identified in microsomal membrane fractions and could be involved in these types of fatty acid turnover (Berger and Schmidt, 1986).

The other major form of fatty acylation involves the covalent attachment of the rare 14-carbon fatty acid, myristate, through amide linkage to proteins. Chemical identification of the amino acid acceptor for myristate has revealed glycine at the amino terminus in all cases. In contrast to the post-translational nature of palmitoylation, myristoylation is tightly coupled to protein synthesis and is abolished immediately upon inhibition of protein synthesis (Magee and Courtneidge, 1985; McIlhinney et al., 1985; Olson and Spizz, 1986; Buss et al., 1984). This observation provided an initial indication that myristoylation was an early covalent modification and indicated that cells do not contain significant pools of nonmyristoylated proteins that can be modified at various times after synthesis. The complete lack of myristoylation in the absense of protein synthesis also indicates that the myristate moiety on proteins is highly stable and does not turnover independently from the polypeptide.

The possibility that myristoylation occurred cotranslationally was examined directly by Wilcox et al. by isolation of peptidyl-tRNA from ^{3}H-myristate-labeled cells. These studies showed that myristate was incorporated into elongating polypeptide chains prior to release from ribosomes (Wilcox, et al., 1987). In order for myristate to be cotranslationally attached to glycine at the amino terminus of polypeptide chains, the initiating methionine, and possibly other amino acids must be removed. Indeed, for the majority of myristoylated proteins examined thus far, glycine has been found to be the penultimate residue. Removal of the amino terminal methionine is a common modification catalyzed by an amino peptidase after elongation of the first 30 to 40 amino terminal amino acids (Palmiter, 1977). The glycine is then available for acylation by N-myristoyltransferase (NMT) which uses myristoyl-CoA as fatty acyl donor (Towler and Glaser, 1986).

Using a series of synthetic peptides as substrates for myristoylation <u>in vitro</u>, Towler and coworkers defined the substrate specificity of NMT (Towler and Glaser, 1986; Towler and Glaser, 1987; Towler et al., 1987). Unlike, N-glycosylation which requires a specific amino acid recognition sequence, the only absolute requirement for myristoylation appears to be glycine as the acyl acceptor. The seven residues immediately following glycine exert positive and negative effects on the ability of a protein to be myristoylated but are not invariant. Peptides with Asn, Gln, Ser, Val or Leu as the residue immediately adjacent to Gly, for example, are effecient substrates for NMT, whereas peptides with Asp, D-Asn, Phe or Tyr at this position are not myristoylated (Towler et al., 1987). Based on rules for myristoylation that have been established from studies using synthetic peptides, it has been possible to predict by analysis of amino acid sequences deduced from cloned cDNA's whether particular proteins are myristoylated <u>in vivo</u>. Considering that every protein that has been found to contain an amino terminal region compatible with myristoylation is myristoylated, it appears that this modification is extremely efficient and is not influenced by other domains of the polypeptide. This is in contrast to N-glycosylation in which only a fraction of potential acceptor sites are modified. Despite the dependence of myristoylation on protein synthesis in vivo, the ability to reconstitute this modification in vitro with artificial peptide substrates demonstrates that this modification is not coupled obligatorily to polypeptide elongation and raises the possibility that post-translational myristoylation could occur in vivo.

NMT has been purified to homogeneity from yeast and partially purified from tissue culture cells (Towler and Glaser, 1986). The enzyme exhibits an apparent molecular weight of 55kD and is highly specific for myristoyl-CoA as an acyl donor. Considering that myristate comprises only 1-3% of cellular fatty acids (Khandwala and Casper, 1971), this high degree of specificity is necessary for the exclusive incorporation of myristate into this class of proteins. NMT is distributed in the soluble and crude membrane fractions from yeast and tissue culture cells (Towler and Glaser, 1986). It has been suggested that the enzyme may associate with ribosomes, thereby facilitating the cotranslational transfer of myristate to nascent polypeptide chains (Olson and Spizz, 1986).

cDNA clones for yeast NMT have recently been isolated (Duronio et al., 1989). The deduced open reading frame encodes a polypeptide of 53 kD and shows no recognizable sequence homology with any protein in currently available databases. Although the amino terminus of the purified protein is blocked, the amino terminus of the deduced open reading frame lacks a concensus for N-myristoylation, suggesting that the blocking group is probably acetate. Over-expression of NMT using a yeast episomal plasmid was without obvious effects on growth kinetics, cell morphology or ^{3}H-myristate-labeled proteins. However, insertional mutagenesis of the NMT locus on yeast chromosome XII caused recessive lethality, indicating that myristoylation is necessary for vegetative cell growth.

In contrast to palmitoylated proteins which are localized almost exclusively to the plasmamembrane, myristoylated proteins exhibit a widespread subcellular distribution (Wilcox and Olson, 1987). The majority of myristoylated proteins are bound tightly to the cytoplasmic surface of the plasmamembrane. However, some myristoylated proteins are restricted to other subcellular organelles. NADH cytochrome b5 reductase, for example, is myristoylated and is associated with the endoplasmic reticulum (Ozols et al., 1984), while the VP2 protein of polyoma virus and Simian virus 40 is localized to the nucleus (Streuli and Griffin, 1987), and the capsid protein VP4 of the picoronaviruses is localized to the cytoplasm and endoplasmic reticulum (Paul et al., 1987). A subset of myristoylated proteins also are found exclusively in the cytosol, while others appear to associate reversibly with membranes (Magee and Courtneidge, 1985; McIlhinney et al., 1985; Olson et al., 1985; Olson and Spizz, 1986). The fact that certain myristoylated proteins reside in the cytosol indicates that myristoylation is not sufficient, by itself, to ensure membrane association. The specific localization of different myristoylated proteins to distinct subcellular membrane systems also suggests that covalent myristate does not function simply as a hydrophobic anchor that partitions into any nearby membrane. If this were the case, myristoylated proteins would be expected to exhibit a more widespread distribution among cellular membranes. An attractive hypothesis that has yet to be proven is that specific myristoyl-protein receptors that participate in sorting of myristoyl-proteins may exist within different intracellular membranes. Such receptors might recognize and bind the myristoylated amino terminal region of specific proteins. Indeed, studies on the binding of normal and non-myristoylated $p60^{src}$ to phospholipid vescicles suggests that the fatty acyl moiety alone is insufficient to mediate tight membrane association and may require additional domains within the polypeptide (Resh, 1988).

Functions for fatty acylation

The potential functions of covalent fatty acids on proteins have been the subject of intense interest. The majority of fatty acylated proteins that have been identified to date play important roles in cellular processes that require protein-protein interactions within membranes. In particular, proteins involved in transmembrane signaling pathways, such as cell surface receptors, protein kinases or their substrates, and guanyl nucleotide regulatory proteins are frequently acylated. Viral gag proteins, which are involved in virus assembly, are also myristoylated.

For palmitoylated glycoproteins, it has been difficult to assign a specific role to fatty acylation. The fact that only a small fraction of cell surface glycoproteins are fatty acylated indicates that this modification is nonessential for transport of proteins to the plasmamembrane. Moreover, for the Vescicular Stomatitis virus G-protein, which is normally palmitoylated during transport from the RER to the Golgi apparatus, site-directed mutagenesis of the acylation site has little or no apparent effect on surface expression (Rose and Gallione, 1981).

The β-adrenergic receptor has recently been shown to undergo palmitoylation at a cysteine residue in the carboxy tail of the protein. Mutation of this cysteine abolishes acylation and results in a dramatic decrease in the efficiency of coupling of the receptor to adenylate cyclase (O'Dowd et al., 1989). These results suggest that the lipid moiety plays an important role in mediating interactions of the receptor with effectors within the plasmamembrane.

For many palmitoylated proteins that are nonglycosylated, the role of covalent fatty acid appears to be essential for membrane binding. Studies of the mammalian RAS proteins yielded the first direct evidence that palmitoylation might mediate the association of otherwise soluble proteins with membranes. Cellular RAS proteins are tightly bound to the cytoplasmic surface of the plasmamembrane where they are postulated to couple growth factor receptors to intracellular effectors in a manner analogous to G-proteins in the β-adrenergic-adenylate cyclase system. RAS proteins are synthesized in the cytosol on free polysomes and are transported to membranes soon after synthesis (Sefton et al., 1982; Willumsen et al., 1984; Willumsen et al., 1984b). Membrane association of RAS occurs coincident with palmitoylation of a cysteine four residues from the carboxy terminus. The amino acid recognition sequence for palmitoylation of RAS is Cys-A-A-X, in which A is any aliphatic amino acid and X can be any residue. This C-terminal sequence is conserved among many RAS-related proteins, some of which have not yet been shown to be palmitoylated, but are highly likely to undergo this modification. Mutation or deletion of amino acids within this conserved sequence has been shown to prevent palmitoylation and to result in RAS proteins that are unable to associate with membranes (Willumsen et al.,

1984a; Willumsen et al., 1984b). Of particular interest is the observation that nonacylated v-RAS proteins, also lack transforming activity. These results suggest that palmitylation of RAS is a prerequisite for membrane binding and that association with the plasmamembrane is necessary for v-RAS to exert its biological effects. To determine whether a different form of lipid modification could restore either membrane association or transforming activity, Buss and coworkers fused the myristoylation signal from Ra-sheed leukemia virus (RaLV) gag protein to the amino termini of cellular RAS proteins that were deleted at the carboxy terminal palmitoylation site (Buss et al., 1989). The resulting fusion proteins were myristoylated and were associated with membranes. Interestingly, myristoylation conferred transforming activity on these cellular RAS proteins. These results indicate that either palmitoylation or myristoylation are necessary and sufficient to juxtapose RAS with appropriate membrane effectors necessary for transduction of signals leading to transformation. The ability of myristate to convert a normal RAS protein into a transforming protein also suggests that myristate confers properties on RAS unique from those conferred by palmitate.

A clue as to the mechanism whereby myristoylation "activates" RAS comes from studies on the dynamics of palmitoylation. Stimulation of fibroblasts with mitogens has been shown to result in accelerated turnover of the palmitate group on cellular H-RAS, suggesting that palmitoylation may act transiently during transmembrane signaling to modify the interaction of RAS with membranes or with other components of the signal transduction pathway (Magee et al., 1987). Deacylation of RAS might, therefore, provide a mechanism for termination of transduced signals. Since myristoylation is an irreversible modification, the RAS proteins described above, which contained the myristoylation concensus sequence, might be irreversibly associated with the plasmamembrane and might thereby lead to persistent activation of growth factor pathways.

The possibility that acylation-deacylation may participate in intracellular signaling is suggested further by studies of a 68 kD protein in macrophages that is immunologically related to the acidic 80K protein in fibroblasts which has been shown to be the major cellular substrate for protein kinase C. Aderem and coworkers have shown that stimulation of macrophages with bacterial lipopolysaccharide (LPS) induces the myristoylation and membrane association of the 68 kD protein (Aderem et al., 1986; Aderem et al., 1988). The membrane-bound form is then believed to be available for phosphorylation by protein kinase C which leads to demyristoylation and release of the phosphorylated form of the protein into into the cytosol. Similar protein kinase C-dependent myristoylation of an 82 kD protein has been observed in HL-60 cells (Malvoisin, 1987). If this type of post-translational myristoylation-demyristoylation does in fact occur for these proteins it would be unique among myristoylated proteins, which are acylated cotranslationally and do not exhibit turnover of the lipid moiety.

Other examples in which protein acylation is modulated in response to cell growth signals include palmitoylation of vinculin, which has been shown to decrease following transformation by SRC (Burn and Burger, 1987). Myristoylation of the 36 kD substrate for SRC-dependent phosphorylation is also reduced in Rous Sarcoma Virus-transformed fibroblasts (Soric and Gordon, 1985). We have recently identified a 64 kD palmitoylated protein, referred to as p64, that is deacylated following mitogenic stimulation of quiescent cells. While the majority of p64 is localized to the cytosol and is nonacylated, the palmitoylated form of the protein is found exclusively in the plasmamembrane where it is resistant to removal by treatments that release peripheral membrane proteins. The specific localization of the fatty acylated form of p64 to the plasmamembrane, combined with the observation that the membrane-bound form of the protein is deacylated in response to mitogenic stimulation, suggests that palmitoylation participates in membrane association of p64 and raises the possibility that this protein may participate in one or more of the plieotropic actions of mitogens on quiescent cells.

Studies of $p60^{src}$ have also shown that myristoylation is sufficient to direct proteins from the cytosol to membranes. SRC is synthesized and myristoylated in the cytosol and migrates to the plasmamembrane within 15 min after completion of translation (Buss et al., 1984). Deletion or mutation of the 8 amino terminal amino acids of SRC prevents myristoylation and membrane association and abolishes the transforming activity of the protein, despite the fact that nonacylated SRC mutants retain tyrosine kinase activity (Cross et al., 1984; Kamps et al., 1985). These results indicate that myristoylation is essential for association of SRC with membranes and that membrane binding is necessary for phosphorylation of critical cellular substrates that participate in induction of transformation. Interestingly, nonmyristoylated mutants of SRC which are unable to transform cells retain the ability to stimulate cell proliferation (Calothy et al., 1987) suggesting that the substrates involved in mitogenesis are distinct from those involved in transformation and that the proteins involved in SRC-dependent stimulation of proliferation are accessible to the soluble form of the kinase.

The ability of a myristoylation concensus sequence to target a cytoplasmic protein to membranes was elegantly demonstrated by Hanafusa and coworkers, who fused the amino terminal 14 amino acids of SRC to chimpanzee α-globin, which is normally soluble (Pellman et al., 1985). The resulting SRC-globin fusion protein associated with membranes. Together, the above studies indicate that either palmitoylation or myristoylation are sufficient to direct proteins to membranes and that these lipid moieties are required for appropriate interactions of acylproteins with other components of the membrane and cytosol.

Modification of proteins with glycosyl-PI anchors

A third type of lipid modification of proteins involves the covalent attachment of GPI. Initial evidence for this type of modification was obtained from studies that showed that treatment of intact cells with PI-specific phospholipase C (PI-PLC) caused release of several cell surface proteins (Low and Finean, 1978). Subsequent studies with the variant surface glycoprotein (VSG) of the parasitic protozoan Trypanosoma brucei led to the identification of GPI as a membrane anchor (Ferguson et al., 1985). Numerous proteins have been shown in recent years to be modified by GPI anchors; however, in only a few cases has the actual structure of the complete membrane anchoring domain been determined. The central features of the GPI anchors that have been characterized thus far include: (1) an ethanolamine that is attached through amide linkage to the carboxy terminus, (2) a phosphodiester bond between the hydroxyl of the ethanolamine and the 6-hydroxyl of mannose, (3) a glycan that can exhibit considerable heterogeneity among different proteins is linked to PI through a glycosidic linkage with glucosamine.

Biogenesis

PI-linked glycoproteins are synthesized in the RER and are translocated cotranslationally across the membrane into the lumen. Pulse-chase experiments indicate that the PI anchor is attached to the newly synthesized polypeptide within 60 seconds following termination of translation (Bangs et al., 1985). Attachment of the anchor is preceeded by proteolytic removal of a hydrophobic carboxy terminal segment, the length of which may vary in different proteins. Amino acid sequence comparisons of a limited number of proteins that acquire PI anchors have failed to reveal exact concensus sequences either for proteolytic processing or for attachment of GPI. Cleavage of the carboxy terminal tail and subsequent modification must be tightly coupled events or the polypeptide would be released into the lumen of the RER where it would be unavailable for modification. At present, it is unclear whether attachment of a GPI anchor involves a single transaminidase which cleaves the acceptor protein near the carboxy terminus and simultaneously transfers the GPI precursor to form the amide linkage. Alternatively, a carboxy-terminal protease and the a GPI transferase may act sequentially. Thus far, neither of these activities have been purified or characterized extensively. The exact nature of the lipid donor for this modification is also unknown, but is presumed to be a preassembled glycosyl phospholipid precursor within the membrane of the RER.

Recent studies with decay accelerating factor (DAF) indicate that the 37 carboxy terminal amino acids serve as a recognition signal for attachment of a GPI anchor. Fusion of this sequence to a truncated form of the Herpes simplex virus glycoprotein D yields a lipid-anchored form of this normally secreted protein (Caras et al., 1987). Conversely, truncation of DAF to remove the carboxy terminal 17 residues is sufficient to convert DAF into a secreted protein (Caras et al., 1989).

Functions for PI anchors

GPI anchors play an obvious role in anchoring proteins to the plasmamembrane that would otherwise be secreted. In addition to mediating the association of these proteins with membranes, GPI anchors may serve more subtle functions. Since proteins linked to the membrane by GPI anchors do not contain transmembrane domains that interact with other membrane proteins or with the cytoskeleton, these proteins are relatively unrestricted in their mobility. Fluorescence photobleaching studies indicate that alkaline phosphatase, for example, exhibits a lateral mobility that is closer to the rate of diffusion of membrane lipids than most integral membrane proteins (Noda et al., 1987). A GPI anchor may thus provide a mechanism for rapid redistribution of proteins at the cell surface.

An additional function for a GPI anchor might be to allow rapid release of specific proteins from the cell surface in response to external stimuli. Activation of a membrane-associated PI-PLC, for example, would release GPI-linked proteins with their oligosaccharide tail, ethanolamine and inositol phosphate, leaving behind diacylglycerol in the membrane. The portion of the GPI anchor retained by the protein might participate in cell-cell recognition or intercellular signaling events, while diacylglycerol could activate protein kinase C, resulting in the generation of an intracellular signal.

Summary and Future Directions

Studies of protein lipidation represent an exciting area for future research because they bring together several diverse fields. While numerous proteins that contain covalent lipids have been identified, relatively little is known of the enzymes involved in these modifications or of the full range of functions of these unique modifications. Among the important questions remaining to be addressed are: (1) What are the mechanisms that underlie the selective association of myristoylated proteins with specific intracellular membranes? (2) How do covalent fatty acids modulate the interactions of proteins with bilayers and with other membrane components? (3) What are the enzymes and precise intermediates responsible for attachment of GPI anchors to proteins? (4) Do specific fatty acids impart unique functions to proteins, or is specificity contained solely in the polypeptide chain? Given the importance of fatty acylation in cellular transformation, understanding the enzymology of these modifications may also permit certain interventions into the process of transformation.

ACKNOWLEDGEMENT:

We thank Ellen Madson for her preparation of the paper.

REFERENCES:

Aderem AA, Albert KA, Keum MM, Wang JKT, Greengard P, Cohn ZA (1988) Stimulus-dependent myristoylation of a major substrate for protein kinase C. Nature 332:362-364.

Aderem AA, Keum MM, Pure E, Cohn ZA (1986) Bacterial lipo-polysaccharides, phorbol myristate acetate, and zymosan induce the myristoylation of specific macrophage proteins. Proc Natl Sci USA 83:5817-5821.

Bangs JD, Hereld D, Krakow JL, Hart GW, Englund PT (1985) Rapid processing of the carboxyl terminus of a trypanosome variant surface glycoprotein. Proc Natl Acad Sci USA 82:3207-3211.

Berger M, Schmidt MF (1986) Characterization of a protein fatty acylesterase present in microsomal membranes of diverse origin. J Biol Chen 261:14912-14918.

Bizzozero OA, McGarry JF, Lees MB (1987) Acylation of endogenous myelin proteolipid protein with different acyl-CoAs. J Biol Chem 262:2138-2145.

Burn P, Burger MM (1987) The cytoskeletal protein vinculin contains transformation-sensitive, covalently bound lipid. Science 235:476-479.

Buss JE, Kamps MP, Sefton BM (1984) Myristic acid is attached to the transforming protein of Rous sarcoma virus during or immediately after synthesis and is present in both soluble and membrane-bound forms of the protein. Mol Cell Biol 4:2697-2704.

Buss JE, Solski PA, Schaeffer JP, MacDonald MJ, Der CJ (1989) Activation of the cellular proto-oncogene product p21Ras by addition of a myristylation signal. Science 243:1600-1603.

Calothy G, Laugier D, Cross FR, Jove R, Hanafusa T, Hanafusa H (1987) The membrane-binding domain and myristylation of $p60^{v\text{-}src}$ are not essential for stimulation of cell proliferation. J Virol 61:1678-1681.

Caras IW, Weddel GN, Davitz MA, Nussenzweig V, Martin DW Jr (1987) Signal for attachment of a phospholipid membrane anchor in decay accelerating factor. Science (Wash DC) 238:1280-1283.

Caras IW, Weddell GN, Williams SR (1989) Analysis of the signal for attachment of a glycophospholipid membrane anchor. J Cell Biol 108:1387-1396.

Cross FR, Garber EA, Pellman D, Hanafusa H (1984) A short sequence in the $p60^{src}$ N terminus is required for p60^{src} myristoylation and membrane association and for cell transformation. Mol Cell Biol 4:1834-1842.

Duronio RJ, Towler DA, Heuckeroth RO, Gordon JI (1989) Disruption of the yeast N-myristoyl transferase gene causes recessive lethality. Science 243:796-800.

Ferguson MAJ, Low MG, Cross GAM (1985) Glycosyl-sn-1,2-dimyristylphosphatidyl-inositol is covalently linked to Trypanosoma brucei variant surface glycoprotein. J Biol Chem 260:14547-14555.

Ferguson MAJ, Williams AF (1988) Cell surface anchoring of proteins via glycosyl-phosphatidylinositol structures. Ann Rev Biochem 57:285-320.

Glick BS, Rothman JE (1987) Possible role for fatty acyl-coenzyme A in intracellular protein transport. Nature 326:309-312.

Grand RJA, Smith KJU, Gallimore PH (1987) Purification and characterization of the protein encoded by the activated human N-ras gene and its membrane localization. Oncogene 1:305-314.

Kamps MP, Buss JE, Sefton BM (1985) Mutation of NH_2-terminal glycine of p60^{src} prevents both myristoylation and morphological transformation. Proc Natl Acad Sci USA 82:4625-4628.

Khandwala AS, Casper CB (1971) The fatty acid composition of individual phospholipids from rat liver nuclear membrane and nuclei. J Biol Chem 246:6242-6246.

Low MG, Finean JB (1978) Specific release of plasma membrane enzymes by a phosphatidylinositol-specific phospholipase. C Biochim Biophys Acta 508:565-570.

Mack D, Berger M, Schmidt MFG, Kruppa J (1987) Cell-free fatty acylation of microsomal integrated and detergent-solubilized glycoprotein of vesicular stomatitis virus. J Biol Chem 262:4297-4302.

Magee AI, Courtneidge SA (1985) Two classes of fatty acid acylated proteins exist in eukaryotic cells. EMBO J 2:1137-1144.

Magee AI, Gutierrez L, McKay I, Marshall C, Hall A (1987) Dynamic fatty acylation of p21^{N-ras}. EMBO J 6:3353-3357.

Malvoisin E, Wild F, Zwingelstein G (1987) 12-O-tetradecan-oyl phorbol 13-acetate stimulates the myristylation of an approximately 82 kDa protein in HL-60 cells. FEBS Lett 215:175-178.

McIlhinney RAJ, Pelley SJ, Chadwick JK, Cowley GP (1985) Studies on the attachment of myristic and palmitic acid to cell proteins in human squamous carcinoma cell lines: evidence for two pathways. EMBO J 4:1145-1152.

Noda M, Yoon K, Rodan GA, Koppel DE (1987) High lateral mobility of endogenous and transfected alkaline phosphatase: a phosphotidylinositol anchor membrane protein. J Cell Biol 105:1671-1677.

O'Dowd BF, Hnatowich M, Caron MG, Lefkowitz RJ, Bouvier M (1989) Palmitoylation of the human β_2-adrenergic receptor. J Biol Chem 264:7564-7569.

Olson EN (1988) Modification of proteins with covalent lipids. Prog Lipid Res 27:177-197.

Olson EN, Glaser L, Merlie JP (1984) α and β subunits of the nicotinic acetylcholine receptor contain covalently bound lipid. J Biol Chem 258:5364-5367.

Olson EN, Spizz G (1986) Fatty acylation of cellular proteins. Temporal and subcellular differences between palmitate and myristate acylation. J Biol Chem 261:2458-2466.

Olson EN, Towler DA, Glaser L (1985) Specificity of fatty acid acylation of cellular proteins. J Biol Chem 260:3784-3790.

Omary MB, Trowbridge IS (1981a) Biosynthesis of the human transferrin receptor in cultured cells. J Biol Chem 256:12888-12892.

Omary MB, Trowbridge IS (1981b) Covalent binding of fatty acid to the transferrin receptor in cultured human cells. J Biol Chem 256:4715-4718.

Ozols J, Carr SA, Strittmatter P (1984) Identification of the NH_2-terminal blocking group of NADH-cytochrome b5 reductase as myristic acid and the complete amino acid sequence of the membrane-binding domain. J Biol Chem 259:13349-13354.

Palmiter RD (1977) Prevention of NH_2-terminal acetylation of proteins synthesized in cell-free systems. J Biol Chem 252:8781-8783.

Paul AV, Schultz A, Pincus S, Oroszlan S, Wimmer E (1987) Capsid protein VP4 of poliovirus is N-myristoylated. Proc Natl Acad Sci USA 84:7827-7831.

Pellman D, Garber EA, Cross FR, Hanafusa H (1985) An N-terminal peptide from p60^{src} can direct myristylation and plasma membrane localization when fused to heterologous proteins. Nature 314:374-377.

Resh MD (1988) Reconstitution of the Rous sarcoma virus transforming protein pp60^{v-src} into

phospholipid vesicles. Mol Cell Biol 8:1896-1905.
Rose JK, Gallione CJ (1981) Nucleotide sequences of the mRNA's encoding the vesicular stomatitis virus G and M proteins determined from cDNA clones containing the complete coding regions. J. Virol. 39:519-528.
Schmidt MFG, Schlesinger MJ (1980) Relation of fatty acid attachment to the translation and maturation of vesicular stomatitis and Sindbis virus membrane glycoproteins. J Biol Chem 255:3334-3339.
Schultz AM, Henderson LE, Oroszlan S (1988) Fatty acylation of proteins. Ann. Rev. Cell Biol. 4:611-647.
Sefton BM, Trowbridge IS, Cooper JA, Scolnick EM (1982) The transforming proteins of Rous sarcoma virus. Harvey sarcoma virus and Abelson virus contain tightly bound lipid. Cell 31:465-474.
Slomiany A, Liau YH, Takogi A, Laszewicz W, Slomiany BL (1984) Characterization of mucus glycoprotein fatty acyl-transferase from gastric mucosa. J Biol Chem 259:13304-13308.
Soric J, Gordon JA (1985) The 36-kilodalton substrate of pp60$^{v\text{-}src}$ is myristylated in a transformation-sensitive manner. Science 230:563-566.
Staufenbiel, M (1987) Ankyrin-bound fatty acid turns over rapidly at the erythrocyte plasma membrane. Mol Cell Biol 7:2981-2984.
Stoffel W, Hillen H, Schroeder W, Deutzmann R (1983) The primary structure of bovine brain myelin lipophilin (proteolipid apoprotein). Hoppe Seylers Z. Physiol. Chem. 364:1455-1466.
Streuli CH, Griffin BE (1987) Myristic acid is coupled to a structural protein of polyoma virus and SV40. Nature 326:619-621.
Towler D, Glaser L (1986) Acylation of cellular proteins with endogenously synthesized fatty acids. Biochemistry 25:884-889.
Towler D, Glaser L (1986) Protein fatty acylation: enzymatic synthesis of an N-myris-toylglycyl peptide. Proc Natl Acad Sci USA 83:2812-2816.
Towler D, Glaser L (1987) Amino terminal processing of proteins by N-myristoylation. J Biol Chem 262:1030-1036.
Towler DA, Adams SP, Eubanks SR, Towery DS, Jackson-Machelski E, Glaser L, Gordon JI (1987) Purification and characterization of a yeast myristoyl CoA:protein N-myristoyl-transferase. Proc Natl Acad Sci USA 84:2708-2712.
Towler DA, Gordon JI (1988) The biology and enzymology of eukaryotic protein acylation. Ann. Rev. Biochem. 57:69-99.
Wilcox CA, Hu JS, Olson EN (1987) Acylation of proteins with myristic acid occurs cotranslationally. Science 26:1275-1278.
Wilcox CA, Olson EN (1987) The majority of cellular fatty acid acylated proteins are localized to the cytoplasmic surface of the plasma membrane. Biochemistry 26:1029-1036.
Willumsen BM, Christensen A, Hubbert NL, Papageorge AG, Lowy DR (1984a) The p21^{ras} C-terminus is required for transformation and membrane association. Nature 310:583-586.
Willumsen BM, Norris K, Papageorge AH, Hubbert NL, Lowy DR (1984b) Harvey murine sarcoma virus p21 ras protein: biological and biochemical significance of the cysteine nearest the carboxy terminus. EMBO J 3:2581-2585.
Yoshimura Y, Agrawal D, Agarwal HC (1987) Cell-free acylation of rat brain myelin proteolipid protein and DM-20. Biochem J 246:611-617.

PREFERENTIAL RELEASE OF GLYCOSYL-PHOSPHATIDYL-INOSITOL ANCHORED PROTEINS FROM HUMAN RED BLOOD CELLS DURING VESICULATION

Peter Bütikofer, Samuel T. Test, Daniel T.Y. Chiu,
Bertram Lubin and Frans A. Kuypers
Children's Hospital Research Institute
747-52nd Street, Oakland, California 94609
U.S.A.

Gycosyl-phosphatidylinositol as an anchor for membrane proteins

Most membrane proteins are attached to the lipid bilayer by means of one or more stretches of relatively hydrophobic amino acid residues. Until recently, it was assumed that the interactions between proteins and lipids that generate the structural integrity of a membrane were non-covalent. In the past few years, a number of proteins were discovered containing covalently attached lipids, such as fatty acids. In most cases the attached lipids did not seem to function as the sole attachment of the proteins to the membrane, but assisted in the binding of these proteins to the lipid bilayer (Olson, 1985; Wilcox and Olson, 1987; Agre et al. 1988). A new group of membrane proteins has been recently described that is attached to the lipid bilayer solely via covalently bound glycosyl-phosphatidylinositol (GPI) (for recent reviews see Low, 1989; Low and Saltiel, 1988; Low, 1988).

The involvement of phosphatidylinositol in anchoring proteins to membranes was first suggested when several investigators reported that certain cell surface proteins could be enzymatically released from the membrane by the action of bacterial phosphatidylinositol-specific phospholipases C (PI-PLC). Subsequent work revealed that phosphatidylinositol was entirely responsible for anchoring those proteins to the membrane, and that it was linked to the protein via a glycan structure. The detailed compositions of the GPI anchors of several proteins have now been reported and suggest that, in spite of their complexity, they may be conserved. The basic structure of the membrane anchors found in most GPI anchored proteins studied so far contains an ethanolamine residue

NATO ASI Series, Vol. H 40
Dynamics and Biogenesis of Membranes
Edited by J.A.F. Op den Kamp

which is amide linked to the carboxy-terminal amino acid of the protein (Figure 1A). The ethanolamine, in turn, is linked via a phosphodiester bond to a short glycan structure consisting of three hexose residues. This structure is attached to the inositol phospholipid through a glucosamine (Low, 1989).

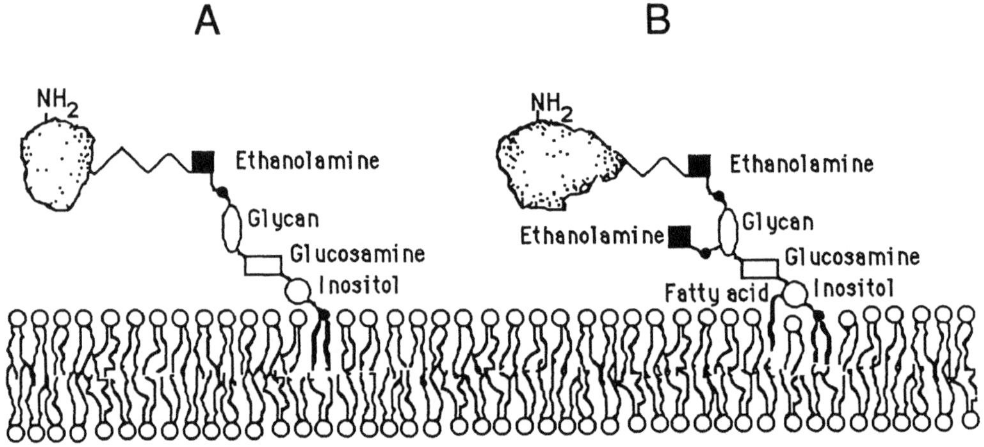

Figure 1: Structure of glycosyl-phosphatidylinositol anchors.

Several phosphatidylinositol-specific phospholipases that were isolated and purified were found to be highly specific for GPI anchors, whereas their action against phospholipids, including phosphatidylinositol, was poor (Davitz et al., 1988, and references therein). GPI anchor-specific phospholipases C and D are now very useful tools in identifying new GPI-anchored proteins and, in addition, suggest a role for phosphatidylinositol in anchoring proteins to membranes. It can be speculated that these phospholipases release GPI-anchored proteins, thereby regulating their expression or action at the cell surface (Braun-Breton et al., 1988). Concomittant with the release of the proteins, biologically active lipids, such as diacylglycerol and phosphatidic acid, would be generated that could affect intracellular metabolism.

However, it has been known for some time that not all GPI anchored proteins are susceptible to hydrolysis by PI-PLC. For example, only 5% to 10% of total AChE is released from human erythrocytes upon treatment with PI-PLC (Silman and Futerman, 1985). Recent analysis of the membrane anchor of human erythrocyte AChE revealed two major variations in its structure as compared to the anchors identified in most other GPI anchored proteins (Figure 1B) (Roberts et al., 1988a & b). An additional phosphoryl-ethanolamine moiety was found to be attached to the glycan core, and a palmitoyl group was acylated to the inositol ring of the phosphatidylinositol. This acylation at the inositol ring seemed responsible for the partial resistence of human erythrocyte AChE to bacterial PI-PLC (Roberts et al., 1988a). It is worth mentioning that from the data available on decay-accelerating factor (DAF) of human erythrocytes one would expect to find that DAF has a very similar GPI structure to that of human erythrocyte AChE, since it also contains an ethanolamine residue with an unblocked amino group, and is partially resistant to hydrolysis by PI-PLC (Medof et al., 1986).

The unique way by which GPI anchored proteins are attached to the bilayer is not only responsible for their specific characteristic resulting in a release from the membrane by phospholipases, but may also result in a relatively high lateral mobility of these proteins. The interaction between GPI anchored proteins and the lipid bilayer is probably comparable to lipid-lipid interactions. In studies using a variety of techniques, the lateral mobility of lipids has been reported to be much faster than that of proteins in membranes (Cherry, 1979; Jacobson et al., 1982). Furthermore, interactions between the cytoskeleton and the cytoplasmic domain of the membrane appear to reduce the lateral mobility of of proteins (Tsuji and Ohnishi, 1986; Tsuji et al., 1988). Thus, it seems likeley that GPI anchored proteins, being attached to the membrane by a phospholipid moiety, would be relatively more mobile in the bilayer than integral proteins. Indeed, several GPI anchored proteins were recently found to have an increased lateral mobility as compared to transmembrane proteins (Ishihara et al., 1987; Noda et al., 1987). Furthermore, it has been speculated that GPI anchored proteins may be involved in transducing signals across a membrane (Low, 1989). After being activated by an external stimulus, this high lateral mobility may allow GPI anchored proteins to rapidly interact with transmembrane proteins, which would then transduce the signal across the membrane to activate second messenger systems.

Vesiculation of human red blood cells

Vesicle release from human red blood cells (RBC) can be induced *in vitro* by various manipulations, such as calcium loading (Allan and Thomas, 1981), ATP-depletion (Lutz et al., 1977), and incubation in presence of dimyristoylphosphatidylcholine (Ott et al., 1981). Although these procedures induce vesiculation by different mechanisms, the sequence of events leading to vesicle release is very similar. In an initial step RBC change their morphology from the normal discoid shape to spiculated cells. Subsequently, the intrinsic domain of the RBC membrane dissociates from the cytoplasmic protein skeleton, which is demonstrated by the absence of spectrin in the released vesicles. The final step involves a fusion event between the membrane bilayers that approach each other during the formation of the membrane spicules. The events leading to RBC vesiculation can be modulated by amphiphilic drugs like chlorpromazine and indomethacin which are known to affect RBC shape and cause the cells to become stomatocytic or echinocytic, respectively (Bütikofer et al., 1988 & 1989a).

Spontaneous vesiculation of RBC has been observed *in vivo* in blood from patients with several hemolytic anemias. These include sickle cell anemia (Allan et al., 1981), hereditary spherocytosis, protein 4.1 deficiency, and hemoglobin H disease (Wagner et al., 1986). Furthermore, release of vesicles from RBC occurs in erythrocytes stored under blood bank conditions (Rumsby et al., 1977), and it is believed that the loss of membrane as microvesicles is a factor that may contribute to the shortened life time of stored RBC. Vesicles can also be detected in freshly drawn blood from normal individuals (Dumaswala and Greenwalt, 1984), and it is possible that the vesiculation which occurs *in vivo* during RBC senescence contributes to the heterogeneous density distribution of erythrocytes (reviewed by Clark, 1988).

Vesicles found in the blood of patients with the above mentioned hemolytic anemias, as well as vesicles shed during RBC storage, have similar protein and lipid compositions as compared to those released from RBC during *in vitro* manipulations. In addition, red cells from dense fractions of density separated RBC (the presumably "older" cells), as well as remnant RBC after vesiculation, have decreased cellular deformability when measured in an ektacytometer (Clark et al., 1983; Wagner et al., 1987; Bütikofer et al., 1989a). Thus, the study of vesicle release from RBC *in vitro* may contribute to the understanding of *in vivo* vesiculation processes and may provide insights into membrane alterations which occur during cell aging.

Release of GPI anchored proteins from RBC during vesiculation

During vesiculation of human RBC *in vitro*, AChE was found to be enriched in the released vesicles relative to RBC membrane phospholipid or protein (Allan et al., 1981; Lutz et al., 1977; Shukla et al., 1978). In addition, it has been reported that between 20% and 80% of total RBC AChE is released from RBC during vesiculation without significant hemolysis (Lutz et al., 1977; Ott et al., 1981; Wagner et al., 1987; Bütikofer et al., 1989b). This preferential release of AChE into the vesicles could be related to the GPI anchor, which could allow AChE to redistribute laterally on the cell surface. In contrast, integral proteins, such as band 3 or glycophorin which may not have such lateral mobility, are present but not enriched in the released vesicles. To test our hypothesis that enrichment of AChE in vesicles released from RBC represents an example of a general phenomenon characteristic for GPI anchored proteins, we studied the release of both AChE and DAF during RBC vesiculation.

We found that the ratio of AChE to phospholipid was increased two- to three-fold in vesicles released from RBC by Ca^{2+}-loading, ATP-depletion, or storage as compared to native RBC (Figure 2). Accordingly, the AChE to phospholipid ratio of remnant RBC after vesiculation, induced by these three conditions, was decreased as compared to untreated RBC (Figure 2). These results clearly demonstrate that, relative to membrane phospholipid, AChE is preferentially released from RBC during these vesiculation processes and that it is enriched in the released vesicles. By contrast, the AChE to phospholipid ratio in vesicles released from RBC by heat treatment was unchanged as compared to heat treated (remnant) RBC after vesiculation, or native RBC (Figure 2).

By immunoblotting with monoclonal antibodies to DAF, we identified DAF in remnant RBC during Ca^{2+}-loading and found that the amount of DAF in these cells decreased as vesiculation progressed (Bütikofer et al., 1989b). At the same time, the released vesicles showed an increased DAF to phospholipid ratio when compared with untreated RBC. A similar observation was made for vesicles released from RBC during ATP-depletion as well as during storage (Figure 3). By contrast, vesicles released during heat treatment showed no

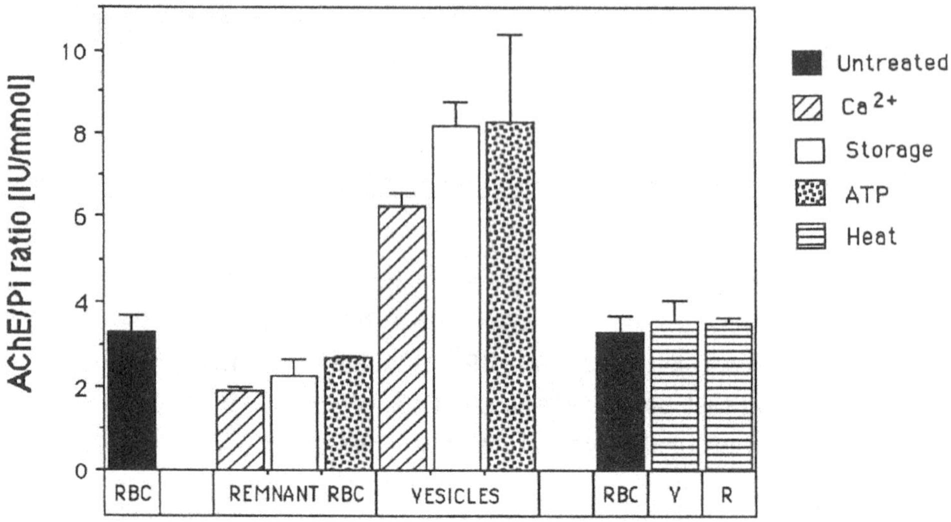

Figure 2: AChE to phospholipid (Pi) ratio in remnant RBC and vesicles. The AChE to phospholipid ratio was determined in ghosts prepared from remnant (vesiculated) RBC obtained after 1 hour of Ca^{2+}-loading, 5-7 weeks of storage, 46 hours of ATP depletion, 40 min of heat treatment, and in vesicles released from these cells.

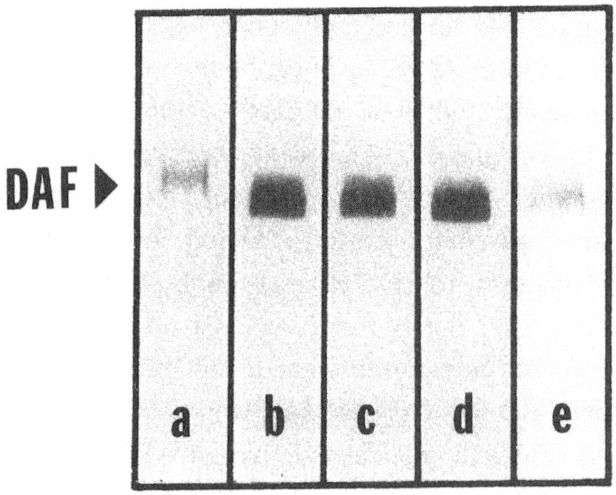

Figure 3: Identification of DAF in RBC and vesicles. Vesiculation was induced by Ca^{2+}-loading, ATP depletion, storage, and heat treatment of RBC. Vesicles were isolated by centrifugation, proteins were separated by SDS-PAGE and subsequently transferred to nitrocellulose membranes. DAF was detected with murine monoclonal anti-DAF antibodies followed by horseradish peroxidase-conjugated anti-mouse IgG. The lanes represent ghosts from untreated RBC (a), and Ca^{2+}- (b), ATP depletion- (c), storage- (d), and heat-induced (e) vesicles, respectively. Each lane contains the same amount of membrane phospholipid.

substantial increase in the DAF to phospholipid ratio as compared to untreated RBC (Figure 3). Thus, the mechanism involving DAF release from RBC and its enrichment in vesicles appears to be similar to that observed for AChE.

Since the ratio of band 3 to phospholipid is similar in vesicles and native RBC (Allan et al., 1980; Shukla et al., 1978), our results suggest that the GPI anchor may be responsible for the preferential enrichment of AChE and DAF in RBC vesicles. For this to occur, these proteins must undergo lateral redistribution on the surface of the RBC prior to vesiculation. We believe that changes in the bilayer architecture occur during the process of vesiculation and that GPI anchored proteins preferentially redistribute into the membrane domains of the echinocytic spicules.

A major component determining the bilayer architecture of the RBC is its membrane skeleton. Detachment of the membrane skeletal protein network from the intrinsic protein domain as well as the lipid bilayer occurs prior to vesicle release. This is demonstrated by depletion of spectrin, the major RBC membrane skeletal protein, in the vesicles. This detachment will likely affect the normal packing of lipids in the bilayer. Furthermore, since the interaction of the membrane skeleton with the intrinsic domain of the bilayer limits the lateral mobility of integral proteins (Tsuji and Ohnishi, 1986; Tsuji et al., 1988), detachment of skeletal proteins from the membrane could generate membrane domains with an increased lateral diffusion constant (Jacobson et al., 1982; Gawrisch et al., 1986). Since GPI anchored proteins have a very large headgroup as compared to their membrane anchor, the relatively high curvature of the RBC membrane at the tips of the echinocytic spicules may favour their accumulation in these areas. All together, this may result in enrichment and retention of GPI anchored proteins in those membrane domains that will eventually be released from the cells as vesicles. A schematic view of our proposed model of membrane rearrangement during vesiculation is given in Figure 4.

Our experiments using heat treatment of RBC to induce vesiculation support the hypothesis that skeletal proteins are involved in the observed rearrangement of the membrane. We found that neither AChE nor DAF were enriched in heat-induced vesicles and that remnant RBC after vesiculation were not depleted of either protein (Figures 2 and 3). The sequence of events leading to vesicle release during heat treatment differs from that of the other vesiculation procedures we used. The spectrin content of vesicles released from heat treated cells was 20-40% of that in the RBC membrane (Calvez et al., 1988),

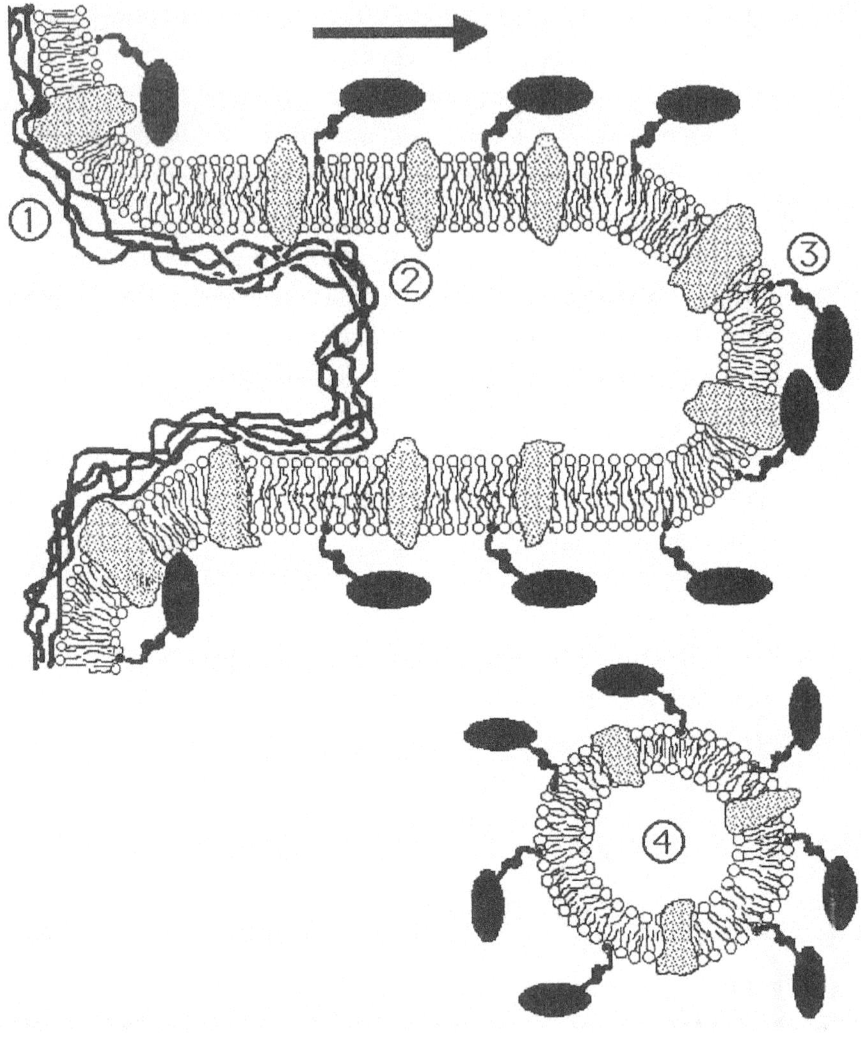

Figure 4: Membrane rearrangement during vesiculation. (1) Interactions between the protein skeleton and the membrane bilayer restrict the lateral mobility of membrane proteins, including that of GPI anchored proteins. (2) A (transient) dissociation of the skeletal network from the membrane is a prerequisite for vesiculation. This dissociation increases the lateral mobility of membrane proteins. Because GPI anchored proteins are attached to the membrane by a phospholipid moiety, these proteins are presumably more mobile on the cell surface and (3) redistribute into those domains of the membrane with increased lateral mobility, where they will be retained in areas of high curvature. These domains at the tip of the echinocytic spicules, which act as a sink for GPI anchored proteins, are subsequently released from the cell as membrane vesicles (4).

whereas vesicles released by the other procedures were largely devoid of spectrin (Allan et al., 1981; Lutz et al., 1977; Ott et al., 1981). Furthermore, vesicles are released from heat treated cells by fragmentation (Wagner et al., 1986) rather than by the typical membrane budding process observed for the other vesiculation procedures. These observations suggest that only limited dissociation of the protein skeleton from the intrinsic domain of the membrane occurs during heat-induced RBC shape change. Therefore, if our hypothesis is correct, membrane domains necessary for accumulation of GPI anchored proteins would not be generated.

Possible physiologic implications

The role of AChE in RBC membranes remains unexplained and there is no evidence that its depletion during the life span of the RBC has physiologic significance. In contrast, DAF serves a critical role in the protection of the RBC from lysis by complement (Nicholson-Weller et al., 1983; Pangburn et al., 1983), and RBC from patients with paroxysmal nocturnal hemoglobinuria, which are deficient in DAF, are highly sensitive to the hemolytic action of complement (Medof et al., 1985). Hence, RBC depleted of DAF during vesiculation can be expected to be more susceptible to complement-induced lysis. Indeed, preliminary results from our laboratory indicate that remnant RBC after release of DAF-enriched vesicles are more susceptible to lysis induced by complement as compared to control RBC (Table 1). It is interesting to note that cells from the dense fraction of normal human RBC were reported to have decreased amounts of DAF per cell (Kinoshita et al., 1985). In addition, RBC with higher density (older cells) have decreased amounts of AChE on their surface as compared to younger cells (Trautsch et al., 1981). This could have resulted from the release of vesicles, enriched in both DAF and AChE, as the cells aged. As a consequence, these older RBC may be more susceptible to complement-induced lysis and this, in turn, may contribute to their ultimate removal from the circulation.

Table 1: Lysis of human RBC by purified anti-I IgM and complement.

	Percent Lysis	
Anti-I IgM added (ug)	Control RBC	Remnant RBC
0.00	0.0	0.0
0.63	1.7	62.6
1.25	8.6	80.8
2.50	26.2	93.0

Untreated RBC and remnant RBC after Ca^{2+}-induced vesiculation were incubated with serial dilutions of purified anti-I IgM and freshly prepared autologous human serum for 30 min at 0 °C, followed by a 60 min incubation at 37 °C. Subsequently, the percent hemolysis was measured in the cell free supernatant after pelleting the RBC.

Our observation that AChE and DAF are depleted in remnant RBC after vesiculation and are enriched in the released vesicles may be representative for all glycosyl-phosphatidylinositol anchored proteins. Since vesiculation is an important physiologic membrane event occurring in many cells, we believe that the release of GPI anchored proteins is not limited to RBC but may have general biological significance. Our findings provide a new hypothesis for the role GPI anchors might play in the attachment of proteins to membranes.

Acknowledgements

The work was supported by grants from The National Institutes of Health HL27059, AM 32094, HL 20985, HL 36255. The authors are grateful to the excellent technical assistance of Ms. Caimin Xu, a visiting scientist from Union Medical College of Beijing, China.

References

Allan D, Limbrick AR, Thomas P, Westerman MP (1981) Microvesicles from sickle erythrocytes and their relation to irreversible sickling. Br. J. Haematol. 47: 383-390

Allan D, Thomas P (1981) Ca^{2+}-induced biochemical changes in human erythrocytes and their relation to microvesiculation. Biochem. J. 198: 433-440

Allan D, Thomas P, Limbrick AR (1980) The isolation and characterization of 60 nm vesicles produced during ionophore A23187-induced budding of human erythrocytes. Biochem. J. 188: 881-887

Braun-Breton C, Rosenberry TL, Pereira da Silva L (1988) Induction of the proteolytic activity of a membrane protein in *Plasmodium falciparum* by phosphatidyl inositol-specific phospholipase C. Nature 332: 457-459

Bütikofer P, Brodbeck U, Ott P (1987) Modulation of erythrocyte vesiculation by amphiphilic drugs. Biochim. Biophys. Acta 901: 291-295

Bütikofer P, Lin ZW, Kuypers FA, Scott MD, Xu C, Wagner GM, Chiu DTY, Lubin B (1989a) Chlorpromazine inhibits vesiculation, alters phosphoinositide turnover and changes deformability of ATP-depleted RBCs. Blood 73: 1699-1704

Bütikofer P, Kuypers FA, Xu C, Chiu DTY, Lubin B (1989b) Enrichment of two glycosylphosphatidylinositol anchored proteins, acetylcholinesterase and decay accelerating factor, in vesicles released from human red blood cells. Submitted for publication.

Calvez JY, Zachowski A, Herrmann A, Morrot G, Devaux PF (1988) Asymmetric distribution of phospholipids in spectrin-poor erythrocyte vesicles. Biochem. 27: 5666-5670

Cherry RJ (1979) Rotational and lateral diffusion of membrane proteins. Biochim. Biophys. Acta 559: 289-327

Clark MR (1988) Senescence of red blood cells: Progress and problems. Physiol. Rev. 68: 503-554

Clark MR, Mohandas N, Shohet SB (1983) Osmotic gradient ektacytometry: Comprehensive characterization of red cell volume and surface maintenance. Blood 61: 899-910

Coakley WT, Bater AJ, Deeley JOT (1978) Vesicle production on heated and stressed erythrocytes. Biochim. Biophys. Acta. 512: 318-330

Davitz MA, Hereld D, Shak S, Krakow J, Englund PT, Nussenzweig V (1987) A glycan-phosphatidylinositol-specific phospholipase D in human serum. Science 238: 81-84

de Vetten MP, Agre P (1988) The Rh polypeptide is a major fatty-acylated erythrocyte membrane protein. J. Biol. Chem. 263: 18193-18196

Dumaswala UJ, Greenwalt TJ (1984) Human erythrocytes shed exocytic vesicles in vivo. Transfusion 24: 490-492

Gawrisch K, Stibenz D, Mops A, Arnold K, Linss W, Halbhuber KJ (1986) The rate of lateral diffusion of phospholipids in erythrocyte microvesicles. Biochim. Biophys. Acta 856: 443-447

Ishihara A, Hou Y, Jacobson K (1987) The Thy-1 antigen exhibits rapid lateral diffusion in the plasma membrane of rodent lymphoid cells and fibroblasts. Proc. Natl. Acad. Sci. USA 84: 1290-1293

Jacobson K, Elson E, Koppel D, Webb W (1982) Fluorescence photobleaching in cel biology. Nature 295: 283-284

Kinoshita T, Medof ME, Silber R, Nussenzweig V (1985) Distribution of decay-accelerating factor in the peripheral blood of normal individuals and patients with paroxysmal nocturnal hemoglobinurea. J. Exp. Med. 162: 75-92

Low MG (1987) Biochemistry of the glycosyl-phosphatidylinositol membrane protein anchors. Biochem. J. 244: 1-13

Low MG (1989) Glycosyl-phosphatidylinositol: a versatile anchor for cell surface proteins. FASEB J. 3: 1600-1608

Low MG, Prasad ARS (1988) A phospholipase D specific for the phosphatidylinositol anchor of cell-surface proteins is abundant in plasma. Proc. Natl. Acad. Sci. USA 85: 980-984

Low MG, Saltiel AR (1988) Structural and functional roles of glycosyl-phosphatidylinositol in membranes. Science 239: 268-275

Lutz HU, Liu SC, Palek J (1977) Release of spectrin-free vesicles from human erythrocytes during ATP-depletion. J. Cell. Biol. 73: 548-560

Medof ME, Kinoshita T, Silber R, Nussenzweig V (1985) Amelioration of lytic abnormalities of paroxysmal nocturnal hemoglobinuria with decay-accelerating factor. Proc. Natl. Acad. Sci. USA 82: 2980-2984

Medof ME, Walter EI, Roberts WL, Haas R, Rosenberry TL (1986) Decay accelerating factor of complement is anchored to cells by a C-terminal glycolipid. Biochem. 25: 6740-6747

Nicholson-Weller A, March JP, Rosenfeld SI, Austen KF (1983) Affected erythrocytes of patients with paroxysmal nocturnal hemoglobinuria are deficient in the complement regulatory protein, decay-accelerating factor. Proc. Natl. Acad. Sci. USA 80: 5066-5070

Noda M, Yoon K, Rodan GA, Koppel DE (1987) High lateral mobility of endogenous and transfected alkaline phosphatase: a phosphatidylinositol-anchored membrane protein. J. Cell Biol. 105: 1671-1677

Olson EN, Towler DA, Glaser L (1985) Specificity of fatty acylation of cellular proteins. J. Biol. Chem. 260: 3784-3790

Ott P, Hope MJ, Verkleij AJ, Roelofsen B, Brodbeck U, van Deenen LLM (1981) Release of spectrin-free vesicles without ATP-depletion. Biochim. Biophys. Acta 641: 79-87

Pangburn MK, Schreiber RD, Müller-Eberhard HJ (1983) Deficiency of an erythrocyte membrane protein with complement regulatory activity in paroxysmal nocturnal hemoglobinuria. Proc. Natl. Acad. Sci. USA 80: 5430-5434

Rennie CM, Thompson S, Parker AC, Maddy A (1979) Human erythrocyte fraction in "Percoll" density gradients. Clin. Chim. Acta 98: 119-125

Roberts WL, Myher JJ, Kuksis A, Low MG, Rosenberry TL (1988) Lipid analysis of the glycoinositol phospholipid membrane anchor of human erythrocyte acetylcholinesterase. J. Biol. Chem. 263: 18766-18775

Roberts WL, Santikarn S, Reinhold VN, Rosenberry TL (1988) Structural characterization of the glycosyl phospholipid membrane anchor of human erythrocyte acetylcholinesterase by fast atom bombardment mass spectrometry. J. Biol. Chem. 263: 18776-18784

Rosenberry TL, Roberts WL, Haas R (1986) Glycolipid membrane-binding domain of human erythrocyte acetylcholinesterase. Fed. Proc. 45: 2970-2975

Rumsby MG, Trotter J, Allan D, Michell RH (1977) Recovery of membrane microvesicles from human erythrocytes stored for transfusion: A mechanism for the discocyte-to-spherocyte shape transformation. Biochem. Soc. Trans. 5: 126-128

Shukla SD, Berriman J, Coleman R, Finean JB, Michell RH (1978) Membrane protein segregation during release of microvesicles from human erythrocytes. FEBS Lett. 90: 289-292

Silman I, Futerman H (1987) Modes of attachment of acetylcholinesterase to the surface membrane. Eur. J. Biochem. 170: 11-22

Trautsch C, Tannert C, Maretzki D (1981) Disproportional loss of membrane constituents in the course of erythrocyte ageing. Acta Biol. Med. Germ. 40: 743-746

Tsuji A, Kawasaki K, Ohnishi S (1988) Regulation of band 3 mobilities in erythrocyte ghost membranes by protein association and skeletal meshwork. Biochem. 27: 7447-7452

Tsuji A, Ohnishi S (1986) Restriction of the lateral motion of Band 3 in the erythrocyte membrane by the cytoskeletal network: Dependence on spectrin association state. Biochemistry 25: 6133-6139

Wagner GM, Chiu DTY, Qju JH, Heath RH, Lubin BH (1987) Spectrin oxidation correlates with membrane vesiculation in stored RBCs. Blood 69: 1777-1781

Wagner GM, Chiu DTY, Yee MC, Lubin BH (1986) Red cell vesiculation - a common membrane physiologic event. J. Lab. Clin. Med. 108: 315-324

Wilcox CA, Olson EN (1987) The majority of cellular fatty acylated proteins are localized to the cytoplasmic surface of the plasma membrane. Biochem. 26: 1029-1036

STRUCTURES FORMED BY MEMBRANE LIPIDS – PHYSICOCHEMICAL PROPERTIES AND POSSIBLE BIOLOGICAL RELEVANCE FOR MEMBRANE FUNCTION.

Göran Lindblom and Leif Rilfors
Department of Physical Chemistry
University of Umeå
S-90187 Umeå
Sweden

A very active area of molecular biology is the effort to investigate relationships between molecular structure and biological function. It is now 65 years since the first attempt (Gorter & Grendel, 1925) was made to describe the basic structure of a biological membrane in terms of a lipid bilayer. Then about 20 yeras ago Singer and Nicolson (1972) presented their significant model, where the membrane lipids form a lipid bilayer matrix in which the working proteins are incorporated. Today the view of the role played by the lipids has been drastically changed from this idea that the main task of the lipids is just to create an inert and inactive thin sheet, to a situation where the lipids participate in the membrane functioning in many ways and in particular by their ability to form different aggregate structures.

Membrane lipids, like other amphiphilic molecules, form a variety of different lyotropic liquid crystalline structures, e.g. lamellar, inverted (denoted *reversed* in the surface chemistry literature) hexagonal, and cubic phases. Almost every membrane so far studied contain at least one lipid that by itself in water forms a nonlamellar structure. This, we believe, leads to very important consequences for the physicochemical properties of the biological membrane and of the membrane function. Experimental evidence for this has also been obtained for some membranes. In particular should be mentioned the membrane of the bacterium *Acholeplasma laidlawii,* where the regulation of the lipid composition is determined by a lipid forming nonlamellar phases (Rilfors et al., 1984). An understanding of the physical chemistry concerning formation and structure of these nonlamellar phases is thus of fundamental biological relevance *irrespective* of whether such phases, or even their structural elements, actually form as stable structures within the cell membranes or elsewhere in a biological system.

NATO ASI Series, Vol. H 40
Dynamics and Biogenesis of Membranes
Edited by J.A.F. Op den Kamp

In this communication we will try to do the impossible, namely, to take the reader through the lipid marsh, while feeding her or him with some basic physical chemistry of association amphiphiles, i.e. self-assembly and aggregation, cubic and inverted hexagonal phase structures, solubilization of hydrophobic molecules, water- and ion-binding, and eventually we will briefly touch upon the fascinating biological implications which this divine physical chemistry lead us into.

Phase equilibria and structural polymorphism

Amphiphilic molecules can be divided into two groups depending on their behaviour in water solution: 1) water-soluble micelle-forming amphiphiles; and 2) water-insoluble swelling amphiphiles. Thus, for example lysophosphatidylcholines (LPC´s) form micelles (1) while phosphatidylcholines (PC´s) only form lamellar liquid crystalline phases (2). However, there is not always a sharp boundary between the two groups since an insoluble swelling lipid at room temperature may be soluble and form micelles at a higher temperature.

Cubic and inverted hexagonal phases form in several membrane lipid systems. As a first step in the investigations of the physicochemical properties of membrane lipids it is desired to determine the phase diagram of the lipid-water system. So far a rather limited number of phase diagrams of membrane lipids are availabe in the literature. However, such diagrams have been collected for a large number of simple surfactants (see e.g. Tiddy, 1980; Ekwall, 1975; Laughlin, 1978). The determinations of phase diagrams are usually a very tedious work using classical methods, and the need for more rapid methods led to the development of NMR techniques for such investigations (Ulmius et al., 1977). These convenient methods, usually based on ^{2}H and ^{31}P NMR, are frequently utilized by us (Brentel et al., 1985; Arvidson et al., 1985; Eriksson et al., 1985; Lindblom et al., 1986; Sjölund et al., 1987). The most abundant phases found for the amphiphilic systems of interest here are: 1) the micellar solution with normal (L_1) or inverted (L_2) aggregate structures; 2) lamellar (L_α) liquid crystalline phases; 3) normal (H_I) or inverted (H_{II}) hexagonal liquid crystalline phases; and 4) a number of different cubic liquid crystalline phases. The structures of the aggregates in groups 1-3 are well established (cf Fig. 1).

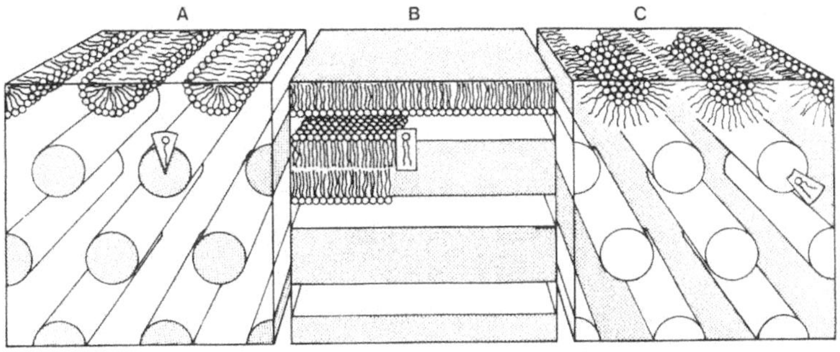

Figure 1. Structure of the most common anisotropic liquid crystalline phases that membrane lipids can form. (A) Normal hexagonal (H_I) phase; (B) lamellar (L_α) phase; and (C) inverted hexagonal (H_{II}) phase. From Rilfors et al., 1984.

The positions in the phase diagram of the various phases occurring in a lipid-water system often give very helpful information about the physicochemical properties and, in particular for the many possible different cubic phases, about the aggregate structure building up the phases. Hitherto, a cubic phase has been observed between almost every other phase in the phase diagram. Cubic phases of membrane lipids often form between the L_α and H_{II} phases. It has also been observed that for some systems isotropic phases, probably of cubic structure, form as metastable intermediates at the thermal phase transition between the L_α and H_{II} phases (Sjölund et al., 1989; Gruner et al., 1988). A cubic phase has so far never been observed between the H_{II} and L_2 phases.

A cubic liquid crystalline lipid-water phase is one in which the lipid aggregates form a three-dimensional lattice. The lipid aggregate units can have different shapes like spheres, rods, or lamellae. In contrast to the cubic phases, lamellar liquid crystalline phases exhibit a one-dimensional periodicity, i.e. lamellar units of infinite extension are stacked regularly (Fig. 1B), and hexagonal liquid crystalline phases exhibit a two-dimensional periodicity with rod-like aggregates of infinite length packed into a hexagonal lattice (Figs. 1A and 1C). Thus, the cubic phases are optically isotropic while the lamellar and hexagonal phases are anisotropic. However, the cubic phases, like other liquid crystalline phases, have no short range order, i.e. the hydrocarbon chains are disordered.

From a methodological point of view it is convenient to divide the cubic phases into two fundamentally different groups: 1) cubic tures having regions which are continuous with respect to both polar (water) and nonpolar (hydrocarbon) components (Fig. 2A); and 2) cubic structures built up either with discontinuous hydrocarbon regions but with continuous water regions (for example normal micellar aggregates in water, Fig. 2B), or with discontinuous water regions but with continuous hydrocarbon regions (for example inverted micellar aggregates in hydrocarbon). Cubic phases belonging to the first group are usually called bicontinuous. Most of the cubic phases formed by membrane lipids have been shown to have such a structure (Rilfors et al., 1986). A cubic phase built up of inverted micelles has, however, not been experimentally observed.

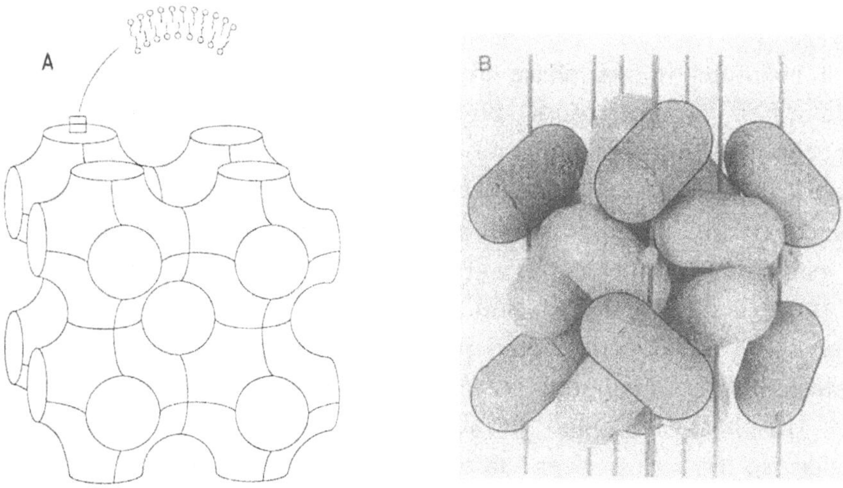

Figure 2. (A) Schematic of a bicontinuous cubic phase belonging to the space group *Im3m*. This infinite periodic minimal surface is called a Schwarz's surface. Lipid monolayers are draped on each side over the minimal surface. (B) A scaled model of a unit cell of a cubic phase structure built up of short rod-like micelles. From Eriksson et al., 1985a.

There is a remarkable difference in viscosity between the two types of structures - all the bicontinuous phases are extremely viscous while the phases built up of micellar aggregates are less viscous. Furthermore, the bicontinuous phases usually exhibit a more hysteretic behaviour. Further details on the different structures of cubic phases, and of the different methods for their study (in particular NMR techniques) can be found in a recent review (Lindblom & Rilfors, 1989).

Theoretical aspects on lipid self-assembly

It is well-known that the amphiphilic property of lipid molecules leads to self-aggregation in water solution, so that the hydrocarbon chains are segregated away from contact with water, while the hydrophilic headgroups are hydrated. The main driving force behind this self-assembly of lipids is the so called *hydrophobic effect*, which on its own would lead to a macroscopic phase separation (one polar and one apolar phase), but the requirement that the polar headgroups must be in contact with water prevents this. Instead certain packing restrictions of the lipid molecules in the aggregates formed have to be fulfilled (Tanford, 1980; Israelachvili et al.,1980). Thus, one of the most useful concepts for a qualitative understanding of the phase behaviour in amphiphile systems is based on a consideration of the shape of the lipid molecules (cf Fig. 1), suggested already by Tartar (1955). According to Israelachvili et al. (1980) the self-assembly of lipid molecules depends on a dimensionless packing parameter defined by the ratio $v(al)^{-1}$, where v is the molecular volume, l the molecular length, and a is the molecular area at the hydrocarbon-water interface. When the packing parameter is equal to unity (cylindrical-like molecules, cf Fig. 1B) there are optimal conditions for the formation of a bilayer structure, while for $v(al)^{-1}>1$ (Wieslander et al., 1980) the molecules are wedge-shaped and the lipid monolayer prefers to curve towards the water region, i.e. an H_{II} phase may form (Fig. 1C). Although this simple approach is very useful for qualitative considerations, it is difficult to use for more quantitative calculations, mainly due to the complex dependence of the area a on temperature and concentration. Furthermore, a change in the molecular shape does not explain why an H_{II} phase is formed at increasing water contents in PC-alkane systems (Sjölund et al., 1987; 1989; Lindblom et al., 1988). This problem was nicely solved by Gruner (1985), who used a concept related to the packing parameter $v(al)^{-1}$, that has a more general character, namely the so called *spontaneous curvature*, H_0, of the lipid monolayer[1]. This concept was first introduced by Helfrich (1973)

---

[1] There are two fundamental types of curvature used in differential geometry that characterize each point on a surface: the Gaussian curvature $K=R_1^{-1}\cdot R_2^{-1}$ and the mean curvature $H={}^1/_2(R_1^{-1}+R_2^{-1})$, where R_1 and R_2 are the principal radii. The surfaces of a so called minimal surface satisfy $H=0$, so that every point is a balanced saddle point with $R_1=-R_2$. We have used the convention that for a lipid monolayer $H>0$ when the layer curves towards the hydrocarbon chain region and $H<0$ when

for PC bilayer systems. Gruner and coworkers have shown that the energy of curvature plays a dominant role in the formation of H_{II} phases. The total curvature free energy, G_{curve}, per area, A, is given by (Anderson et al., 1988)

$$G_{curve} = K_m A \langle(H-H_o)^2\rangle$$

where H is the local mean curvature of the monolayers, and K_m is the elastic bendning constant. It can thus be seen that it is advantageous to have H close to H_o to minimize the free energy of curvature. However, for a phospholipid monolayer to form a cylinder of radius $R_o = 1/H_o$ there will also be a cost of nonzero packing energy (Gruner, 1985). In particular, the acyl chains of the PC molecules must stretch to fill the interstitial regions between the cylindrical aggregates building up the

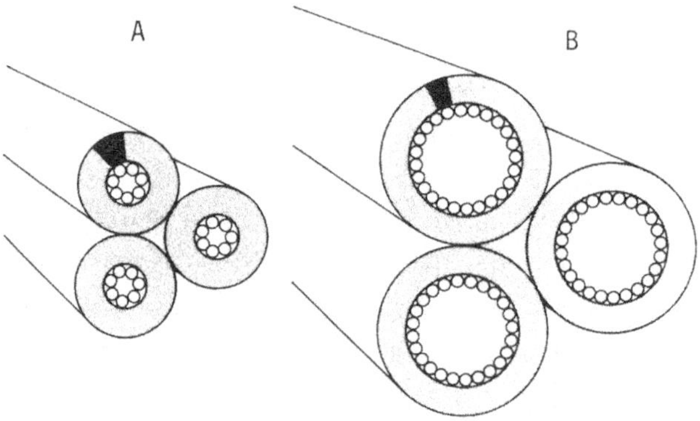

Figure 3. Schematic illustration of an H_{II} phase having acyl chains with (A) a high degree of unsaturation, and (B) a low degree of unsaturation. From Sjölund et al., 1989.

H_{II} phase (Fig. 3). Note that the region in question (shaded) will increase with the size of the water cylinders. Thus, the forces of curvature and packing oppose each other, and such a situation is referred to as a "frustration" (Charvolin & Sadoc, 1987). It can therefore be concluded that the smaller the radius of the water cylinder is, the smaller are the hydrophobic interstices and the easier it is for the lipid acyl chains to

---

the layer curves towards the aqueous region. It has been shown (Hyde, 1989) that the effective packing parameter is related to the mean and Gaussian curvatures by

$$(v/al)_{eff} = 1 + Hl + Kl^2/3$$

It was first suggested by Scriven (1976) that bicontinuous cubic phase structures can be described by periodic minimal surfaces (for a review see Lindblom & Rilfors, 1989).

elongate to fill the volume. This explains why the H_{II} phase tends to form at low water contents. Similarly, it can be understood why increasing acyl chain length also favours the H_{II} phase formation. It has been shown in several studies (Sjölund et al., 1987; 1989; Gruner, 1985; Tate & Gruner, 1987) that this "frustration" can also be eliminated by the addition of hydrophobic molecules. NMR studies (Sjölund et al., 1987) strongly indicate that these molecules preferentially partition into the interstitial region between the cylinders building up the H_{II} phase structure. Thus, we showed by investigating ^2H NMR spectra of perdeuterated *n*-dodecane solubilized in the dioleoyl-PC (DOPC)-water system that the deuteron quadrupole splitting disappeared and changed over to a single narrow peak on going from an L_α phase to the water-rich H_{II} phase. Furthermore, it was found that the translational diffusion of dodecane in the H_{II} phase was almost as rapid as in pure dodecane. The diffusion coefficients were measured with the NMR method (Lindblom & Wennerström, 1977), which was feasible since the static dipolar couplings of dodecane in this H_{II} phase were very close to zero. This can be taken as a further strong indication of that the dodecane molecules preferentially reside in the interstitial regions of the H_{II} phase.

It can be expected that the more saturated the acyl chains of the phospholipids are, i.e. the greater the packing constraints for formation of the H_{II} phase are, the larger the water cylinders and the water content in the H_{II} phase must be. Thus, if a phosphoplipid-dodecane-water system is able to form an H_{II} phase at all, this ability will increase in the following order: dipalmitoyl-PC<1-palmitoyl-2-oleoyl-PC<DOPC<

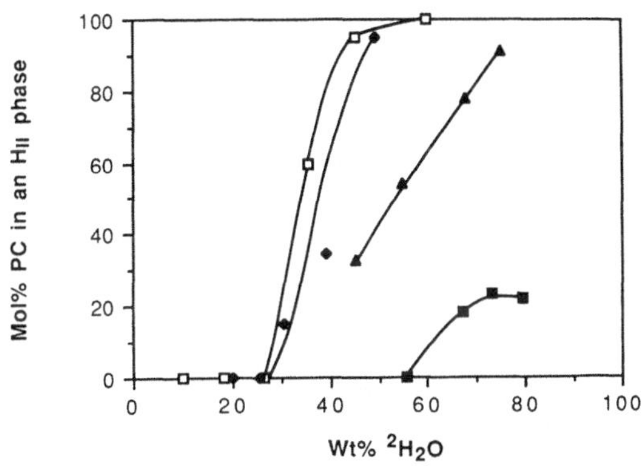

Figure 4. Phase equilibria at 45°C in different PC-*n*-dodecane-water systems. The molar ratio of dodecane/PC = 2. (◆) DLiPC; (□) DOPC; (▲) POPC; (■) DPPC. From Sjölund et al., 1989.

<dilinoleoyl-PC. In recent experiments by Sjölund et al. (1989) this was found to be the case (Fig. 4). It has also been shown in a preliminary X-ray diffraction study that the water cylinders in the DOPC-dodecane-water system may grow to very large sizes having diameters of the order of 20 nm (S. Gruner, G. Lindblom, L. Rilfors & M. Sjölund, to be published).

Obviously, the size of the polar headgroup plays an important role for the formation of an H_{II} phase; smaller headgroups favouring H_{II} phases. Thus, it can be expected that phospholipids having *charged* headgroups should promote the formation of the lamellar phase. Recently, we found (M. Sjölund, G. Lindblom & L. Rilfors, to be published) that this behaviour is not always observed. NMR and X-ray diffraction investigations clearly showed that the divalent dioleoylphosphatidic acid (DOPA) forms an H_{II} phase up to a water content of about 60 wt%, then at higher water concentrations an L_α phase is stable. This latter phase was found to be very well ordered (macroscopically aligned in the test tube) so that ^2H and ^{23}Na NMR quadrupole splittings (Lindblom et al., 1976) of water and counterions could be studied up to a water content of *98* wt% (Fig. 5).

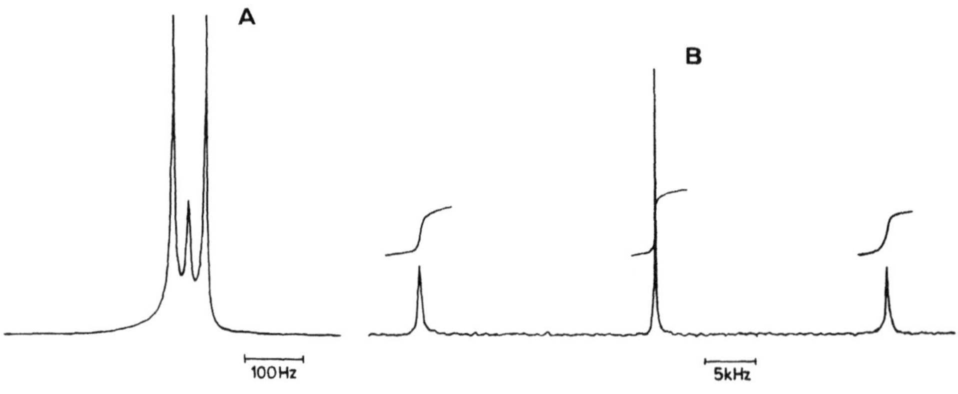

Figure 5. NMR spectra from an L_α phase composed of DOPA and 95 wt% water showing quadrupole splittings of (A) ^2H-water and (B) ^{23}Na-counterions. From M. Sjölund, G. Lindblom & L. Rilfors, to be published.

From the water splittings it was concluded that the bilayers in the L_α phase are fully hydrated, and addition of water results in an increase of "free" water between the lamellae. The decrease in the water splitting with increasing water content was treated by applying a two-site model developed previously (see for example Arvidson et al., 1985), where free and bound water molecules are assumed to exchange rapidly between the sites. The observed quadrupole splitting within this model then follows the equation

$$\Delta^2H=nX_L/X_w\,|\chi S|$$

where X_L and X_w are the mole fractions of lipid and water, respectively, χ is the quadrupole coupling constant, S is the order parameter, and n is the average number of water molecules bound to each group. By plotting Δ^2H against X_L/X_w a straight line is obtained strongly indicating that water molecules added to the L_α phase are trapped between the bilayers as free water (i.e. one-dimensional swelling). It is interesting to note that the L_α phase in this system can swell to very large water quantities, above 98 wt%. This is a typical behaviour for ionic surfactants (Ekwall, 1975) which can be explained by the electrostatic forces acting over long distances between the charged lamellar surfaces (Jönsson & Wennerström, 1987). On the other hand, it is found that the nonionic glucolipids monoglucosyldiacylglycerol (MGluDG) and diglucosyldiacylglycerol (DGluDG) have a proportionally low ability to take up water, whether the phase structure is lamellar (DGluDG) or inverted hexagonal (MGluDG) (Lindblom et al., 1986). This behaviour was recently shown by FT-IR (A. Nilsson, A. Holmgren, L. Rilfors & G. Lindblom, to be published) to be mainly due to the great possibility for these lipids to form strong hydrogen bonds between the sugar moities in the headgroups. Consequently, the so called hydration force (Le Neveu et al., 1976) is much weaker for the nonionic glucolipids than for the phospholipids (Marra, 1986; Lindblom et al., 1986)

The experimental discovery, almost 20 years ago, of quadrupole splittings of sodium counterions in lyotropic liquid crystalline systems (Lindblom, 1971) has made it possible to very conveniently determine the degree of ion binding in lamellar and hexagonal phases. In particular, this NMR method is probably still the only one with which so called "ion condensation" (Manning, 1972) in lyotropic liquid crystals can be directly investigated. Here, we will briefly report some of our results of sodium counterion binding obtained in the lamellar phase of the DOPA-water system (Sjölund et al., to be published). In the ion condensation model the counterions are assumed to bind to the charged lamellar surfaces to reduce the effective charge density to a constant critical value (Wennerström et al., 1979)

$$(|\sigma|a)_{crit} = 1.31\pi\varepsilon\varepsilon_0 k_B T|z|/e$$

where σ is the surface charge density, and the distance between the lamellae is 2a. The other symbols have their usual meaning. The ion condensation behaviour implies a number of non-trivial properties of the counterion distribution in lyotropic liquid crystalline phases. One of the most important and striking one is that the number of bound ions is unaffected by a dilution of the system, i.e. it is independent of the water content as long as $|\sigma|a$ is sufficiently large. Such a situation is actually observed for the DOPA-water system, where it is found that the ^{23}Na quadrupole splitting is constant, equal to about 25 kHz over the whole interval of water concentration studied (60-98 wt%). Note that this consequence of the ion condensation model is in sharp contrast to what is expected from a simple chemical equilibrium model with a single type of binding site characterized by a single binding constant, K. In such a model a dilution of the system would lead to a decrease in the number of bound ions, which would be reflected in a decrease in the measured ^{23}Na quadrupole splitting.

Lipid-water systems forming nonlamellar phases

A large body of data has been produced on the phase structures formed by a variety of membrane lipids. The knowledge of the properties of lipids has expanded greatly during the past ten years, and several reviews have been published (Cullis & De Kruijff, 1979; Cullis et al., 1985, 1986; De Kruijff et al., 1981, 1985; Lindblom & Rilfors, 1989; Rilfors et al., 1984). Only a very brief summary of these publications will be presented.

The phase structure formed by membrane lipids is influenced by several factors: 1) the chemical structure of the lipid; 2) the surrounding conditions like temperature, hydrostatic pressure, hydration of the lipid, and pH; and 3) the interaction with other molecules present in the aqueous region, in the hydrocarbon-water interfacial region, or in the hydrocarbon region of the lipid aggregates.

Chemical structure. The structure of the polar headgroup is of great importance for the phase structure formed by lipid molecules. Under physiological conditions, lipids with small and/or poorly hydrated headgroups (PE, MGluDG, MGalDG) often form H_{II} and cubic phases, while lipids with large and/or well hydrated headgroups (PC, PS, PG, DPG, PI, PA, DGluDG, DGalDG) form an L_α phase.

The structure of the hydrocarbon chains of the lipid also affects the phase structure. Generally, the temperature for the formation of an inverted nonlamellar phase is decreased when the chain length is increased, when double bonds or methyl branches are introduced into the chains, and when an ester linkage to the glycerol backbone is replaced by an ether linkage.

An important group of lipids in mammalian tissues are the sphingolipids, which have the long-chain base sphingosine, or a derivative of it, as a backbone. When glycerolipids and sphingolipids with identical or similar polar head groups are compared, it has been found that the sphingolipids have a greater capability to stabilize a lamellar phase. Some anionic sphingolipids even form H_I phases and micellar solutions, and in this respect behave like ordinary soaps despite that the sphingolipids contain two hydrocarbon chains.

Surrounding conditions. If an increased temperature affects the phase equilibria of a membrane lipid-water system, it generally causes a transition from an L_α phase to a cubic or an H_{II} phase. An increase in the hydrostatic pressure has a very small, but reversed, effect.

Although cell membranes are surrounded by water, differences in hydration of the polar headgroups are of relevance to biological systems. The effects of varying the degree of lipid hydration is rather complex. A transition from an L_α to a cubic or an H_{II} phase is often observed when the water content is decreased in a lipid-water mixture. However, in the systems monoolein-water and MGluDG-water, an increase in the water content cause transitions from an L_α to a cubic phase, and from a cubic phase to an H_{II} phase, to occur, respectively.

The phase structure formed by a lipid containing ionic groups can be affected by the pH of the surrounding water phase. When the pH is lowered, a transition from an L_α to an H_{II} phase may take place. For the zwitterionic lipid PE, the transition occurs when going from an alkaline to a neutral pH, and for the anionic lipids PS, DPG, and PA, a change from a neutral to an acidic pH promotes the transition.

Interaction with other molecules. Cations and water-soluble molecules are able to induce phase transitions in membrane-lipid water systems. Ca^{2+}, Mg^{2+}, and Mn^{2+} induce a transition from an L_α to an H_{II} phase for the anionic lipids DPG and PA. Monovalent cations can also promote the formation of an H_{II} phase. Li^+ has this effect on PS, and Na^+ has this effect on PE and DPG at molar concentrations. The fusogenic substance polyethylene glycol has been reported to increase the temperature for the L_α-H_{II} phase transition (T_{LH}) of PE. Cryoprotectants like trehalose also affect the T_{LH}, but conflicting results have been obtained (Wistrom et al., 1988; Bryszewska & Epand, 1988).

The influence of hydrocarbons like *n*-alkanes, benzene, and cyclohexane on the phase equilibria of above all PE, PC, and PE-PC mixtures has been investigated (Sjölund et al., 1989, and references therein). In general, hydrocarbons promote or facilitates the formation of cubic and H_{II} phases.

The effects of a number of amphiphilic molecules have been investigated. A discussion of all these molecules in various lipid-water systems would be too comprehensive in this context, and therefore just some examples are given.

The shorter-chain *n*-alcohols (ethanol, butanol) stabilize the L_α phase, while the longer-chain *n*-alcohols promote the formation of an H_{II} phase in PE-water systems. Detergents and lyso-PC´s form micellar solutions at high water contents, and they have the ability to form an L_α phase when mixed in certain ratios with lipids forming an H_{II} phase. Mono- and diacylglycerols are formed during fat digestion in the small intestine in vertebrates; both molecules induce the formation of cubic and H_{II} phases when mixed with PE or PC.

The influence of cholesterol on the phase equilibria of membrane lipid-water systems has been intensively studied. In general, the addition to these systems of up to approx. equimolar amounts of cholesterol induce the formation of cubic and H_{II} phases. However, above these concentrations it has been found that the effect of cholesterol reverses, leading to a progressive stabilization of the L_α phase (Epand & Bottega, 1987). If cholesterol is mixed in equimolar concentrations with lyso-PC an L_α phase is formed. The effect of cholesterol in some cases depend on the acyl chain composition of the lipids involved. For example, for PE/PC mixtures, addition of the sterol promotes the formation of an H_{II} phase if PC has unsaturated acyl chains, but stabilize the L_α phase if PC has saturated acyl chains; moreover, cholesterol decreases the T_{LH} for PE with monounsaturated acyl chains, whereas the opposite effect is observed for human erythrocyte PE with a large fraction of polyunsaturated acyl chains.

Some steroids in which the the hydroxyl group has been replaced by a keto group have a stronger ability than cholesterol to induce the formation of an H_{II} phase.

Peptides and proteins constitute a very heterogeneous group of molecules regarding size, shape, and charge. Contrary to molecules like alkanes, alcohols, and detergents, it is therefore impossible to predict the effect of peptides and proteins on the phase equilibria of lipid-water mixtures. The hydrophobic peptide gramicidin has been extensively studied, and it has been found to induce the formation of cubic or H_{II} phases in PE-water and PC-water systems. The amphiphilic peptides cardiotoxin and mellitin interact both electrostatically and hydrophobically with lipid bilayers, and they form an H_{II} phase together with DPG (Batenburg et al., 1985, 1987). Cytochrome *c*, a highly basic protein, induces nonlamellar phase structures in a total lipid extract from mitochondria, in DPG systems, and in mixtures of DPG and PE or PC. On the other hand, glycophorin, an integral glycoprotein, has a bilayer stabilizing effect in PE and DPG systems. Such an effect has also been obtained with cytochrome oxidase and chlorophyllase.

Biological relevance of cubic phases and isotropic structures

Membrane fusion. Membrane fusion is a very important phenomenon in all cells. In particular it is involved in processes of transport in which membranes encapsulate different cell substances (membrane traffic). Other examples are the sperm-egg fusion, and virus-cell fusion. Several recent reviews on membrane fusion are available (Verkleij, 1984; Siegel, 1987; Chernomordik et al. 1987).

The molecular mechanisms behind the fusion process are poorly understood, and a number of factors are discussed in the reviews cited above. Here we will confine ourselves to the fusion occurring in lipid systems, in particular systems where isotropic structures probably are involved. In the course of fusion structural reorganizations have to occur in the lipid bilayers, so that for example two lipid vesicles can form one new larger vesicle. Thus, the essential step in fusion is the rearrangement of the lipid molecules of the two apposed membranes to form a single, continuous membrane. The lipid phase structure, and correlation of non-leaky fusion with phase transitions of macroscopical systems, plays an important role in this respect. Using vesicles made of

DOPE-Me Ellens et al. (1986) found increased fusion to occur in the temperature range in which isotropic structures were formed in dispersions of this lipid. This isotropic structure probably corresponds to a cubic liquid crystalline phase (Gruner et al. 1988). When the temperature was above the transition temperature for formation of the H_{II} phase, lysis of the vesicles developed. A prerequisite for the membrane fusion to take place is the formation of intermembrane structures.

The theoretical works by Siegel (1986; 1987) strongly indicate that a sequence of intermembrane structures are involved in the mechanism of the L_α/H_{II} phase transitions. Siegel suggested that several lipid systems exhibiting L_α/H_{II} phase transitions should form a type of intermediate that would produce isotropic structures in a temperature interval starting tens of degrees below the T_{LH}. Fluctuations lead to the formation of intermembrane micellar intermediates (IMIs), so that the lipids now can exchange in the outer monolayers. The initial step in a phase transition is the formation of these very short-lived IMIs. These micelles may then aggregate in the plane of the two apposed membranes and transform either to an H_{II} phase or to a new intermediate called interlamellar attachment (ILA), which is very long-lived compared with the IMI. The ILA is suggested to have a structure similar to the unit cell in some bicontinuous reversed cubic phases (Fig. 2A). The conditions for the formation of ILAs are given a very simple form in Siegels´ theory. The probability for an ILA to form largely depends upon an experimental parameter closely related to the spontaneous radius of curvature (R_0) in Gruners´ theory (Gruner, 1985) and to the packing parameter introduced by Israelachvili et al. (1980). Below a certain value of this parameter the IMIs can form ILAs, which in turn results in membrane fusion with non-leaky mixing of the aqueous contents.

Regulation of lipid composition in biological membranes. Cubic and isotropic phases can be formed by total lipid mixtures isolated from biological membranes, and lipid structures allowing isotropic motional averaging are formed even in intact membranes. It can therefore be assumed that membrane lipids, at least in metabolically active membranes, must have the possibility to rearrange to isotropic structures in order to fulfil certain functions. However, nonbilayer structures are not allowed to occur permanently in a biological membrane since this would destroy its barrier properties. Instead, the membrane lipids should be in the "vicinity" of a phase transition under physiological conditions so that local and transient nonbilayer structures can form easily in response to some triggering mechanism. It can thus be assumed that the balance between bilayer and nonbilayer lipid structures is actively regulated in cells growing under different environmental conditions. Experimental results (Wieslander et al., 1980, 1986; Rilfors et al., 1982; Lindblom et al., 1986; Goldfine et al., 1987) show that this assumption seems to be true.

The regulation of membrane lipid composition and the physico-chemical properties of the membrane lipids have been intensively studied in *A. laidlawii*. The membrane of this organism contains six amphiphilic lipids, three nonionic and three ionic. MGluDG and DGluDG are the major lipids (55-75 mol%). The polar headgroup composition of the lipids is affected by several factors: the growth temperature; incorporation of different fatty acids into the lipids; and incorporation of sterols, hydrocarbons, alcohols, and detergents into the membrane (Wieslander et al., 1980, 1986; Rilfors et al., 1987). All these factors affect the molar ratio MGluDG/DGluDG. MGluDG forms cubic and H_{II} phases above 10^oC while DGluDG and the other lipids form an L_α phase (Lindblom et al., 1986). The following conclusion was drawn: when the growth conditions of the cells are changed in such a way that the phase equilibria of the membrane lipids are shifted towards nonbilayer structures, the cells respond by changing the ratio MGluDG/DGluDG in order to stabilize the bilayer structure. The formation of the nonlamellar phases by *A. laidlawii* lipids depends critically upon the concentration of MGluDG.

The regulation of the "bilayer stability" is exemplified in Fig. 6. When the proportion of the oleoyl chains in the membrane lipids is increased, *without* changing the polar headgroup composition, the phase equilibria are shifted towards the cubic and H_{II} phases (Wieslander et al., 1981; Khan et al., 1981). However, by lowering the molar ratio MGluDG/DGluDG the transition to the nonlamellar phases begins at approx. the same temperature. A cubic phase appears in all the lipid extracts. Accordingly, two kinds of restriction seem to determine the membrane lipid composition in the *A. laidlawii* cells: Firstly, a large proportion of the lipids are not allowed to enter the gel phase since this will inactivate the membrane-bound enzyme and transport activities; and secondly,

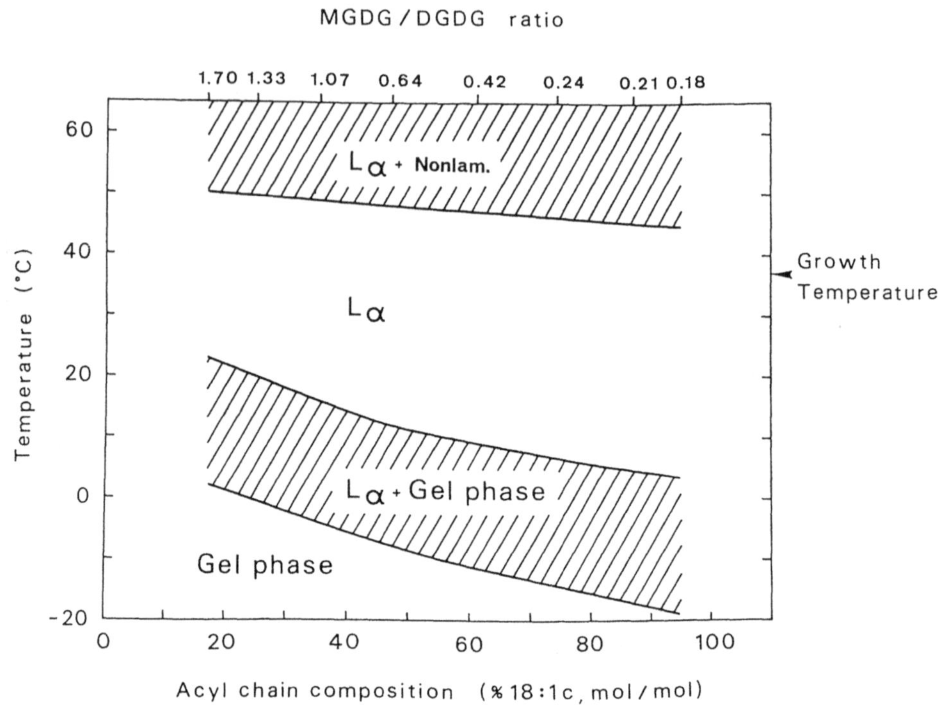

Figure 6. Phase equilibria of total lipid mixtures, containing different amounts of palmitoyl and oleoyl chains, from membranes of *A. laidlawii* grown at 37°C. The metabolically obtained MGluDG/DGluDG ratios are also shown. From Lindblom et al., 1986.

nonbilayer structures are not allowed to form permanently since this will break down the barrier properties of the membrane.

Similar experiments as those described above have recently been performed with *Clostridium butyricum* (Goldfine et al., 1987). The major membrane lipids in this organism are PE, plasmaenylethanolamine (PlaE), and the glycerol acetal of PlaE (GAPlaE). The ratio (PE + PlaE)/GAPlaE is decreased when the degree of acyl chain unsaturation is increased. The PE plus PlaE fraction forms an H_{II} phase above 0°C while GAPlaE forms an L_α phase up to at least 50°C. Similar regulatory mechanisms thus seem to operate in *A. laidlawii* and *C. butyricum* although the membrane lipids involved differ completely.

The membrane lipids in *B. megaterium* are PE, PG, and DPG, and the lipids contain mainly iso and anteiso methyl-branched, saturated acyl chains. When the growth temperature is increased the fraction of PE remains fairly constant (55-65 mol%), while the molar ratio iso/anteiso acyl chains is markedly increased. PE isolated from *B. megaterium* grown at 20°C have a low ratio of iso/anteiso acyl chains and begins to form a cubic phase at 50°C at low water contents. PE isolated from cells grown at 55°C have a 10-fold higher value of the ratio iso/anteiso acyl chains and form an L_α phase up to at least 65°C (Rilfors et al., 1982). Since an increase in the temperature will increase the tendency of PE to form cubic and H_{II} phases, the temperature-induced regulation of the acyl chain composition in *B. megaterium* membranes is most probably necessary in order to maintain a stable lipid bilayer.

Many organisms, both prokaryotic and eukaryotic, regulate their membrane lipid composition in response to various changes in the environmental conditions, such as the ambient temperature, supply of fatty acids, ionic conditions, and the presence of sterols, hydrocarbons, and alcohols in the membrane (Rilfors et al., 1984). Nearly all biological membranes that have been investigated contain at least one lipid species that can form a cubic and/or an H_{II} phase, and the formation of these phases is dependent on all the factors mentioned above. It is therefore possible that one aim with the lipid regulation mechanisms is to keep the balance between the bilayer and nonbilayer lipid structures on an appropriate level. Recently, Gruner et al. (1988) proposed that the membrane is striving towards a constant spontaneous radius of curvature of the lipid monolayer, an idea also partly based on our experimental findings on *A. laidlawii* membranes. The generality of this hypothesis is currently investigated in our laboratories.

Membrane structures. Etioplasts are found in leaves of plants grown in the dark and after the exposure to light they transform into chloroplasts. The etioplast contains a highly ordered, branched tubular membrane structure called the prolamellar body (PLB) which develops into the thylakoid membranes of the chloroplast in the presence of light. The three-dimensional arrangement of the tubules in the PLB has been investigated by several authors (Murakami et al., 1985 and references therein). The basic unit of most PLB:s is a tetrapodal structure which has four short tubular arms (30-35 nm in diameter) meeting at one point with tetrahedral angles. This is similar to the arrangement of the structural elements in a bicontinuous cubic phase belonging to the space group *Pn3* or *Pn3m*. A fact favouring the possibility that the PLB has a bicontinuous cubic structure is its lipid composition: the galactolipids MGalDG and DGalDG account for approximately 75-80 mol% and the ratio

MGalDG/DGalDG is high (1.6-1.8) (Selstam & Sandelius, 1984). Galactolipid mixtures with this ratio can form a cubic phase (Brentel et al., 1985). Moreover, one pigment-protein complex dominates the protein content of the PLB, the NADPH-protochlorophyllide oxidoreductase (Murakami et al., 1985). The accumulation of this pigment-protein complex is unique for the PLB and may facilitate or induce the formation of this structure (Selstam & Sandelius, 1984).

The second intracellular membrane network discussed will be the endoplasmic reticulum (ER). Two functionally distinct regions of ER exist: 1) the rough ER, which in most cell types forms wide, flattened sacs arranged in a compact, parallel array; however, in some cell types the rough ER forms a three-dimensional network of tubular elements; and 2) the smooth ER which as a rule forms a three-dimensional network of fine tubules with a diameter of 30-60 nm. It may be speculated that the networks of tubules resemble a bicontinuous cubic phase, in which proteins are embedded (H. Wennerström, personal communication).

Fat digestion. In most vertebrates the fat digestion occurs in the upper small intestine through the action of pancreatic lipase, colipase, and bile. Colipase binds to the surface of fat droplets in the presence of bile salts and provides an attachment site for lipase (Patton, 1981). Lipase attacks two of the ester bonds of the triacylglycerol molecules, producing first a diacylglycerol and one fatty acid and then a monoacylglycerol and another fatty acid. The only one of these substances able to form a liquid crystalline phase by itself is monoacylglycerol.

During the fat digestion a number of different lipid phases are formed, among others lamellar and cubic liquid crystalline phases (Patton & Carey, 1979). It can be speculated that the cubic phase is preferred in front of a micellar solution or any other liquid crystalline phase, in order to effectively carry out the work of fat digestion. Lipases perform their enzymatic reaction with highest efficiency at the water-hydrocarbon interface of a lipid bilayer. A bicontinuous cubic phase formed by lipid bilayers might therefore constitute a suitable matrix for the enzymatic process: 1) the substrates can be incorporated into the cubic phase and will thus have the proper location for the lipolytic attack; 2) large interfacial surfaces for the lipases are created in a small volume by the cubic phase; and 3) the products are effectively "stored" in the cubic phase. A cubic phase of monoacylglycerol-water can be dispersed by sodium cholate, but it is not known if the cubic phase is directly involved in the mucosal uptake of lipids.

Conclusions

Several arguments make it probable that cubic and isotropic lipid structures are the nonbilayer structures that are of greatest biological relevance: 1) Low amounts of energy are required for the small structural reorganizations of lipids and proteins that occur during the transformations between a lamellar phase and a bicontinuous cubic phase represented by a minimal surface or a constant-mean-curvature surface; 2) the nonbilayer phases and structures formed by the total lipids extracted from biological membranes, and by intact membranes, are predominantly cubic phases or isotropic lipid structures; and 3) membrane fusion, regulation of membrane lipid composition, the structure of the prolamellar body, and fat digestion seem to involve cubic and/or other isotropic structures.

References

Anderson DM, Gruner SM, Leibler S (1988) Geometrical aspects of the frustration in the cubic phases of lyotropic liquid crystals, Proc Natl Acad Sci 85:5364-5368

Arvidson G, Brentel I, Khan A, Lindblom G (1985) Phase equilibria in four lyso-phosphatidylcholine/water systems. Exceptional behaviour of 1-palmitoyl-glycerophosphocholine, Eur J Biochem 152:753-759

Batenburg AM, Bougis PE, Rochat H, Verkleij AJ, De Kruijff B (1985) Penetration of a cardiotoxin into cardiolipin model membranes and its implications on lipid organization, Biochemistry 24:7101-7110

Batenburg AM, Hibbeln JCL, Verkleij AJ, De Kruijff B (1987) Melittin induces H_{II} phase formation in cardiolipin model membranes, Biochim Biophys Acta 903:142-154

Brentel I, Selstam E, Lindblom G (1985) Phase equilibria of mixtures of plant galactolipids. The formation of a bicontinuous cubic phase, Biochim Biophys Acta 812:816-826

Bryszewska M, Epand RM (1988) Effects of sugar alcohols and disaccharides in inducing the hexagonal phase and altering membrane properties: implications for diabetes mellitus, Biochim Biophys Acta 943:485-492

Charvolin J, Sadoc JF (1987) Periodic systems of frustrated fluid films and "bicontinuous" cubic structures in liquid crystals, J Phys (Paris) 48:1559-1569

Chernomordik LV, Melikyan GB, Chizmadzhev YA (1987) Biomembrane fusion: a new concept derived from model studies using two interacting planar lipid bilayers, Biochim Biophys Acta 906:309-352

Cullis PR, De Kruijff B (1979) Lipid polymorphism and the functional roles of lipids in biological membranes, Biochim Biophys Acta 559:399-420

Cullis PR, Hope MJ, De Kruijff B, Verkleij AJ, Tilcock CPS (1985) In: Kuo, JF (ed) Phospholipids and cellular recognition, CRC Press, Boca Raton, Florida, pp 1-59

De Kruijff B, Cullis PR, Verkleij AJ, Hope MJ, Van Echteld CJA, Taraschi TF (1985) Lipid polymorphism and membrane function. In: Martinosi AN (ed) The enzymes of biological membranes. Plenum Press, New York, pp 131-204

Ekwall P (1975) Composition, properties and structures of lipid crystalline phases in systems of amphiphilic compounds, Adv Liq Cryst 1:1-142

Ellens H, Bentz J, Szoka FC (1986) Fusion of phosphatidylethanolamine-containing liposomes and mechanism of the L_α-H_{II} phase transition, Biochemistry 25:4141-4147

Epand RM, Bottega R (1987) Modulation of the phase transition behavior of phosphatidylethanolamine by cholesterol and oxysterols, Biochemistry 26:1820-1825

Eriksson P-O, Lindblom G, Arvidson G (1985a) NMR Studies of 1-palmitoyllysophosphatidylcholine in a cubic liquid crystal with a novel structure, J Phys Chem 89, 1050-1053

Eriksson P-O, Rilfors L, Lindblom G, Arvidson G (1985) Multicomponent spectra from ^{31}P NMR studies of the phase equilibria in the system dioleoylphosphatidylcholine-dioleoylphosphatidylethanolamine-water, Chem Phys Lipids 37, 357-371

Goldfine H, Johnston NC, Mattai J, Shipley GG (1987) Regulation of bilayer stability in *Clostridium butyricum*: Studies on the polymorphic phase behavior of the ether lipids, Biochemistry 26:2814-2822

Gorter E, Grendel F (1925) On biomolecular layers of lipid on the chromacytes of the blood, J Exp Med 41:439-443

Gruner SM (1985) Intrinsic curvature hypothesis for biomembrane lipid composition: A role for nonbilayer lipids, Proc Natl Acad Sci USA 82:3665-3669

Gruner SM, Tate W, Kirk GL, So PTC, Turner DC, Keane DT, Tilcock CPS, Cullis PR (1988) X-ray diffraction study of the polymorphic behavior of N-methylated dioleoylphosphatidylethanolamine, Biochemistry 27:2853-2866

Helfrich W (1973) Elastic properties of lipid bilayers. Theory and possible experiments, Z Naturforsch Teil C 28:693-703.

Hyde ST (1989) Microstructure of bicontinuous surfactant aggregates, J Phys Chem 93:1458-1464

Israelachvili JN, Marcelja S, Horn RG (1980) Physical principles of membrane organization, Q Rev Biophys 13:121-200

Jönsson B, Wennerström H (1987) Phase equilibria in a three component water-soap-alcohol system. A thermodynamic model, J Phys Chem 91:338-352

Khan A, Rilfors L, Wieslander Å, Lindblom G (1981) The effect of cholesterol on the phase structure of glucolipids from *Acholeplasma laidlawii* membranes, Eur J Biochem. 116, 215-220

Laughlin RG (1978) Solvation and structural requirements of surfactant hydrophilic groups, Adv Liq Cryst 3:41-148

Le Neveu DM, Rand RP, Parsegian VA (1976) Measurement of forces between lecithin bilayers, Nature 259:601-603

Lindblom G (1971) Ion binding in liquid crystals studied by NMR III. ^{23}Na quadrupolar effects in a model membrane system, Acta Chem Scand 25:2767-2768

Lindblom G, Persson N-O, Arvidson G (1976) Ion binding and water orientation in lipid model membrane systems studied by NMR, Adv Chem Ser 152:121-141

Lindblom G, Wennerström H (1977) Amphiphile diffusion in model membrane systems studied by pulsed NMR, Biophys Chem 6:167-171

Lindblom G, Brentel I, Sjölund M, Wikander G, Wieslander Å (1986) Phase equilibria of membrane lipids from *Acholeplasma laidlawii*. The importance of a single lipid forming non-lamellar phases, Biochemistry 25:7502-7510

Lindblom G, Sjölund M, Rilfors L (1988) Effect of *n*-alkanes and peptides on the phase equilibria in phosphatidylcholine-water systems, Liq Cryst 3:783-790

Lindblom G, Rilfors L (1989) Cubic phases and isotropic structures formed by membrane lipids - possible biological relevance, Biochim Biophys Acta 988:221-256

Manning GS (1972) Polyelectrolytes, Ann Rev Phys Chem 23:117-140

Marra J (1986) Direct measurements of attractive van der Waals and adhesion forces between uncharged lipid bilayers in aqueous solutions, J Colloid Interface Sci 109:11-20

Murakami S, Yamada N, Nagano M, Osumi M (1985) Three-dimensional structure of the prolamellar body in squash etioplasts, Protoplasma 128:147-156

Patton JS, Carey MC (1979) Watching fat digestion, Scince 204:145-148

Patton JS (1981) In: Johnson LR (ed) Physiology of the gastrointestinal tract, Raven Press, New York, pp 1123-1154

Rilfors L, Khan A, Brentel I, Wieslander Å, Lindblom G (1982) Cubic liquid crystalline phase with phosphatidylethanolamine from *Bacillus megaterium* containing branched acyl chains, FEBS Letters 149:293-298

Rilfors L, Lindblom G, Wieslander Å, Christansson A (1984) Lipid bilayer stability in biological membranes. In: Manson LA, Kates M (eds) Biomembranes Vol. 12, Plenum Press, New York, pp 205-245

Rilfors L, Eriksson P-O, Arvidson G, Lindblom G (1986) Relationship between three-dimensional arrays of "lipidic particles" and bicontinuous cubic lipid phases, Biochemistry 25:7702-7711

Rilfors L, Wikander G, Wieslander Å (1987) Lipid acyl chain-dependent effects of sterols in *Acholeplasma laidlawii* membranes, J Bacteriol 169:830-838

Selstam E, Sandelius AS (1984) A comparison between prolamellar bodies and prothylakoid membranes of etioplasts of dark-grown wheat concerning lipid and polypeptide composition, Plant Physiol 76:1036-1040

Siegel DP (1987) In: Sowers AE (ed) Cell fusion. Plenum Press, New York, pp 181-207

Siegel DP (1986) Inverted micellar intermediates and the transitions between lamellar, cubic, and inverted hexagonal amphiphile phases, Chem Phys Lipids 42:279-301

Singer SJ, Nicolson GL (1972) The fluid mosaic model of the structure of cell membranes, Science 175:720-731

Sjölund M, Lindblom G, Rilfors L, Arvidson G (1987) Hydrophobic molecules in lecithin-water systems I. Formation of reversed hexagonal phases at high and low water contents, Biophys J. 52:145-153

Sjölund M, Rilfors L, Lindblom G (1989) Reversed hexagonal phase formation in lecithin-alkane-water systems with different acyl chain unsaturation and alkane length, Biochemistry 28:1323-1329

Tanford C (1980) The hydrophobic effect, 2nd edn. Wiley, New York

Tartar HV (1955) A theory of the structure of micelles of normal paraffin chain salts in aqueous solution, J Phys Chem 59:1195-1199

Tate MW, Gruner SM (1987) Lipid polymorphism of mixtures of dioleoylphosphatidylethanolamine and saturated and monounsaturated phosphatidylcholines of various chain lengths, Biochemistry 26:231-236

Tiddy GJT (1980) Surfactant-water liquid crystal phases, Phys Rep 57:1-46

Ulmius J, Wennerström H, Lindblom G, Arvidson G (1977) Deuteron NMR studies of phase equilibria in a lecithin-water system, Biochemistry 16:5742-5745

Verkleij AJ (1984) Lipidic intramembraneous particles, Biochim Biophys Acta 779:43-63

Wennerström H, Lindman B, Lindblom G, Tiddy GJT (1979) Ion condensation model and NMR studies of counterion binding in lyotropic liquid crystals, J Chem Soc Faraday Trans I 75:663-668

Wieslander Å, Christiansson A, Rilfors L, Lindblom G (1980) Bilayer stability in membranes. Regulation of lipid composition in *Acholeplasma laidlawii* as governed by molecular shape, Biochemistry 19:3650-3655

Wieslander Å, Rilfors L, Johansson LB-Å and Lindblom G (1981) Reversed cubic phase with membrane glucolipids from *Acholeplasma laidlawii*. ^{1}H, ^{2}H and diffusion nuclear magnetic resonance measurements, Biochemistry 20:730-735

Wieslander Å, Rilfors L, Lindblom G (1986) Metabolic changes of membrane lipid composition in *Acholeplasma laidlawii* by hydrocarbons, alcohols, and detergents. Arguments for effects on lipid packing, Biochemistry 25:7511-7517
Wiström CA, Crowe LM, Crowe JH (1988) Prevention of hexagonal phase transitions in phosphatidylethanolamine, Biophys J 53:129a

STUDIES OF THE METABOLISM AND FUNCTION OF SN-1-STEAROYL-2-ARACHIDONOYL LIPIDS IN ANIMAL CELLS

J.A. Glomset
Howard Hughes Medical Institute
SL-15
University of Washington
Seattle, Washington 98195
U.S.A.

Animal cell membranes contain an impressive diversity of lipids. Each lipid presumably contributes to the general structure of the membrane bilayer, but certain lipids may also have specific functions. For example, phosphoinositides play a key role in transmembrane signalling (Berridge, 1987), anchor specific proteins to the outer leaflet of the plasma membrane (Low, 1989), and regulate the interactions of intracellular cytoskeletal proteins (Lassing and Lindberg, 1988). Other lipids also may be required for specific membrane functions, but remain for the most part to be identified. Individual molecular species of lipids as well as lipid classes may be involved. As an illustration, the molecular species of phosphatidylinositol (PI)[1] that anchor proteins to the outer leaflet of the plasma membrane generally differ from those that predominate inside cells. The latter mainly consist of the sn-1-stearoyl-2-arachidonoyl species, for reasons that have yet to be fully explained. In the studies that are summarized below my colleagues and I focused attention on this species of PI and on corresponding species of other lipids in order to identify mechanisms that might influence their formation and function in animal cell membranes.

---

1. Abbreviations: PI=phosphatidylinositol, PA=phosphatidic acid, PC=phosphatidylcholine, PE=phosphatidylethanolamine, PS=phosphatidylserine, DG=diacylglycerol, MG=monoacylglycerol, PDGF=platelet-derived growth factor, Pi=inorganic phosphate.

NATO ASI Series, Vol. H 40
Dynamics and Biogenesis of Membranes
Edited by J.A.F. Op den Kamp

Effects of PDGF on the lipids of Swiss 3T3 cells

Our interest in the metabolism and function of sn-1-stearoyl-2-arachidonoyl lipids was stimulated by a series of studies that Andreas Habenicht initiated several years ago in our laboratory. He labeled quiescent Swiss 3T3 cells with ^{3}H-glycerol and ^{14}C-arachidonic acid, stimulated the cells with PDGF, and observed both a rapid disappearance of radioactive PI and a concomitant appearance of radioactive DG, MG, and free arachidonic acid (Habenicht et al, 1981). It seemed likely at the time that most of the cell phosphoinositides contained an sn-1-stearoyl-2-arachidonoyl DG backbone, that they were being hydrolysed by a PDGF-activated phospholipase C, and that some of the sn-1-stearoyl-2-arachidonoyl DG that was generated by the phospholipase C reaction was being further hydrolysed by a DG lipase. Indeed, subsequent experiments showed that about 80% of the PI of Swiss 3T3 cells consists of the sn-1-stearoyl-2-arachidonoyl species and that the relative content of this species of PI does not change when the phosphoinositides of intact cells are hydrolysed in response to PDGF (Brian Williams, unpublished results).

Nevertheless, subsequent analyses of the cell DG by reverse-phase high performance chromatography showed that PDGF was stimulating not only the accumulation of sn-1-stearoyl-2-arachidonoyl DG, but also the accumulation of other DG species (Fig. 1). Furthermore, analyses of the cell and medium MG showed that PDGF was stimulating both the accumulation of arachidonoyl MG and the accumulation of other species of MG (Fig. 2). This suggested that several additional phospholipids were being hydrolysed in company with sn-1-stearoyl-2-arachidonoyl phosphoinositides. This possibility would be compatible with the recent results of other investigators who have shown that vasopressin and other agonists can induce a phospholipase C-mediated degradation of PC and PE in other cells (for a review, see Exton, 1988).

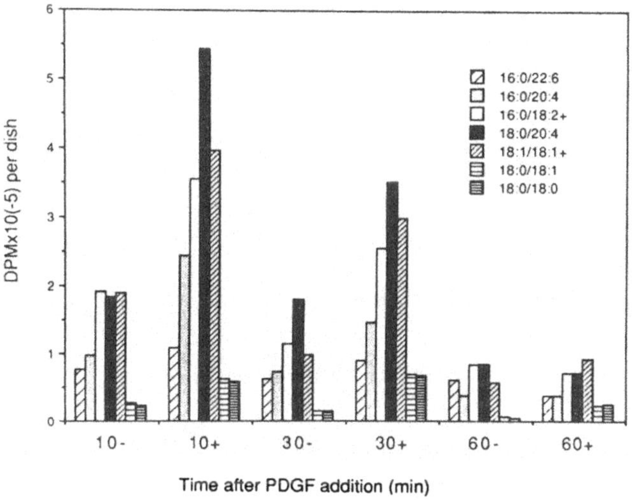

Fig.1 Molecular species of radioactive DG in quiescent or PDGF-stimulated cells that had been prelabeled with radioactive glycerol. Cells were prelabeled with ^{3}H-glycerol for 48 h and then rinsed with medium containing unlabeled glycerol. At 10, 30, or 60 min after the addition of PDGF (+) or vehicle (-), cells were harvested, lipids were extracted, DGs were isolated and derivatized, and molecular species were separated by high performance liquid chromatography on a reverse-phase column. In the figure, the terms 16:0/22:6, etc. refer to the respective sn-1-acyl-2-acyl DGS. (From MacDonald et al, 1988a, with permission)

If several different species of DG transiently accumulate in response to PDGF, is there a mechanism that might selectively channel species that contain an arachidonoyl moiety into the phosphoinositide cycle? Recent experiments in our laboratory by Marnie MacDonald appear to have partially answered this question. While studying the influence of PDGF on the DG kinase reaction in Swiss 3T3 cells, she discovered a membrane-associated form of the kinase that preferentially phosphorylates arachidonic acid-containing DGs (MacDonald et al, 1988 a,b). Subsequent studies in the laboratory by Rozenn Lemaitre have shown that many animal tissues contain a comparable enzyme which can be chromatographically separated from nonspecific DG kinases. It seems reasonable to postulate that one function of this enzyme may be to promote the conversion of arachidonic acid-containing DGs into phosphoinositides (MacDonald et al, 1988a).

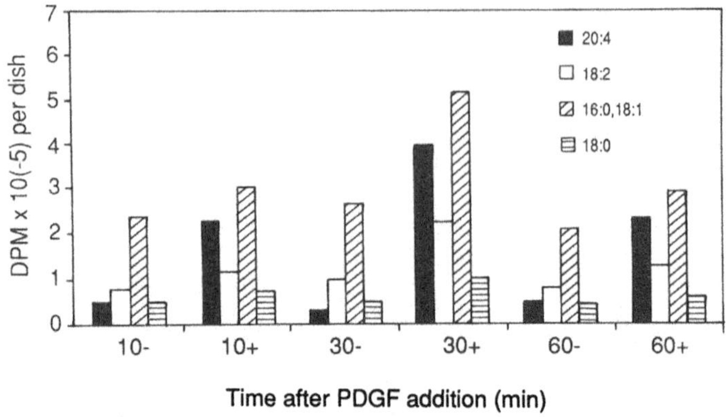

Fig.2 Molecular species of radioactive MG in quiescent or PDGF-stimulated Swiss 3T3 cells that had been prelabeled with radioactive glycerol. The data are from the same experiment shown in Fig. 1. MGs were isolated from the cells and medium, combined, and then analysed by reverse-phase HPLC. A fraction containing both arachidonoyl MG and linoleoyl MG was further subfractionated by thin layer chromatography on silver nitrate plates. In the figure, the terms 20:4, etc. refer to MGs containing the corresponding fatty acids.

Additional experiments in our laboratory may have identified a second mechanism that might channel DGs into phosphoinositides. They showed that PDGF blocks both the conversion of labeled ethanolamine into PE and the conversion of labeled serine into PS (Table 1) without affecting the cellular uptake of either precursor (not shown). If these effects reflect the inhibition of an sn-1,2-DG ethanolamine-phosphotransferase, channeling of DG into the phosphoinositide cycle might be the predicted result.

Formation of sn-1-stearoyl-2-arachidonoyl lipids in quiescent Swiss 3T3 cells

The above experiments provided information about mechanisms that might influence the recycling of arachidonic acid-containing DGs into phosphoinositides in PDGF-stimulated Swiss 3T3 cells, but did not explain why quiescent Swiss 3T3 cells and most other animal cells show a high relative content of sn-1-stearoyl-2-arachidonoyl PI. Previous investigators had concluded that neither this species of PI nor comparable species of other phospholipids are formed <u>de novo</u> , but that they are formed instead by the "remodeling" or "retailoring" of other phospholipids (for reviews, see Holub and Kuksis, 1978, Thompson and Martin, 1984, and Lynch and Thompson, 1988). For

<u>Table 1. Effects of PDGF on the incorporation of radioactive serine, ethanolamine, or glycerol into 3T3 cell phospholipids.</u>

Parallel dishes of quiescent Swiss 3T3 cells were incubated for 60 min with ^{3}H-serine, ^{14}C-ethanolamine, or ^{3}H-glycerol, in the presence or absence of PDGF. Then the lipids were extracted and chromatographed in triplicate on thin layer plates, and the radioactivity determined.

Precursor	PE	PS	PC	PI
	Mean radioactivity per dish (counts/min)			
^{3}H-serine				
- PDGF	916	7832	2103	n.d.
+ PDGF	363	2108	1157	n.d.
% change	-57	-73	-44	
^{14}C-ethanolamine				
- PDGF	1383	n.d.	224	n.d.
+ PDGF	764	n.d.	203	n.d.
% change	-44		-9	
^{3}H-glycerol				
- PDGF	17967	3540	93589	5119
+ PDGF	11607	2174	73796	14582
% change	-35	-38	-21	+85

n.d. = not determined

example, Holub and Kuksis injected labeled ^{32}Pi and ^{14}C-glycerol into rats and found that the liver rapidly incorporated both precursors into monoenoic and dienoic species of PI, but only slowly incorporated the precursors into tetraenoic PI. This led them to conclude that monoenoic and dienoic species of PI were being formed <u>de novo</u> and to suggest that deacylation-reacylation mechanisms were subsequently converting these species into tetraenoic PI (Holub and Kuksis, 1971). Other investigators had previously reached similar conclusions with regard to the formation of tetraenoic species of rat liver PA, PC, and PE (Hill et al.,1968, Kanoh, 1969, and Åkesson et al, 1970).

Though the results of these investigators suggested that some tetraenoic species of phospholipids may not be formed _de novo_, the postulated deacylation-reacylation mechanism of lipid "remodeling" seemed to require further study. For example, the precise number of enzymes that would be required to "remodel" the various lipid classes remained to be determined, and mechanisms that might control these enzymes remained to be identified. More importantly, alternative mechanisms of phospholipid "remodeling" remained to be excluded. Catriona Simpson, Cindy Richardson, Weiling King, and Brian Williams and I therefore decided to use quiescent Swiss 3T3 cells as a model in a further investigation of mechanisms that might lead to the formation of sn-1-stearoyl-2-arachidonoyl species of PI and other lipids.

One of our first goals was to determine how many lipid classes in these cells might show evidence for acyl chain "remodeling". Accordingly, we pulsed the cells for 1 h with ^{3}H-glycerol, then chased the label for varying periods of time, and measured the changes in radioactivity in molecular species of several different lipid classes. We found that most of the radioactive glycerol that was incorporated into lipids during the 1 h pulse was distributed among monoenoic and dienoic lipid species. This was in good agreement with the results that previous investigators had obtained with other cell systems. We also found that the radioactivity in arachidonic acid-containing species increased during the chase in the following lipid classes: PC, PI, PS, DG, PE, and PA. Each class showed an increase in radioactivity in the sn-1-stearoyl-2-arachidonoyl species. In addition, there was an increase in radioactivity in sn-1-palmitoyl-2-arachidonoyl PC and in several species of PS. The changes in PS, which presumably reflected head group exchange reactions, led to a more than twenty fold increase in the total radioactivity of this lipid class.

Because our previous experiments with PDGF-treated cells had demonstrated that the mitogen can induce a transient accumulation of sn-2-arachidonoyl MG (Fig. 2), we wondered whether a pathway for "recycling" this species of MG might exist in Swiss 3T3 cells and be contributing to the formation of sn-1-stearoyl-2-arachidonoyl lipids. To investigate this possibility we incubated quiescent Swiss 3T3 cells with ^{3}H-glycerol-labeled sn-2-arachidonoyl MG in the presence of excess unlabeled glycerol, and followed the incorporation of radioactivity into PI and other cell lipids. We found that radioactivity rapidly appeared in PI, PC, PE, PS, PA, DG, and TG, and that the radioactivity was mainly distributed among sn-1-stearoyl-2-arachidonoyl and sn-1-palmitoyl-2-arachidonoyl species of the different diacyl lipid classes. Each class contained more radioactivity in the sn-1-stearoyl species than in the sn-1-palmitoyl species. Interestingly, however, a much higher proportion of the sn-1-stearoyl species was present in PI and PS than in the other lipids. There was an intermediate proportion of this species in PE and DG, and a low proportion in PC and generally also in PA. This led us to formulate the working hypothesis that is shown in Fig. 3.

This hypothesis assumes that sn-2-arachidonoyl MG can be formed by several different mechanisms including the hydrolysis of plasma TG by lipoprotein lipase, the hydrolysis of cell phospholipids by the combined action of phospholipase Cs and a neutral lipase, or the hydrolysis of cell TG by a TG lipase. It postulates that the sn-2-arachidonoyl MG can be converted to sn-1-acyl-2-arachidonoyl DG by an MG acyltransferase (Coleman and Haynes, 1986) or to sn-1-acyl-2-arachidonoyl PA by the successive action of an MG kinase and an acyltransferase. The key feature of the hypothesis, however, is that it postulates that a special branched pathway exists that specifically forms sn-1-stearoyl-2-arachidonoyl PI, PE, and PS. The existence of such a pathway would account for the high contents of radioactive sn-1-stearoyl-2-arachidonoyl PI and PS that were observed in our experiments with ^{3}H-glycerol-labeled sn-2-arachidonoyl MG and would shed light on the significance of the earlier-mentioned effects of PDGF on the incorporation of labeled ethanolamine into PE and labeled serine into PS (see Table 1).

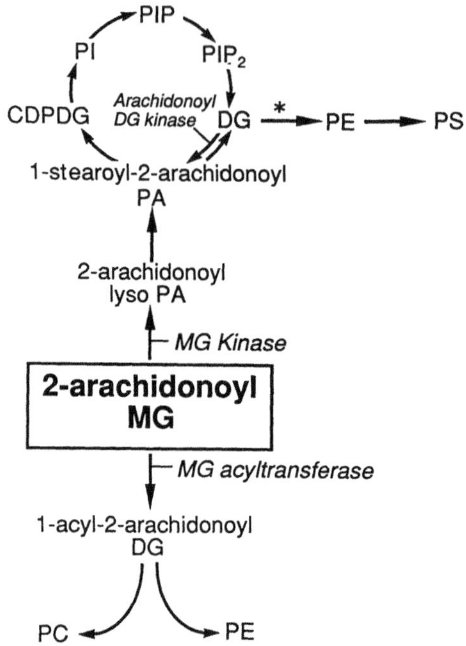

Fig. 3. Postulated pathway for recycling of sn-2-arachidonoyl MG into phospholipids. The asterisk indicates a postulated site of control by PDGF.

Role of sn-1-stearoyl-2-arachidonoyl phospholipids in animal cells

Arachidonoyl lipids are of special interest because some of them can be hydrolysed in response to agonists (see, for example, Kaya, et al., 1989, and Chilton, 1988). The arachidonic acid that is released by the hydrolysis reaction can then be converted into eicosanoids that have powerful autocrine and paracrine effects. However, Ken Applegate and I have been exploring the possibility that arachidonoyl lipids may also function in other ways. In particular, we have been using computer-based molecular modeling programs to obtain clues concerning the potential structural role of these lipids in animal cell membranes. We showed previously (Applegate and Glomset, 1986) that the methylene-interrupted double bond domains of both docasahexaenoic acid and arachidonic acid can adopt an "angle iron-shaped" conformation that has unusual packing properties. In this conformation the individual methylene carbons that separate pairs of cis double bonds are aligned to form a straight axis. The double bond carbons project outward from this axis, forming double bond planes that are alternately aligned in two perpendicular, molecular planes. The molecular planes of two or more polyenoic fatty acids can come into close apposition, thus optimizing attractive, intermolecular van der Waals forces. Furthermore, the "V grove" between the molecular planes of each polyenoic fatty acid can come into close contact with a saturated fatty acyl chain.

To determine how an angle iron-shaped conformation of arachidonic acid might affect the interactions of sn-1-stearoyl-2-arachidonoyl lipids in a membrane bilayer, we have recently been using the Cray computer in San Diego to investigate the packing properties of this species of DG. We have found that sn-1-stearoyl-2-arachidonoyl DGs can form regular, tightly packed arrays that show packing energies that are much more favorable than those of corresponding monoenoic DGs. This raises the potentially interesting possibility that sn-1-stearoyl-2-arachidonoyl lipids might be able to form microdomains within cell membranes (Fig. 4).

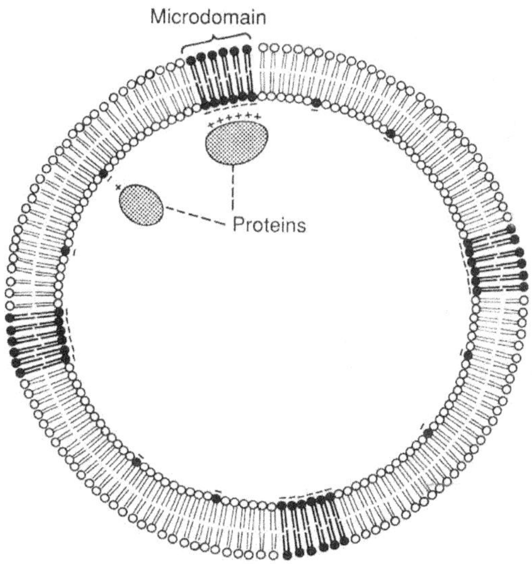

Fig. 4 Microdomains of sn-1-stearoyl-2-arachidonoyl PS and other lipids in a cell membrane are postulated to form binding sites for peripheral membrane proteins.

The probability that such microdomains might exist in membranes would probably be a function of many additional variables including temperature, the nature of the phospholipid head groups, the nature of the membrane proteins that interact with these head groups, and the nature of the lipids that are located on the opposing leaflet of the membrane bilayer. Nevertheless, the possibility that microdomains of sn-1-stearoyl-2-arachidonoyl lipids might exist in membranes seems worth exploring in view of their potential functional importance. For example, as Fig. 4 suggests, the content of PS in a microdomain of sn-1-stearoyl-2-arachidonoyl lipids might be metabolically programmed to form a cluster of negative charges that could act as an attachment site for specific membrane proteins.

Summary

Studies of the effects of PDGF on Swiss 3T3 cells have yielded evidence concerning two different mechanisms that might channel arachidonic acid-containing DGs into the PI cycle. In addition, studies of the formation of sn-1-stearoyl-2-arachidonoyl lipids in quiescent Swiss 3T3 cells have shown that these cells can selectively convert sn-2-arachidonoyl MG into sn-1-stearoyl-2-arachidonoyl PI and PS. Finally, computer-based molecular modeling studies have shown that sn-1-stearoyl-2-arachidonoyl DGs can form tightly packed arrays in simulated membranes. These results provide new information about the biosynthesis of arachidonoyl lipids in animal cells and raise the possibility that one role of these lipids may be to form functionally important microdomains in cell membranes.

Acknowledgements

The research that has been summarized here was supported by the Howard Hughes Medical Institute and by U.S. Public Health Service Grant RR-00166.

References

Åkesson B, Elovson J, Arvidson G (1970) Initial incorporation into rat liver glycerolipids of intraportally injected [^{3}H]-glycerol. Biochim Biophys Acta 210:15-27

Applegate KR, Glomset JA (1986) Computer-based modeling of the conformation and packing properties of docosahexaenoic acid. J Lipid Res 27:658-680

Berridge MJ (1987) Inositol trisphosphate and diacylglycerol: two interacting second messengers. In: Richardson CC (ed) Annual review of biochemistry, vol 56. Annual Reviews, Palo Alto, California, p 159

Chilton FH (1989) Potential phospholipid source(s) of arachidonate used for the synthesis of leukotrienes by the human neutrophil. Biochem J 258:327-333

Coleman RA, Haynes EB (1986) Monoacylglycerol acyltransferase. J Biol Chem 261:224-228

Exton JH (1988) Mechanisms of action of calcium-mobilizing agonists: some variations on a young theme. FASEB J 2:2670-2676

Hill EE, Husbands DR, Lands WEM (1968) The selective incorporation of ^{14}C-glycerol into different species of phosphatidic acid, phosphatidylethanolamine, and phosphatidylcholine. J Biol Chem 243:4440-4451

Holub BJ, Kuksis A (1978) Metabolism of molecular species of diacylglycerophospholipids. In: Paoletti R, Kritchevsky D (eds) Advances in lipid research, vol 16. Academic Press, New York San Francisco London, p 1

Holub BJ, Kukis A (1971) Differential distribution of orthophosphate-^{32}P and glycerol-^{14}C among molecular species of phosphatidylinositols of rat liver in vivo. J Lipid Res 12:699-705

Kanoh H (1969) Biosynthesis of molecular species of phosphatidyl choline and phosphatidyl ethanolamine from radioactive precursors in rat liver slices. Biochem Biophys Acta 176:756-763

Kaya H, Patton GM, Hong SL (1989) Bradykinin-induced activation of phospholipase A_2 is independent of the activation of polyphosphoinositide-hydrolyzing phospholipase C. J Biol Chem 264:4972-4977

Lassing I, Lindberg U (1988) Evidence that the phosphatidylinositol cycle is linked to cell motility. Exp Cell Res 174:1-15

Lynch DV, Thompson GA (1988) Enzymatic pathways in the regulation of membrane fluidity. In: Aloia RC, Curtain CC, Gordon, LM (eds) Physiological regulation of membrane fluidity. Alan R. Liss, New York, p 323 (Advances in Membrane Fluidity, vol 3)

MacDonald ML, Mack KF, Richardson CN, Glomset JA (1988) Regulation of diacylglycerol kinase reaction in Swiss 3T3 cells. J Biol Chem 263:1575-1583

MacDonald ML, Mack KF, Williams BW, King WC, Glomset JA (1988) A membrane-bound diacylglycerol kinase that selectively phosphorylates arachidonoyl-diacylglycerol. J Biol Chem 263:1584-1592

Thompson GA, Jr., Martin CE (1984) Regulation of phospholipid fatty acid composition. In: Shinitzky M (ed) Physiology of membrane fluidity, vol I. CRC Press, Boca Raton, Florida, p 99

INVOLVEMENT OF THE INTRACELLULAR LIPID TRAFFIC DURING PLASMA MEMBRANE BIOGENESIS IN HIGHER PLANTS

Patrick MOREAU, Pascal BERTHO and Claude CASSAGNE

IBCN-CNRS, Université de Bordeaux II, 1 rue Camille Saint-Saëns,
33077 Bordeaux-cedex, France

Introduction

In this paper , we present different results obtained by the *in vivo* approach on the intracellular transport of lipids and fatty acids to the plasma membrane of seven-day-old etiolated leek seedlings. The complexity of the processes involved in plasma membrane biogenesis needs further investigations at both the molecular and cellular levels, but, at the cellular level, this complexity probably exceeds the possibilities of the *in vivo* analysis.Accordingly, a new approach,based on *in vitro* cell-free systems, which has proved to be suitable for studies on protein transport, was developped recently to study the membrane transfer of lipids. Some preliminary results concerning this new approach are presented and discussed.

Lipid and Fatty Acid Transfer to the Plasma Membrane

Saturated very long chain fatty acids (VLCFA) are characteristic components of the plasma membrane of leek epidermis (Cassagne & Lessire, 1978). It was demonstrated that these molecules are synthesized in the endoplasmic reticulum (Cassagne & Lessire, 1978) and, partially, in the Golgi complex (Moreau et al., 1988a). Moreover, the plasma membrane, isolated by phase partition in an aqueous two-polymer phase system, is unable to synthesize these molecules (Moreau et al., 1988a and 1988b).

A methodology has been devised to investigate intermembrane transfers of lipids and VLCFA by seven-day-old etiolated leek seedlings *in vivo* (Moreau et al., 1986). *In vivo* pulse-chase experiments, followed by membrane fractionation of microsomal pellets on linear density sucrose gradients,demonstrated a tranfer of lipids and VLCFA between a light membrane fraction (1.08-1.09 g/ml) and an heavier one (1.16-1.18 g/ml).The latter fraction was found to contain all the membrane vesicles originating from the plasma membrane but was relatively heterogeneous. In order to study the transfer of fatty acids to the plasma membrane, the latter was purified by phase partition following *in vivo* pulse-chase experiments (Moreau et al., 1988b). Lipid and fatty acid radioactivities were then determined after analysis of lipids according to Heape et al.,1985, and fatty acid analysis was performed as described previously (Moreau et al.,1986).

NATO ASI Series, Vol. H 40
Dynamics and Biogenesis of Membranes
Edited by J.A.F. Op den Kamp

The total label of the crude microsomes was unchanged during the various chase times (405 ± 34 dpm/mg prot. x 10^{-3}). However, an increase of the label in the plasma membrane was observed during the chase (table I).

After a two-hour labeling period, the amounts of total labeled fatty acids found in the plasma membrane increased by 13%, 55% and 140% for 30min, 60min and 120min of chase, respectively. The detailed analysis of the radioactivities of the various plasma membrane fatty acids (table I) indicates that, after a two-hour chase, labeled C16, C18 and VLCFA are increased 2-, 2.9- and 3.5-fold, respectively.

Table I. Fatty acid transfer to the plasma membrane during the chase following a two-hour pulse

	DPM / mg Proteins x 10^{-3}					
	Microsomes			Plasma Membrane		
Chase time (min)	C16	C18	VLCFA	C16	C18	VLCFA
0	221	186	10.5	81	38	9.5
30	223	218	12	98	36	11
60	206	142	11	135	46	18.5
120	202	175	12	165	110	33

These results clearly demonstrated the transfer of fatty acids to the plasma membrane of a higher plant and show that, during the chase, there is also an enrichment of the labeled VLCFA, compared to labeled C16 and C18 fatty acids, in the plasma membrane. Lipid analysis revealed that the transferred lipids were essentially neutral lipids, phosphatidylcholine and phosphatidylethanolamine (Moreau et al.,1988b).

Monensin-induced accumulation of lipids at the level of the Golgi complex

The above results and others resulting from double-labeling experiments and pulse-chase experiments using short chase times (Moreau et al.,1988a), suggest that an ER-Golgi-Plasma Membrane pathway could be one of the routes used by the cell to transfer lipids and VLCFA to the plasma membrane.The involvement of the Golgi complex was further investigated using monensin, a carboxylic ionophore known to disturb the vesicular transport of glycoproteins at the level of the Golgi apparatus (Farquhar,1985; Robinson 1985), due to a swelling of the Golgi stacks (Tartakoff,1983; Morré et al.,1983; Morré et al.,1985).

In order to study the penetration of monensin into leek seedlings, we first devised techniques to extract, isolate and quantify the drug (Bertho et al.,1989). The assay of monensin allowed, for the first time, to determine the amount of the drug that really enters into the cells and the proportion of the drug that is incorporated into the microsomal membranes. It was demonstrated that only a very small amount of monensin is associated with the membranes, the bulk of the drug being recovered in the supernatant. The weak binding of monensin to microsomal membranes, compared to other amphipathic molecules (Conrad & Singer,1979), led to the conclusion that monensin behaves as a classical amphipathic molecule (Bertho et al.,1989).

Table II. Effect of monensin on the intracellular distribution of lipids

Membrane fractions	Relative spec. radioact. of total fatty acids Control	+ Monensin (μM) 0.5	 1	 2
Light membranes (1.08-1.09 g/ml)	1	0.84	0.63	0.71
ER membranes (1.12 g/ml)	1	1.1	1.1	1.2
Golgi membranes (1.13-1.14 g/ml)	1	1.3	1.67	1.77

Relative specific radioactivities were determined as explained in the text and those obtained without monensin were taken as equal to 1.

The effect of monensin on the overall lipid synthesis was studied in order to determine the concentrations of the drug having no, or only slight, effects on the level of lipid synthesis. We showed that, using short incubation periods (30min) and monensin concentrations of up to 2 μM, the overall fatty acid synthesis from (1-^{14}C)acetate is unchanged, thus allowing to study the effect of the drug on the intracellular distribution of the labeled lipids (Bertho et al.,1989). The analysis of the effect of monensin was carried out using three concentrations of the drug (0.5, 1 and 2 μM). After incubation with (1-^{14}C)acetate,with or without monensin, different batches of seedlings were homogenized for subcellular fractionation. The specific radioactivities of the total fatty acids of the different fractions were then determined. The *spec. label of fraction (cpm/mg prot.) / spec. label of gradient (cpm/mg prot.)* ratios were calculated for each fraction and these values (relative specific radioactivities) were compared for the various conditions (with or without monensin). We observed a dose-dependent increase of labeled fatty acids in the 1.13-1.14 g/ml Golgi-enriched fraction (table II). During the same time, the amount of labeled fatty acids decreased in the light membranes (1.08-1.09 g/ml) and remained more or less constant at the site of lipid synthesis (1.12 g/ml).The increase of labeled fatty acids in the Golgi fraction reaches 67% and 77% for monensin concentrations of 1 and 2 µM, respectively. For the same concentrations, the decreases in the light membranes were 37% and 29%.

Using pulse-chase experiments, the involvement of the intracellular lipid transport by membrane flow has been studied in several systems (Moreau et al.,1988a; Mills et al.,1984; De Silva & Siu,1981), and the participation of the Golgi complex in chinese hamster lung fibroblasts (Lipsky & Pagano,1985), brain cells (Townsend et al.,1984) and epithelial cells (van Meer et al.,1987), has been analysed. The results shown in table II demonstrate, for the first time in plant cells,that monensin induces an intracellular redistribution of the acetate-labeled lipids and in particular, their accumulation in the Golgi complex. Such studies will able to determine which lipids are concerned and in what proportion.

Are the lipids accumulated in the Golgi complex,upon monensin addition, destined to be transported to the plasma membrane ?

To answer this question, the effect of monensin on the lipid distribution between the plasma membrane and the intracellular membranes was studied. After *in vivo* incubations carried out in the presence or absence of monensin, membrane fractionation was performed and the plasma membrane was prepared by phase partition as described earlier (Moreau et al.,1988b).Two different experimental conditions were used: the seedlings were either incubated for 30 min with or without 1 µM monensin in the presence of labeled acetate or, were preincubated for 30 min with or without 1 µM monensin prior to the incorporation of labeled acetate for 30 min.The specific radioactivities of total lipids were determined for the whole microsomes and the plasma membrane fraction.The relative specific radioactivities of lipids of the plasma membrane and the whole microsomes were calculated for each experimental condition.

These preliminary experiments showed that the ratio decreases by around 25% when monensin is only present during the incorporation of labeled acetate and, by at least 45% when the seedlings are preincubated for 30 min with the drug.These data show that the arrival of newly synthesized lipids to the plasma membrane is inhibited by monensin and that this effect is more important when the seedlings are preincubated with the drug.Moreover, the fatty acid analysis showed that the VLCFA could be more affected than the C16 and C18 fatty acids, in good agreement with the fact that VLCFA are enriched in the plasma membrane (Cassagne & Lessire,1978; Moreau et al.,1988b).

These results, compared with those obtained for the Golgi fraction (see above), suggest that the decrease of neosynthesized lipid transfer to the plasma membrane could be correlated, at least partly, with the accumulation of these molecules in the Golgi fraction.

The in vitro cell-free system approach

In order to study the mechanisms by which the intermembrane transfer could take place, it is required that an *in vitro* reconstitution of the transfer between donor and acceptor membranes be attempted.The cell-free approach, suitable for the study of protein transport (Nowack et al.,1987; Rothman,1988), could also be useful in the case of lipids.

The first results obtained by this new approach in the case of lipids, using membrane fractions prepared from etiolated leek seedlings, are presented.The methodology used is derived directly from that described by Nowack and coworkers (1987). The acceptor membranes (Golgi membranes) are fixed on nitrocellulose filters and the labeled donor membranes (ER membranes) are incubated with the acceptor membranes with or without ATP and other various cofactors, to produce the transition vesicles which can then fuse with the acceptor membranes.The study of the reconstitution of vesicle formation by the donor membrane and the fusion with the acceptor membrane, allows to determine how the various factors (membrane proteins, cytosol, ATP and other nucleotides, pH, temperature.....) are, or not, implicated in these events (Nowack et al.,1987; Paulik et al.,1988).

In order to reconstitute the transfer of membrane lipids between the ER and the Golgi membranes in leek seedlings, the following steps were realised. Firstly, we determined the best conditions for fixing the Golgi membranes to nitrocellulose filters. Then, before incubation with the labeled donor membranes, filters were saturated by 5% BSA to avoid important non specific binding of donor membranes to the filters. The efficiency of the saturation by BSA was determined and we found that only 1.2 % ± 0.1 % (n=8) of the total label of the donor membranes was fixed to the filters after saturation.

For the reconstitution of membrane transfer , the donor membranes (ER membranes = 1.12 g/ml fraction) and the acceptor membranes (Golgi membranes = 1.13-1.14 g/ml fraction) were incubated together under the following conditions: **i.** The Golgi membranes, before being fixed to nitrocellulose filters, were boiled for 5 min at 100°C. The incubation with the donor membranes was carried out in the presence of 1mM ATP and an ATP-regenerating system.

ii. The ER and Golgi membranes were incubated in the absence of ATP. **iii.** The ER and Golgi membranes were incubated in the presence of 1 mM ATP and an ATP-regenerating system.

The results are presented in the table III. In order to compare the data obtained under the various conditions from nine independent experiments, the radioactivities measured on the filters corresponding to the boiled Golgi membranes were given the arbitrary value of 100 and the other values (-ATP and +ATP) were calculated using this value as the reference.

The comparison of the data obtained with or without ATP show that there is a stimulation of the transfer by ATP.The increase of radioactivity found in the absence of ATP (as compared to the boiled controls) can be explained, either by the presence of preformed transition vesicles in the donor membranes, and/or by the presence of endogenous ATP allowing the formation of some transition vesicles.

Table III. Assay of reconstitution of membrane lipid transfer between the ER and the Golgi membranes

Experimental conditions	Lipids transferred (arbitrary units)
Boiled Golgi membranes	100
- ATP	132 ± 3 (n=9)
+ ATP	162 ± 10 (n=9)

The arbitrary units employed were calculated as explained in the text (see above)

We have calculated that the amount of labeled lipids transferred in the assays was about 3000 dpm / mg prot. of Golgi membranes / h ± 900 (n=9). The mean percentage of radioactivity transferred represents 0.61 % ± 0.07 % (n=9) of the total radioactivity of the donor membranes. This percentage is of the same order of magnitude as those (between 0.5 % and 1.5 %) obtained by Nowack and coworkers with membrane fractions prepared from rat liver (Nowack et al.,1987). These authors recently isolated the transition vesicles from rat liver by free-flow electrophoresis (Paulik et al.,1988) and demonstrated that they function as an intermediate compartment in the membrane transfer between the ER and the Golgi complex, the percentage of membrane transfer obtained with the purified transition vesicles reaching 10 to 30 % (Paulik et al.,1988).
Such *in vitro* cell-free systems open new perspectives in the study of the cellular mechanisms and factors that govern membrane biogenesis and membrane inter-relations in animal and plant cells.

References

Bertho P, Moreau P, Juguelin H, Gautier M , Cassagne C (1989) Monensin-induced accumulation of neosynthesized lipids and fatty acids in a Golgi fraction prepared from etiolated leek seedlings. Biochim Biophys acta 978: 91-96.

Cassagne C, Lessire R (1978) Biosynthesis of saturated very long chain fatty acids by purified membrane fraction from leek epidermal cells. Arch Biochem Biophys 191: 146-152.

Conrad MJ, Singer SJ (1979) Evidence for a large internal pressure in biological membranes. Proc Natl Acad Sci 76: 5202-5206.

De Silva NS, Siu CH (1981) Vesicle-mediated transfer of phospholipids to plasma membrane during cell aggregation of *Dictyostelium discoideum*. J Biol Chem 256: 5845-5850.

Farquhar MG (1985) Progress in unraveling pathways of Golgi traffic. Ann Rev Cell Biol 1: 447-488.

Heape AM, Juguelin H, Boiron F, Cassagne C (1985) Improved one-dimensional thin-layer chromatographic technique for polar lipids. J Chromatogr 322: 391-395.

Lipsky NG, Pagano RE (1985) Intracellular translocation of fluorescent sphingolipids in cultured fibroblasts: endogenously synthesized sphingomyelin and glucocerebroside analogues pass through the Golgi apparatus en route to the plasma membrane. J Cell Biol 100: 27-34.

Mills JT, Furlong ST, Dawidowicz EA (1984) Plasma membrane biogenesis in eukariotic cells: translocation of newly synthesized lipid. Proc Natl Acad Sci 81: 1385-1388.

Moreau P, Bertho P, Juguelin H, Lessire R (1988a) Intracellular transport of very long chain fatty acids in etiolated leek seedlings. Plant Physiol Biochem 26: 173-178.

Moreau P, Juguelin H, Lessire R, Cassagne C (1988b) Plasma membrane biogenesis in higher plants: *in vivo* transfer of lipids to the plasma membrane. Phytochem 27: 1631-1638.

Morré DJ, Boss WF, Grimes H, Mollenhauer HH (1983) Kinetics of Golgi apparatus membrane flux following monensin treatment of embryogenic carrot cells. Eur J Cell Biol 30: 25-32.

Morré DJ, Minnifield N, Mollenhauer HH (1985) Kinetics of monensin-induced swelling of Golgi apparatus cisternae of H-2 hepatoma cells. Eur J Cell Biol 37: 107-110.

Nowack DD, Morré DM, Paulik M, Keenan TW, Morré DJ (1987) Intracellular membrane flow: reconstitution of transition vesicle formation and fusion in a cell free system. Proc Natl Acad Sci 84: 6098-6102.

Paulik M, Nowack DD, Morré DJ (1988) Isolation of a vesicular intermediate in the cell-free transfer of membrane from transitional elements of the endoplasmic reticulum to Golgi apparatus cisternae of rat liver. J Biol Chem 263: 17738-17748.

Robinson DG (1985) Plant membranes: endo- and plasma membranes of plant cells. John Wiley & Sons, Inc., New York.

Rothman JE (1988) Reconstruction of organelle function in cell-free systems. In: Morré DJ, Howell KE, Cook GMW, Evans WH (eds) Cell-free analysis of membrane traffic. Alan R. Liss, Inc., New York, p 311.

Tartakoff AM (1983) Perturbation of the structure and function of the Golgi complex by monovalent carboxylic ionophores. Meth Enzymol 98: 47-59.

Townsend LE, Benjamins JA, Skoff RP (1984) Effects of monensin and colchicine on myelin galactolipids. J Neurochem 43: 139-145.

van Meer G, Stelzer EHK, Wijnaendts-van-Resandt RW, Simons K (1987) Sorting of sphingolipids in epithelial (Madin-Darby canine kidney) cells. J Cell Biol 105: 1623-1635.

EVIDENCE FOR AN INTERMEMBRANE TRANSPORT OF LIPIDS TO PERIPHERAL NERVE MYELIN *IN VIVO* BY A NOVEL *PULSED WAVE* PROCEDURE.

Anthony HEAPE, Françoise BOIRON and Claude CASSAGNE.

Institut de Biologie Cellulaire et Neurochimie du CNRS.
1, rue Camille Saint-Säens,
33077 Bordeaux cedex.
FRANCE

Three principal methods have been employed for the study of intracellular transport in nervous tissues : autoradiography associated with electron microscopy, time-staggered double-isotope labelling and isotopic pulse-chase experiments. Although the autoradiographical (Gould and Dawson, 1976; Gould et al., 1987; Kumara-Siri and Gould, 1980; Rawlins, 1973) and the double-labelling methods (Konat et al., 1985; Pereyra et al., 1983; Shimomura et al., 1984) can provide useful complementary information, only the isotopic pulse-chase experiments (Benjamins and Iwata, 1979; Burkart et al., 1982; Konat, 1981; Konat et al., 1985), associated with the use of intracellular transport inhibitors, would be able to demonstrate unambiguously the existence of intermembrane transfer events implicated in membrane biogenesis. No *in vivo* studies of peripheral nerves have been carried out using this method, most probably because of the difficulty in controlling the experimental conditions.

Recently, we carried out an *in vivo* study of palmitate metabolism in the sciatic nerves of one-month-old mice (Heape et al., 1989). The results of this study showed that the administration of 1-4 pmoles of [^{3}H]palmitate to mouse sciatic nerves by intraneural injection leads to the synthesis of labeled polar lipids (phospholipids + galactosphingolipids), the quantity of which increased linearly with the amount of substrate injected, indicating that the enzyme systems synthesizing polar lipids were not saturated by the exogenous substrate. Furthermore, the physiological disturbance created by the presence of the exogenous substrate was of short duration, since 90% of the administered [^{3}H]palmitate were metabolized in less than 30 minutes. Thereafter, the labeled free fatty acid pool remained constant. After only 2 hours *in vivo* incubation, nearly 70% of the total labeled products (complex lipids + free fatty acids) recovered from the nerves were polar lipids, and a further 20% were associated with the neutral lipids, excluding free fatty acids. Given the nature of the radioactive products present in the nerves, which indicate that the experimental conditions employed are compatible with the functioning of practically all the major enzyme systems implicated in polar lipid formation, and the high, reproducible level of metabolism of the administered substrate, our conditions can be considered to be close to those encountered in the physiological situation.

NATO ASI Series, Vol. H 40
Dynamics and Biogenesis of Membranes
Edited by J.A.F. Op den Kamp

Finally, the natural elimination of the radioactive substrate by its complete metabolization within a few minutes of its injection, opens up the possibility for the *in vivo* study of intracellular lipid transport implicated in peripheral nerve myelin assembly, by means of a novel method for which we have proposed the name *"pulsed wave"* . This method, although based on the same principle, has important advantages over the isotopic pulse-chase method, since, in the absence of an isotopic chase (whose sole purpose is to "drown" the labeled substrate), it is no longer necessary to submit each nerve to a second traumatizing injection, whose consequences are unknown, and the administration of large quantities of potentially toxic substrates (e.g. fatty acids) to the nerves can be avoided.

Here, we have used the *pulsed wave* procedure, associated with membrane fractionation by isopycnic centrifugation on linear sucrose gradients, to investigate the possibility of demonstrating, *in vivo* , the existence of an intermembrane movement of lipids to the myelin of mouse sciatic nerves.

Experimental Procedure for Pulsed Wave Experiments

Anaesthetized one-month-old normal mice received an intraneural injection (Heape et al.,1986) of [^{3}H]palmitate (46Ci /mmole ; 284000 ± 11600 dpm /nerve) into their left sciatic nerve as described (Heape et al.,1989). After *in vivo* incubation, both sciatic nerves of each mouse were dissected out, pooled (8 mice / experiment) and homogenized in ice-cold distilled water. The sciatic nerve membranes were then fractionated as shown in Fig.1A. After sucrose (refractometry) and protein determination (Bradford, 1976), $CHCl_3$ / CH_3OH / H_2O lipid extraction was performed on all the resulting fractions. Lipids were resolved (Heape et al., 1985) and their radioactivity was determined (Heape et al.,1989). Results are expressed as "dpm / pmol of '[^{3}H]palmitate'(total label) recovered in the sciatic nerve homogenate" (dpm / pmole ; Heape et al.,1989). For the zero time incubation, the sciatic nerves were dissected out and placed in ice-cold water containing 2.27 x 10^6 dpm of [^{3}H]palmitate. All other conditions and treatments were the same as those employed for the *in vivo* experiments.

Results and Discussion

After isopycnic centrifugation on a linear 0 - 1.4 M sucrose gradient, the membranes of myelinated peripheral nerves are distributed all along the gradient. The total protein profile in the gradient (Fig.1B) presents a single peak whose maximum (which is usually considered to represent the myelin) is centred at a position corresponding to a sucrose concentration of 0.533 ± 0.025 M (n=5). A slight shoulder is frequently observed between 0.8 and 1.0 M sucrose. Very little protein is observed at sucrose concentrations lower than 0.3 M. 64% of the total

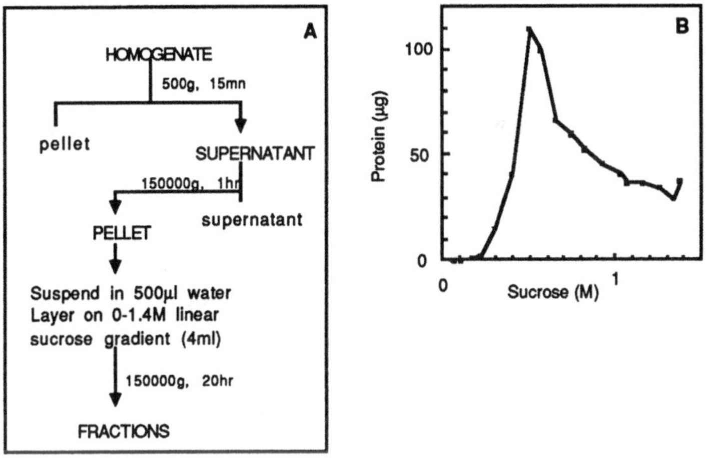

Figure 1

A) Experimental protocol for sciatic nerve membrane fractionation.

B) Typical distribution of total proteins in the linear 0 - 1.4M sucrose gradient after isopycnic centrifugation.

activity of the endoplasmic reticulum (ER) marker, CDP-choline : diacylglycerol choline phosphotransferase (estimated as described by Radominska-Pyrek et al.,1977), was present in the membranes sedimenting between 0.6 and 0.8 M sucrose, while only 12% was associated with the maximum of the adjacent myelin peak. Similarly, the highest stearoyl-CoA elongase activity (a microsomal enzyme probably associated with the ER), measured using a modification of the method of (Boiron et al.,1984), was observed in the membrane fraction (0.6 M sucrose) adjacent to the maximum of the protein peak. In view of these observations, and despite an inevitable cross-contamination of the two types of membrane (myelin and ER), we can consider that these two membranes are reasonably well resolved on the sucrose gradients.

Hence, if the majority of complex lipid acylation takes place in the ER of the sciatic nerves (not yet proven), short *in vivo* incubation periods in the presence of [^{3}H]palmitate should lead to a distribution of the radioactive lipids whose maximum would be situated in the membrane fractions slightly heavier than the myelin. If these lipids are subsequently transfered to, and accumulated in the myelin, then incubation periods much longer than the time required to metabolize all of the available labeled substrate administered to the nerves (30 minutes ; Heape et al.,1989) should lead to a label distribution presenting a maximum at the summit of the myelin peak. For membrane fractions through which the different lipids pass on their way to the myelin, one should be able to observe a transitory increase (wave) of the radioactivity of each of the concerned lipids in these fractions, by analyzing their distribution in the gradients after *in vivo* incubation periods of varying duration. In order to visualize such a process, we carried out a short-term kinetics study of [^{3}H]palmitate metabolism by the sciatic nerves of one-month-old mice *in vivo,* employing the experimental procedure described above.

Total radioactivity recovery from the membrane fractions (73720 ± 2520 dpm/pmole) of the sucrose gradients after isopycnic centrifugation and in the 500g pellet (P500 : 20340 ± 1640 dpm/pmole) did not vary as a function of the length of the *in vivo* incubation period (1 pmole = 101200 dpm). The remaining radioactivity (7530 ± 3165 dpm/pmole) was associated with the 150000g supernatant, of which less than 15% were associated with polar lipids, regardless of the incubation time (3min - 2hr).

Table 1 shows the variations of the total radioactivities of the polar lipids in the P500 and the membrane fractions of the sucrose gradients, and of 3 major polar lipids (phosphatidylcholine [PC], phosphatidylethanolamine [PE] and the hydroxycerebrosides [H-CB]) in the sucrose gradients as a function of time.

TABLE 1. POLAR LIPID LABELING IN THE MEMBRANE FRACTIONS AS A FUNCTION OF TIME.

	RADIOACTIVITY (dpm / pmole)					
Incubation time (min)	0	3	5	30	60	120
Polar lipids P500	364	3365	4662	8004	13705	15148
Polar lipids gradient	1730	15278	16381	36734	50741	59203
PC gradient	105	6500	7368	21534	25973	29101
PE gradient	173	1782	2021	6062	8974	10098
H-CB gradient	65	170	292	1030	2891	3234

These results illustrate the rapid initial incorporation of the radioactivity into the polar lipids during the first 30 min., as observed previously for the whole sciatic nerves (Heape et al., 1989), followed by a slower, but continued polar lipid synthesis up to 2hr.

Figure 2 shows the radioactive PC profiles in the gradients during the rapid incorporation phase. It is clear from these results that [^{3}H]PC appears more rapidly in the membrane fractions (enriched in endoplasmic reticulum [ER]?) slightly heavier (0.6 - 0.8 M sucrose) than the peak myelin fraction (0.55 M sucrose), but that this latter fraction subsequently contains more labeled PC than any other fraction.This observation is compatible with the existence of an intermembrane transfer of PC from the ER-enriched fractions to the myelin. This notion is further supported by the slight decrease of label in the 0.6 - 0.8 M region of the gradient between 3 and 5min., and the simultaneous, but larger increase of the label in the lighter fractions. However, the increase in the amount of [^{3}H]PC in the light (myelin) fractions could also result from an intrinsic acylation process in these membranes, the initial "delay" in the the labeling of the PC being due to the time required for the substrate to reach the acylating enzyme

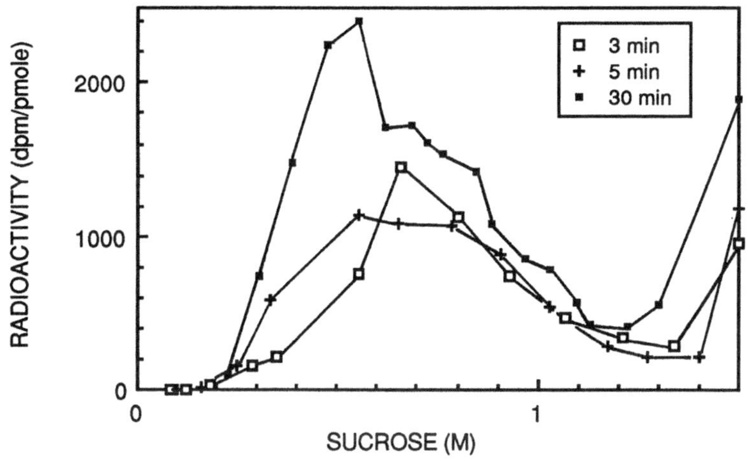

Figure 2 : Distribution of labeled PC in the sucrose gradients during the rapid incorporation phase.

systems in the myelin. Similar evolutions of the radioactivity profiles were observed for PE and H-CB.

The results obtained for the longer incubation periods (1 and 2hr), when the rate of labeled polar lipid synthesis is considerably reduced (see Table 1), allowed to demonstrate the existence of a net transfer of tritiated lipids from the membranes in the 0.6 - 1.0 M region of the gradient to the other membrane fractions. Table 2 summarizes the results obtained for PC, PE and H-CB.

For these lipids, significant increases in the amount of label between 1 and 2hr were observed in the 0.35-0.5 M sucrose region ("pre-myelin"), at the summit of the myelin peak and between 1.0 and 1.35 M sucrose. During the same period, the radioactivity decreased in the 0.75-1.0 M region, but the most important decrease was observed in the 0.6-0.75 M region of the gradient (ER?). The label associated with PC, PE and H-CB in the 0-0.35 M region and in the fractions beyond 1.35 M sucrose did not exhibit any significant changes between 1 and 2hr.

Taken together, these results constitute the first strong biochemical evidence for the existence of an intermembrane transport of lipids occurring during the assembly of the PNS myelin membrane *in vivo*. Although the existence of such events could be expected, and has already been suggested (Benes et al., 1973; Gould and Dawson, 1976; Pleasure and Prockop, 1972; Rawlins, 1973), they have never been demonstrated in this biological system, largely because of the high degree of complexity of the system and the correlated difficulty of separating the different membrane compartments. The refinement of the *pulsed wave* method employed here and the development of membrane fractionation procedures adapted to myelinated nerve tissues, will enable us to obtain more detailed knowledge concerning the biogenesis of the myelin sheath *in vivo*.

TABLE 2. POLAR LIPID SYNTHESIS AND TRANSFER DURING THE SLOW LABEL INCORPORATION PHASE.

LIPID	A (dpm/pmole)	B (dpm/pmole)	C (dpm/pmole)
PC	3128	5317	2189
PE	1124	1794	670
H-CB	343	743	400

A :Total radioactivity incorporated into PC, PE and H-CB in the sucrose gradient membrane fractions between 1 and 2hr.

B :Sum of the radioactivities incorporated into PC, PE and H-CB in the membrane fractions exhibiting an increase in the label of these lipids between 1 and 2hr.

C = B-A : Increase in the radioactivity of PC, PE and H-CB in the "accumulating" membrane fractions due to a net transfer of these lipids from the 0.6 - 1.0M region of the sucrose gradient = MINIMUM TRANSFER ACTIVITY

References

Benjamins JA, Iwata R (1979) Kinetics of entry of galactolipids and phospholipids into myelin. J Neurochem **32**:921-926

Boiron F, Darriet D, Bourre JM, Cassagne C (1984) Decreased biosynthesis of saturated C20-C24 fatty acids by the Trembler mouse sciatic nerve. Neurochem Int **6**:109-116

Bradford MM (1976) A rapid and sensitive method for the quantitation of microgram quantities of protein utilizing the principle of protein-dye binding. Anal Biochem **72**:248-254

Burkart T, Caimi L, Siegrist HP, Herschkowitz NN, Wiesmann UN (1982) Vesicular transport of sulfatide in the myelinating mouse brain. Functional association with lysosomes. J Biol Chem **257**:3151-3156

Gould RM, Dawson RMC (1976) Incorporation of newly formed lecithin into peripheral nerve myelin. J Cell Biol **68**:480-496

Gould RM, Holshek J, Silverman W,Spivak WD (1987) Localization of phospholipid synthesis to Schwann cells and axons. J Neurochem **48**:1121-1131

Heape MA, Juguelin H, Boiron F, Cassagne C (1985) An improved one-dimensional thin-layer chromatographic technique for polar lipids. J Chromat **322**:391-395

Heape A, Boiron F, Cassagne C (1989) High uptake and rapid metabolism of palmitate in peripheral nerves of normal and Trembler mice in vivo : similarities and differences. Neurochem Int : *In press*

Konat G (1981) Intracellular translocation of newly synthesized myelin proteins in the rat brain stem slices. Exp Neurol **73**:254-266

Konat G, Divkaran P, Samorajski T, Wiggins RC (1985) Intracellular translocation of myelin proteolipid protein. J Neurochem **44**: 1500-1510

Kumara-Siri MH, Gould RM (1980) Enzymes of phospholipid synthesis : axonal versus Schwann cell distribution. Brain Res **186**:315-330

Pereyra PM, Braun PE, Greenfield S, Hogan (1983) Studies on subcellular fractions which are involved in myelin assembly : labeling of myelin proteins by a double radio-isotope approach indicates developmental relationships. J Neurochem **41**:974-988

Pleasure DE, Prockop DJ (1972) Myelin synthesis in peripheral nerve in vitro : sulphatide incorporation requires a transport lipoprotein. J Neurochem **19**:283-295

Radominska-Pyrek A, Strosznajder J, Dabrowiecki Z Goracci G, Chojnacki T, Horrocks LA (1977) Enzymic synthesis of ether types of choline and ethanolamine phosphoglycerides by microsomal fractions from rat and liver. J Lipid Res **18**:53-58

Rawlins FA (1973) A time-sequence autoradiographic study of the in vivo incorporation of [1,2-^{3}H]cholesterol into peripheral nerve myelin. J Cell Biol **58**:42-53

Shimomura K, Yahara S, Kishimoto Y, Benjamins JA (1984) Metabolism of cerebroside and sulfatides in subcellular fractions of developing rat brain. Biochim Biophys Acta **795**:265-270

PHOSPHOLIPID LOCALIZATION AND MOBILITY IN PLASMODIUM INFECTED ERYTHROCYTES.

J.A.F. Op den Kamp, G. Moll, A.P. Simões and B. Roelofsen
Centre for Biomembranes and Lipid Enzymology.
State University of Utrecht
Padualaan 8
3584 CH Utrecht
The Netherlands

Upon infection of erythrocytes with Plasmodium, the malaria causing parasite, the red cell membrane undergoes a variety of changes. Proteins, encoded by the parasitic DNA are inserted into the membrane (Howard 1988, Hommel and Semoff, 1988), membrane permeability is increased (Tanabe *et al*, 1983; Sherman, 1979) pores (Ginsburg *et al*, 1986) are formed, the mobility of phospholipids is increased (Taraschi *et al*, 1986; Beaumelle *et al*, 1988; Deguercy *et al*, 1986; van der Schaft *et al*, 1987; Moll *et al*, 1988; Haldar *et al*, 1989) and the ultrastructural appearance of the cell can be modified by the development of knobs on the membrane surface (Howard, 1988). Recently, a series of studies was undertaken to investigate the fate of the phospholipid complement of the host cell membrane and to establish possible, parasite induced, modifications in the composition as well as the organization of these basic membrane constituents. Here we present a short overview of the most relevant observations. Emphasis will be laid on three aspects of our work: (i) the localization of phospholipids in the membrane of the erythrocyte following infection; (ii) the transfer of phospholipids from the environment into the infected red cell, and (iii) the possible alterations in phospholipid species composition of the host erythrocyte membrane. Central to these studies are observations that parasitised erythrocytes contain, at the final stage of infection, about 6 times as much phospholipid as the non-infected cells (van der Schaft *et al*, 1987) and that the parasite, although equipped with a

NATO ASI Series, Vol. H 40
Dynamics and Biogenesis of Membranes
Edited by J.A.F. Op den Kamp

complete set of enzymes to synthesize complex phospholipids, cannot synthesize its own fatty acids (Sherman, 1979).

Phospholipid localization in the membranes of parasitised erythrocytes.

The polar headgroup composition of the phospholipids in the monkey erythrocyte membrane does not change upon infection with *Plasmodium knowlesi*. Using a Percoll/sorbitol density gradient centrifugation technique (Kutner *et al,* 1985) to separate parasitized from non-parasitized cells, followed by the isolation of red cell membranes with Affi-gel beads (Vial *et al,* 1989), it was possible to determine phospholipid compositions of the various fractions. Table I shows that sphingomyelin (SPH), phosphatidylcholine (PC), phosphatidylethanolamine (PE) and phosphatidylserine (PS) are present in similar relative amounts before and after infection.

Table I: Phospholipid compositions (mole %) of erythrocyte membranes of schizont stage infected erythrocytes and normal cells, both isolated on Affigel beads, and of normal erythrocytes subjected directly to lipid extraction are presented. The number of determinations is given between parentheses.

	erythrocyte membrane of schizont infected cells (2)	normal erythrocyte membrane (2)	normal erythrocytes (5)
Sphingomyelin	16.7±1.5	16.2±1.2	15.5±0.3
Phosphatidylcholine	41.1±0.8	39.3±3.0	40.2±3.2
Phosphatidylethanolamine	27.6±0.5	26.3±2.3	25.0±2.6
Phosphatidylserine/-inositol	18.1±1.7	18.2±2.0	15.3±0

In contrast, a comparison with the phospholipid composition of the parasites itself (Table II) shows a remarkable difference: the parasite does not contain any sphingomyelin, which is in agreement with previous observations (Rock e*t al,* 1971; De Zeeuw *et al*, 1972; Mc Clean *et al,*

1976). This fact, of which the physiological implication is not known, appears to be a very usefull property of the system because it allows to calculate -even for each individual class - which fraction of the phospholipids that can be extracted from the infected red cells (red cell membranes + parasite membranes) should be assigned to the host cell membrane.

Table II: Phospholipid content of uninfected and *Plasmodium knowlesi* infected erythrocytes. Uninfected and trophozoite infected erythrocytes were counted in a Neubauer haemacytometer at 0.05 % hematocrit. Cells were pelleted and subjected to lipid extraction. Phospholipid analyses were performed according to Broekhuyse *et al*, (1969) and Rouser *et al*, (1970). Data are the means of three measurements and expressed as nmol phospholipid/ 5.10^8 cells ± SD values.

	uninfected cells	infected cells
Sphingomyelin	55.2 ± 7.6	58.4 ± 4.6
Phosphatidylethanolamine	101.7 ± 6.0	569.9 ± 11.8
Phosphatidylcholine	161.3 ± 19.1	796.6 ± 25.5
Phosphatidylserine/-inositol	63.0 ± 7.8	165.5 ± 18.7
Cardiolipin	0.0 ± 0.0	20.7 ± 8.6
Phosphatidic acid	3.1 ± 3.0	5.7 ± 4.2
Total Phospholipid	384.3	1616.8

Application of the commonly used tools to locate the phospholipids in the membranes of parasitised cells results in a mixed set of data. As indicated in the introduction, the membrane properties, including their permeability, are changed upon infection and as a result the data obtained with phospholipases, chemical probes and exchange proteins have to be interpreted carefully.

Treatment of the cells with sphingomyelinase C is the only approach which gives straight forward results: 80 % of this phospholipid can be hydrolysed in intact cells and, since the parasite does not contain any SPH, the residual 20 % must be present in the inner layer of the erythrocyte membrane. Treatment of infected cells with bee venom and/or *Naja naja* phospholipase A_2 also results in a partial hydrolysis of PC and PE. However, no plateau in the

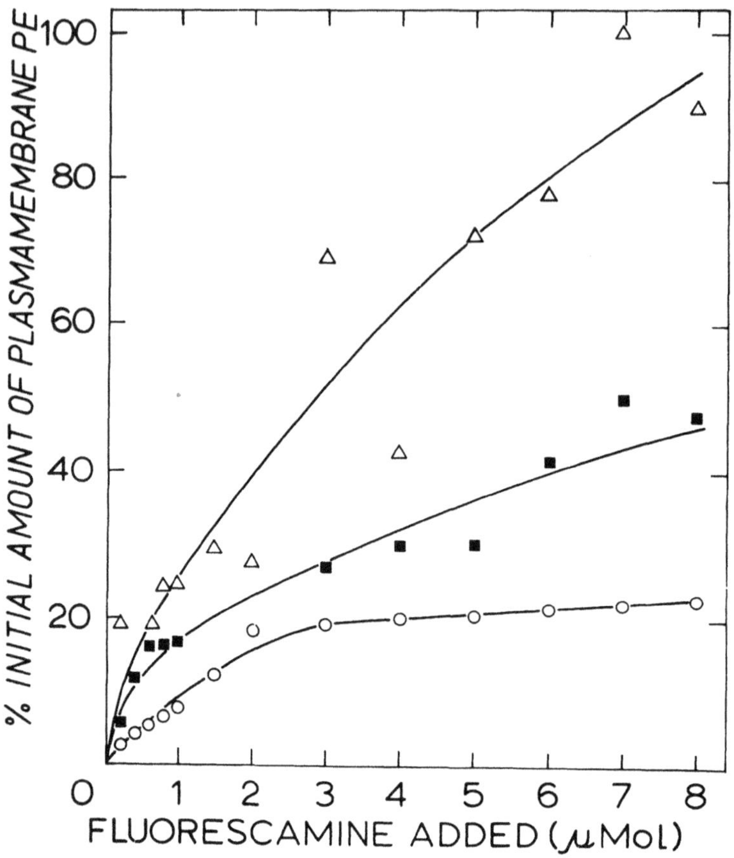

Figure 1. Concentration dependent labeling of PE by fluorescamine in the erythrocyte membrane of normal red blood cells (o) at room temperature and early trophozoit stage infected cells (■) and a mixture of 60% ring stage and 40% schizont stage infected cells (Δ) at 0-4°C. The increase of the fluorescamine derivative of PE is expressed as the increase of the initial amount of PE present in the erythrocyte membranes of the cells calculated by using sphingomyelin (which is not present in the parasite) as an internal standard. Labeling of unparisitized cells from malaria infected blood and erythrocytes from monkeys that have been cured from malaria by chloroquine treatment, shows the same pattern as labeling of normal erythrocytes.

hydrolysis patterns can be observed in these cases. In contrast to incubations with control erythrocytes, hydrolysis continues, eventually resulting in lysis of the parasitized cells. It is impossible therefore to interprete these results in terms of a particular transbilayer distribution of PC and PE. The only conclusion which can be drawn from these experiments is that no PS is hydrolysed under these conditions; that is: before lysis of the parasitized cells starts to occur. Aminogroup containing lipids can also be localized with fluorescamine, and in control monkey erythrocytes 18 % of the PE and none of the PS is labelled by this enzyme. Similar data are found for parasitized cells, although in these experiments the labelling pattern is more complicated due to penetration of the probe into the cells. As is obvious from the results presented in Fig. 1, the conclusion that 18 % of the PE is present in the outer monolayer of the membrane of trophozoite infected cells can be obtained by extrapolation only, whereas in the

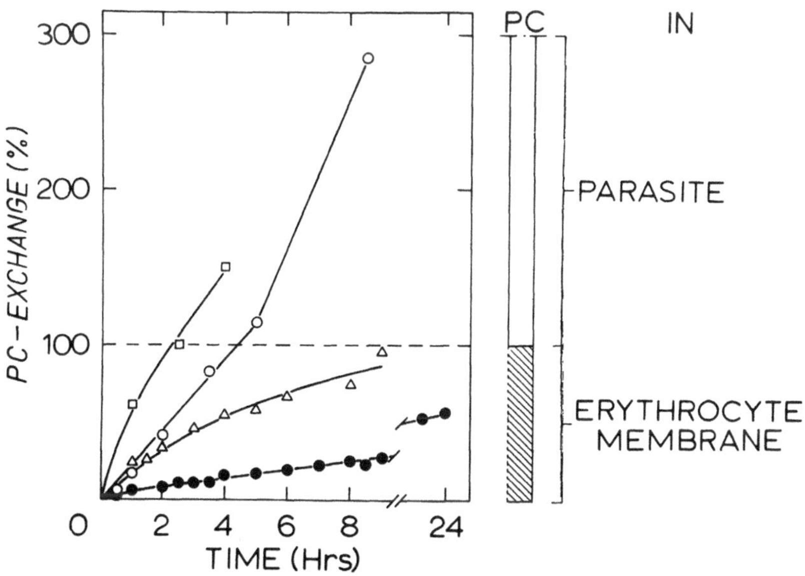

Figure 2. Exchangeability of phosphatidylcholine in intact normal erythrocytes (●), early trophozoite stage infected cells (Δ), a mixture of 60% ring stage and 40% schizont stage infected cells (o) and late trophozoit stage infected cells (□). The PC specific exchange protein was used to exchange PC between (^{14}C-PC) containing PC/PA/cholesterol vesicles and the cells. The exchangeability of PC is measured by determining specific radioactivities of PC in the cells as a function of time. The data are expressed as a percentage of the total amount of PC present in the erythrocyte membrane, calculated by using sphingomyelin (which is absent in the parasite) as an internal standard. The bar indicates the amount of PC present in the erythrocyte membrane and the parasite of parasitized cells.

case of cells which contain parasites in the schizont stage the results are not at all conclusive. Attempts to localize the PC with the help of a PC-specific transfer protein (PC-TP) from bovine liver failed as well.

A first prerequisite for the application of this tool is that transbilayer movement of the PC in the membrane is slow in comparison with the rate of PC exchange. The data presented in Fig. 2 demonstrate that this condition is not fullfilled in case of parasitized erythrocytes.

Already at the trophozoite stage of infection, externally added PC which is introduced into the cells by means of the PC-TP, appears to accumulate in the outer, as well as in the inner layer of the erythrocyte membrane and -at a later stage of infection- vast amounts of the PC even appear to reach the parasites.

In order to circumvent the above difficulties, we have applied the following approaches. In the first place we have used a chemically modified phospholipase with an extremely high penetrating capacity. As a result, this phospholipase can hydrolyse the phospholipids in the outer membrane layer of the erythrocyte within minutes (van der Wiele *et al,* 1988a), and in such short incubation periods the effects of phospholipid transbilayer movements will be minimal. The enzyme used is an acylated derivative of pig pancreatic phospholipase A_2, containing a palmitoyl residue attached to the lysine at position 116 of the otherwise fully amidinated protein (van der Wiele *et al*,1988b). Incubations of trophozoite and schizont infected monkey erythrocytes with this so-called Pal-AMPA resulted in clearly discernible, readily hydrolyzable pools of PC and PE. For PC, this pool accounted for 25-30 % of all of the host cell plasma membrane, which is slightly less than the fraction of PC that can be degraded in intact non-parasitized cells. In agreement with the preliminary results derived from the fluorescamine experiments, about 15 % of the PE in the host cell plasma membrane appeared to be accessible for Pal-AMPA, whereas virtually no PS was hydrolysed.

As a second approach we used a non-destructive and non-invasive method to assess the amount of PS present at the outer surface of the infected erythrocytes. The method is based on the observation that the formation of the so-called prothrombinase complex (composed of prothrombin, activated coagulation factors V and X and Ca^{2+}) and thereby the conversion of prothrombin into thrombin, is critically dependent on the presence of membrane bound PS (Bevers *et al*, 1982). When it is used in a comparative mode with e.g. open ghosts, it can give a fair indication of the percentage of the total amount of PS which is externally located. In agreement with the data obtained with the phospholipases, this assay detected less than 1 % of the PS present in the erythrocyte membrane to be located in the outer layer of the plasma membrane of the infected cell.

In summary, the conclusion can be made that the growth and development of *Plasmodium knowlesi* within the monkey erythrocyte hardly effects the regular distribution of phospholipids in the host cell plasma membrane. All of the PS, and most of the PE are in the inner layer whereas 80 % of the SPH is present in the outer layer. The precise distribution of PC is not known yet, due to the fact that infected cells do not stand the combined treatment of sphingomyelinase C plus phospholipase A_2.

All the above strategies were furthermore used to study the localization of phospholipids in a number of control cells, i.e. (i) cells from a monkey which was not infected with the parasite; (ii) non-parasitised cells from an infected monkey, separated from the parasitised cells by a specific centrifugation technique (Kutner *et al,* 1985); and (iii) cells from a monkey which, after infection, thad been cured from the disease with chloroquin. None of these cells showed any abnormality with respect to the distribution of phospholipids over the two membrane layers.

These observations are not surprising in view of the consideration that it will be only in the interest of the intra-erythrocytic growing parasite to leave the outer membrane surface of its host as much as possible in the original, non-modified state in order to escape the clearing systems which are specialised in removing abnormal cells from circulation. On the other hand, our data are in conflict with results published by others (Gupta *et al,* 1982; Joshi *et al*, 1986, 1987, 1988 ; Schwartz *et al*, 1987). The possible reasons for these discrepancies have recently been discussed (Moll *et al,* submitted). In short: data published by Gupta and coworkers have been obtained with parasitised cells, which may have suffered from a considerable extent of energy deprivation and, as a consequence of the thus caused lack of ATP, had to face a poorly (

if at all) functioning of the aminophospholipid translocase that is present in the red cell plasma membrane (Seigneuret and Devaux, 1984; Tilley *et al,* 1986). In addition it has been shown that malaria infected cells exhibit a reduced spectrin content (Wallach and Conley, 1977, Yuthavong *et al,* 1979). As a consequence, Gupta and colleagues may have performed their experiments using cells in which both mechanisms that are known to be involved in maintaining normal phospholipid asymmetry in the red cell membrane (Middelkoop et al, 1988, 1989) had been strongly impaired. Data published by Schwartz *et al.* (1987) are difficult to interprete, because the way in which they converted percentages of hydrolysis as determined in whole lipid extracts (erythrocytes + parasites), into amounts of phospholipid present in the outer layer of the membrane, is not at all clear.

Transfer of phospholipids into parasitised erythrocytes

It will be clear from the above discussion that much effort has been put in determining the transbilayer distribution of phospholipids in the membrane of *Plasmodium knowlesi* infected erythrocytes. Justification for this effort was found in the fact that the parasite is unable to synthesize fatty acids and therefore requires an import of fatty acids -maybe even of intact phospholipids- for the synthesis of its own membranes, and that this import process might involve (and upset) the native phospholipid organization of the host cell membrane. As the concentration of free fatty acids in the serum is low, it is reasonable to consider the possibility that these essential constituents are imported in the form of di- or triglycerides or as complex phospholipids. The latter possibility was studied in more detail and the basis of these studies was formed by the observations already mentioned above, namely that, in the presence of the PC-TP, exogenous PC can easily be exchanged with the entire PC complement of the erythrocyte membrane and that this exchange process may even involve the PC present in the parasite.

In a series of experiments, the details of which have been described by Moll *et al.*

(1988), the following observations were made.

- When radiolabeled PC containing vesicles are incubated with parasitised cells, radioactive PC can be recovered from the cells, even when cells and vesicles had been incubated in the absence of PC-TP (Table III). This is in contrast to the situation in non-parasitised cells where PC transfer requires the presence of high concentrations of the transfer protein.

Table III. Introduction of phosphatidylcholine into trophozoite infected erythrocytes. [^{14}C choline, ^{3}H oleate] PC with a $^{3}H/^{14}C$ ratio of 2.68 was introduced into the cells using donor vesicles of PC, PA and cholesterol . Incubations, carried out at 37°C, were performed with 60 µl of infected cells , without the transfer protein as well as in the presence of 3 µM PC-TP, in a total volume of 210 µl. In one of the incubations, PC-TP was present for one h. only. Subsequent incubation during 4 h (^{a}) was carried out in the absence of PC donor vesicles. After incubation, cells were washed and lipids were extracted , separated by TLC and analysed. The uptake data are expressed as % of the total PC present in the erythrocyte membranes.

incubation time (h)			
+ PC - TP	- PC - TP	PC transferred	PC : $^{3}H / ^{14}C$
0	1	7.3 %	2.50
0	5	61.8 %	2.50
1	0	36.2 %	2.41
5	0	187.2 %	2.89
1	4^{a}	29.0 %	2.50

When vesicles, containing radiolabeled PC, are incubated with parasitised cells in the presence of the PC-TP, the accumulation of PC is much higher and, when calculated on the basis of the amount of PC present in the plasma membranes of the parasitised cells, amounted to 174 %. This figure demonstrates that exchange is no longer limited to the plasma membrane of the host cell, but that a considerable fraction of the exogenous PC ends up in the parasite as well.

- When double labeled PC is used in these experiments, the ratio of [^{3}H] oleate vs. [^{14}C] choline does not change during or after the uptake of this PC by cells and parasites. Furthermore neither of these two labels can be found in the other phospholipids present in the system.

- When vesicles, containing radiolabeled PE, are incubated with parasitised cells, a substantial part of this PE accumulates in the cells (Fig. 3). When such incubations are carried out in the absence of a lipid transfer protein, none of the radioactivity present in the PE is recovered in other phospholipids. In case a non-specific lipid transfer protein is present in the incubation system, part of the PE appears to be converted into PC.

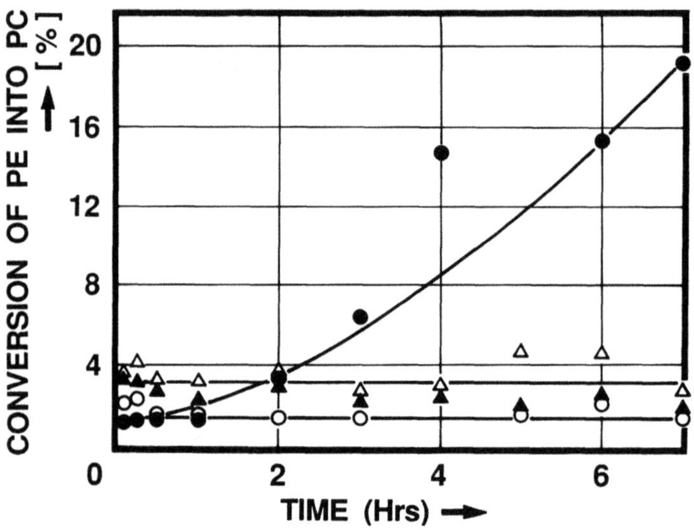

Fig 3.Conversion of PE into PC by intraerythrocytic trophozoites.
[^{14}C] PE was transfered from PC / PA / PE / cholesterol vesicles (0.4 : 0.01 : 0.5 : 0.4, molar ratio) towards trophozoite infected cells and control cells, both in the presence and absence of the nsLTP. Incubations were carried out with 350 µl packed cells in a total volume of 1550 µl. After incubation, the cells were washed, lipids were extracted, separated by TLC and analysed. The conversion of PE into PC is plotted as percentages of the total amount of PE which is incorporated into uninfected cells without nsLTP (Δ); uninfected cells with nsLTP (o); infected cells without nsLTP (▲) and infected cells with nsLTP (●).

- When vesicles, containing radiolabeled PS, are incubated with the parasitised cells, rapid accumulation of this PS can be observed (Fig.4). A large portion of the PS is converted into PE and the rate and extent of this conversion depend on the incubation conditions. PS accumulation in the absence of a transfer protein is accompanied by fast and extensive conversion into PE, whereas this conversion proceeds at a lower rate and to a smaller extent in the presence of the lipid transfer protein.

The conclusions to be drawn from these observations are the following.

The parasitised cell is able, in contrast to the non-parasitised erythrocyte, to accumulate substantial amounts of phospholipids from the environment. Both the imported PC and most of the PE, are used by the parasite as such, whereas PS is largely converted into PE. The latter

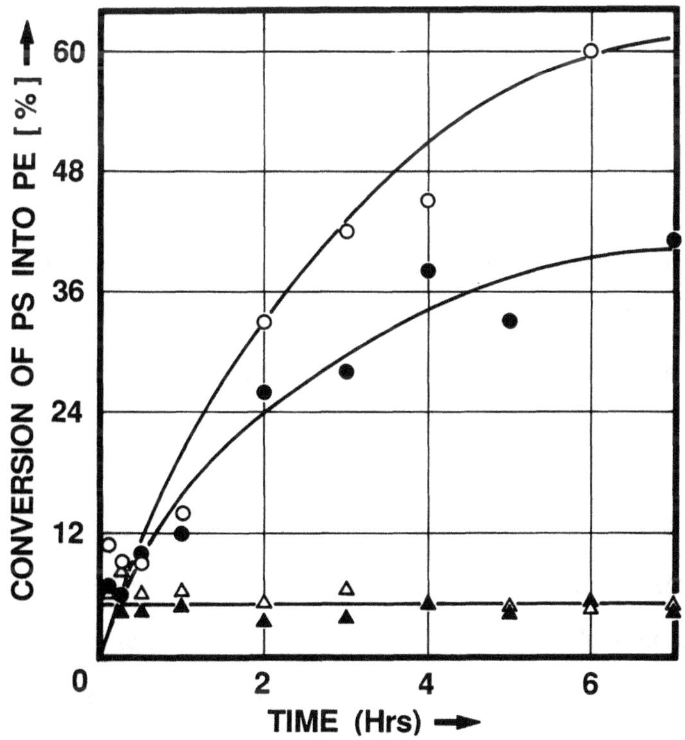

Fig. 4. Conversion of PS into PE by intraerythrocytic trophozoites.
[14C] PS was transfered from PC / PA / PS / cholesterol vesicles (0.4 : 0.01 : 0.78 : 0.4, molar ratio) towards trophozoite infected cells and uninfected cells, both in the presence and absence of the nsLTP. Incubations were carried out with 350 µl packed cells in a total volume of 1465 µl. After incubation, the cells were washed, lipids were extracted, separated by TLC and analysed. The conversion of PS into PE is plotted as percentages of the total amount of PS which is incorporated into uninfected cells without nsLTP (△); uninfected cells with nsLTP (▲); infected cells without nsLTP (○) and infected cells with nsLTP (●).

conversion is in line with the low PS content of the parasites.

At least two mechanisms of phospholipid uptake might be operative. Protein mediated exchange of PC can be followed by a rapid transbilayer movement of the newly introduced PC from the outer towards the inner leaflet of the host cell membrane, subsequently followed by transfer of this PC to the parasite via an unknown pathway. The observation that this accumulation also occurs in the absence of a lipid transfer protein, suggests that mechanisms like pinocytosis or endocytosis are active.

Alterations in phospholipid molecular species of the host erythrocyte membrane

Phosphatidylcholine is the major phospholipid in both unparasitized and *P. knowlesi* parasitized monkey erythrocytes. With the major goal of detecting any eventual alterations in PC molecular species composition associated with parasite development, we have studied its composition in control and parasitized erythrocytes, as well as in the isolated host cell membrane. The characterization of the molecular species composition was accomplished with Reversed Phase High Performance Liquid Chromatography, complemented with Gas Liquid Chromatography, essentially as described by Blank *et al*, (1984). Again using a Percoll density gradient centrifugation technique (Kutner *et al*, 1985) to separate parasitized from unparasitized cells, followed by the isolation of the red cell membrane with Affi-Gel beads (Vial *et al*, 1989), we could show that in trophozoite parasitized cells the PC molecular species composition is identical to that in the isolated host erythrocyte membrane. However, as shown in (Fig. 5), this composition differs from the one of unparasitized erythrocytes that had been recovered from the same batch of infected blood: there is a remarkable increase in 16:0/18:2-PC which seems to be compensated for by a decrease in 18:0/20:4-, 18:0/18:2- and 16:0/16:0- phosphatidylcholines. These differences are exacerbated in the schizont stage of parasite development.

Intriguing is the observation that even unparasitized erythrocytes from sick monkeys exhibit some differences when compared to other controls, such as erythrocytes from monkeys which had never been infected with malaria and erythrocytes from chloroquine treated monkeys. The

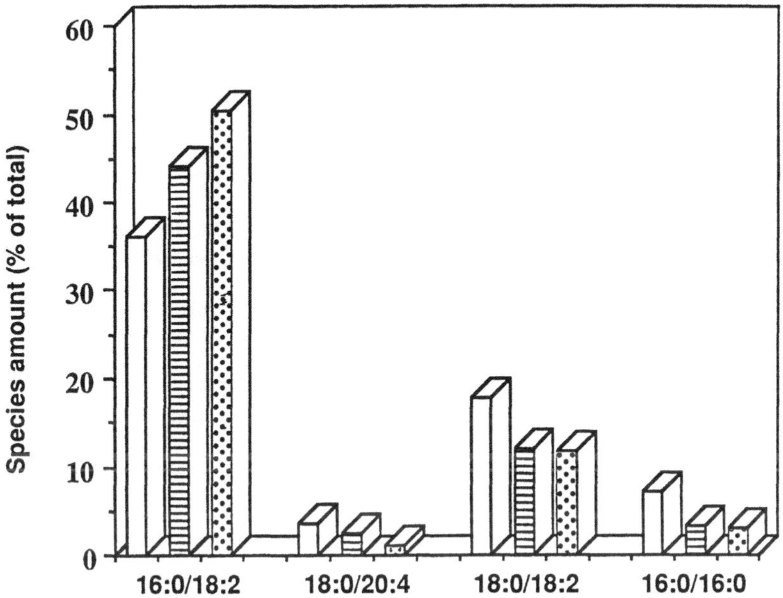

Fig. 5 Main alterations in phosphatidylcholine diacyl species composition of monkey erythrocytes associated with P. knowlesi development.

Cells were obtained from monkeys having a 20-30% parasitemia. Lipids were extracted and PC was isolated by TLC. Diacylglycerobenzoate derivatives were obtained from PC and separated by Reverse Phase HPLC; the amount of each molecular species within multispecies HPLC peaks was determined by GLC (Blank et al, 1984). For each of the four species depicted, the columns correspond, from left to right, to unparasitized, trophozoite-parasitized and schizont-parasitized erythrocytes.

extent by which the PC molecular species composition in the various samples differs from that in the controls, i.e. cells from never infected monkeys can be ordered as follows: erythrocytes from chloroquine treated monkeys ~ unparasitized erythrocytes from infected monkeys < host cell membranes of trophozoite parasitized erythrocytes ~ trophozoite-parasitized erythrocytes < schizont-parasitized erythrocytes. These results demonstrate that *P. knowlesi* can alter the molecular species composition in the parasitized cells and ultimately that in the host cell membrane, what is probably due to a high exchange of PC between the parasite and host cell membrane. Since different molecular species exhibit different fitting properties in the lipid bilayer, and are therefore associated with a specific lipid packing, it is conceivable that by changing the molecular species composition of the host cell membrane, the parasite is manipulating its physiological properties in order to create the conditions that allow the uptake of nutrients and to ensure its survival.

References

Beaumelle BD, Vial HJ, Bienvenue A (1988) Enhanced transbilayer mobility in malaria infected erythrocytes. J Cell Physiol 135 : 94-100

Bevers EM, Comfurius P, van Rijn JLML, Hemker HC and Zwaal RFA (1982) Generation of prothrombin-converting activity and the exposure of phospatidylserine at the outer surface of platelets. Eur J Biochem 122 : 429-436

Blank ML, Robinson M, Fitzgerald V and Snyder F (1984) Novel quantitative method for determination of molecular species of phospholipids and diglycerides.J Chromatogr 298 : 473-482

Broekhuyse RM (1969) Quantitative two dimensional thin-layer chromatogrphy of blood phospholipids. Clinica Chimica Acta 23 : 457-461

Deguercy G, Schrevel J, Duportail G, Laustriat G and Kuhry JS (1986) Membrane fluidity changes in P.berghei-infected erythhrocytes, investigated with a specific plasma membrane fluorescent probe. Biochem Int 12 : 21-32

De Zeeuw RA, Wysbeek J, Rock RC and Mc Cormick G (1972) Composition of phospholipids in Plasmodium knowlesi membranes and in host erythrocyte membranes. Proc Helminth Soc Wash 39 : 412-418

Ginsburg H, Kutner S, Zangwil M and Cabantchik ZI (1986) Selective properties of pores induced in host erythrocyte membrane by Plasmodium falciparum infected cells. Effect of parasite maturation. Biochim Biophys Acta 861 : 194-196

Gupta CM, Alam A, Mathur PN and Dutta GP (1982) A new look at the non parisitized erythrocytes in Plasmodium falciparum infected erythrocytes. Nature 299: 259-261

Haldar K, de Amorim AF, Cross GAM, (1989) Transport of fluorescent phospholipid analogues from the erythrocyte membrane to the parasite in Plasmodium falciparum infected cells. J Cell Biol 108 : 2183-2192

Hommel M and Semoff S (1988) Expression and function of erythrocyte associated surface antigens in malaria. Biology of the Cell 64 : 183-203

Howard RJ (1988) Malarial proteins at the membrane of P.falciparum infected erythrocytes and their involvement in cytoadherence to endothelial cells. Prog Allergy, 41 : 98-147

Joshi P, Dutta GP and Gupta CM (1987) An intracellular simian malarial parasite induces stage dependent alterations in membrane phospholipid organization of its host erythrocyte. Biochem J 246 : 103-108

Joshi P, Alam A, Chandra R, Puri SK, Gupta CM (1986) Possible basis for membrane changes in non parasitized erythrocytes of malaria infected animals.Biochim Biophys Acta 862 : 220-222

Joshi P and Gupta CM (1988) Abnormal phospholipid organization in Plasmodium falciparum infected erythrocytes. Brit J Haematol, 68 : 255-259

Kutner S, Breuer WV, Ginsburg H, Aley SB and Cabantchik ZI (1985) Characterization of permeation pathways in the plasma membrane of human erythrocytes infected with early stages of Plasmodium falciparum : association with parasite development
J Cell Physiol 125 : 521-527

Mc Clean, S, Purdy WC, Kabat A, Sampugna J, de Zeeuw RA and Mc Cormick, G (1976) Analysis of phospholipid composition of Plasmodium knowlesi and Rhesus erythrocyte membranes. Anal Chim Acta 82 : 175-185

Middelkoop E, Lubin BH, Bevers EM, Op den Kamp, JAF, Comfurius P, Chiu DT-Y, Zwaal RFA, van Deenen, LLM and Roelofsen B (1988) Studies on sickled erythrocytes provide evidence that the asymmetric distribution of phosphatidylserine in the red cell membrane is maintained by both ATP dependent translocation and interaction with membrane skeletal proteins. Biochim Biophys Acta 937 : 281-288

Middelkoop E, van der Hoek EE, Bevers EM, Comfurius P, Slotboom AJ, Op den Kamp JAF, Lubin BH, Zwaal RFA and Roelofsen B (1989) Involvement of ATP dependent aminophospholipid translocation in maintaining phospholipid asymmetry in diamide treated human erythrocytes. Biochim Biophys Acta, 981 : 151-160

Moll GN, Vial HJ, Ancelin ML, Op den Kamp JAF, Roelofsen B and van Deenen LLM (1988) Phospholipid uptake by *Plasmodium knowlesi* infected erythrocytes. FEBS Lett 232 : 341-346

Rock RC, Standefer JC, Cook RF, Little W and Sprinz H (1971c) Lipid composition of *Plasmodium knowlesi* membranes : comparison of parasites and microsomal subfractions with rhesus erythrocyte membranes. Comp Biochem Physiol 38 B : 425-437

Rouser G, Fleischer S, Yamamoto A, Fleischer S. (1970) Two dimensional thin layer chromatographic separation of polar lipids and determination of phospholipids by phosphorus analysis of spots. Lipids 5 : 494-496

Van der Schaft PH, Beaumelle B, Vial HJ, Roelofsen B, Op den Kamp JAF and van Deenen LLM (1987) Phospholipid organisation in monkey erytrocytes upon *Plasmodium knowlesi* infection. Biochim Biophys Acta 901 : 1-14

Schwartz RS, Olson JA Raventos-Suarez C, Yee M, Heath RH, Lubin B and Nagel RL (1987) Altered plasma membrane phospholipid organization in Plasmodium falciparum infected human erythrocytes. Blood 60 : 401-407

Seigneuret M and Devaux PF (1984) ATP dependent asymmetric distribution of spin labeled phospholipids in the erythrocyte membrane : relation to shape changes. Proc Nat Acad Sci USA 81 : 3751-3755

Sherman J (1979) Biochemistry of Plasmodium . Microbiol Rev 43 : 453-495

Tanabe K, Mikkelse RB and Wallach DFH (1982) Calcium transport of Plasmodium chabaudi infected erythrocytes. J Cell Biol 93, 680-684

Taraschi TF, Parashar A, Hooks M and Rubin M (1986) Perturbation of red cell membrane structure during intracellular maturation of Plasmodium falciparum. Science 232 : 102-104

Tilley L Cribier S, Roelofsen B, Op den Kamp JAF and van Deenen LLM (1986) ATP dependent aminophospholipid translocation across the human erythrocyte membrane FEBS Lett 194 : 21-27

Vial HJ, van der Schaft PH, Beaumelle BD, Thuet MJ, Op den Kamp JAF (1989) Improved isolation of Plasmodium knowlesi infected erythrocyte host cell membrane on polycationic beads. Parasitol Res 75 : 419-421

Wallach DFH and Conley M (1977) Altered membrane proteins of monkey erythrocytes infected with simian malaria. J Molec Med 2 : 119-136

Van der Wiele FChr, Atsma W, Roelofsen B, van Linde M, van Binsbergen, J Radvanyi F, Raykova D, Slotboom AJ and de Haas GH (1988a) 1 : site-specific epsilon-NH2 monoacylation of pancreatic phospholipase A2. 2 : Transformation of soluble phospholipase A2 into a highly penetrating membrane bound form. Biochemistry 27 : 1688-1694

Van der Wiele FChr, Atsma W, Dijkman R, Schreurs AMM, Slotboom AJ, and de Haas GH, (1988b) Site specific epsilon-NH2 mono-acylation of pancreatic phospholipase A2. Biochemistry 27 : 1683-1688

Yuthavong Y, Wilairat P, Panypan B, Potiwan C and Beale GH (1979) Alterations in membrane proteins of mouse eruthrocytes infected with different species and strains of malaria parasites. Comp Biochem

LIPID TRANSPORT IN *PLASMODIUM FALCIPARUM* -INFECTED ERYTHROCYTE MEMBRANES.

Kasturi Haldar
Dept. Microbiology and Immunology
Stanford University
Stanford CA 94305

During its intraerythrocytic development the malaria parasite induces changes in the transport of phospholipids across the host erythrocyte membrane. In the mature erythrocyte, protein mediated transport of phosphatidylethanolamine (PE) and phosphatidylserine (PS) to the cytoplasmic leaflet is an energy requiring process and appears to serve as the primary mechanism for maintaining phospholipid asymmetry at the plasma membrane (Daleke and Huestis, 1985; Tilley et. al., 1986; Zachowski et. al., 1986) Band 7 has been identified as the putative amino-phospholipid translocator in human erythrocytes (Connor and Schroit 1988). Human erythrocyte membranes do not contain a phosphatidylcholine (PC) transporter and PC transbilayer movement in these membranes is extremely slow (Op den Kamp, 1979). In erythrocytes infected with the malaria parasites *Plasmodium falciparum* and *Plasmodium knowlesi*, there is an increased uptake and exchangeability of exogenously added radiolabelled PS, PE and PC with lipids of the host cell membrane (van der Schaft et. al., 1987; Moll et. al., 1988). However little is known about the molecular basis of these changes in parasitised erythrocytes. We have investigated the movement of fluorescent analogues of phospholipids in parasitised erythrocytes to elucidate mechanisms of lipid transport from the red cell membrane to the intracellular parasite.

Transbilayer movement of C6NBD-PC in parasitised erythrocytes.

In earlier studies (Haldar et. al. 1989) we demonstrated that when the probe 1-acyl-2-(-4-nitrobenzo-2-oxa-1,3-diazole)-aminocaproyl phosphatidylcholine (C6NBD-PC) is incorporated into uninfected and *P. falciparum* infected erythrocyte membranes at 0^oC, it remains at the cell surface. Equal amounts of the fluorescent lipid are delivered to infected and uninfected erythrocyte membranes. At 10^oC rapid internalisation of the lipid is observed in infected erythrocytes, at all stages of parasite growth. Removal of exofacial C6NBD-PC by back extraction indicates that 67.4 +/- 12.4% of the added label was internalised in a population of infected erythrocytes at 60% parasitemia. In contrast, in uninfected erythrocytes only 27.3 +/-9.1% of the label was resistant to back extraction (Table I). Our results indicate that the

NATO ASI Series, Vol. H 40
Dynamics and Biogenesis of Membranes
Edited by J.A.F. Op den Kamp

Internalisation of C6NBD-PS in parasitised erythrocytes.

When parasite cultures were incubated with 1-oleyl-2 - (-4-nitrobenzo-2-oxa-1,3-diazole) - aminocaproyl phosphatidylserine (C6NBD-PS) on ice for 15 min, the lipid analogue was incorporated into the erythrocyte membrane. As shown in Fig. 1A, equal amounts of label were transferred to uninfected cells and those at 69% parasitemia, over a range of 0-2 μg/ml. The amount of label transferred to the erythrocyte membrane did not vary with the degree of parasitism of the cell population. As shown in Fig. 2A, when cells were incubated with C6NBD-PS at 0°C a cell surface labelling pattern was observed. When the cells were warmed to 10°C for 30 min, fluorecence was transported to the parasite in the infected erythrocytes (Fig. 2B). TLC analysis of lipids from parasitised erythrocytes and uninfected erythrocytes indicated that the fluorescent label was not chemically altered by the parasite and could be attributed to the C6NBD-PS originally added to the cells (not shown). The internalisation of fluorescent lipid was not observed at 0°C. In order to determine whether the probe suffered any self quenching in the cells (and consequently remained undetected in the micrographs in Fig 2A and B), we compared the ratio of fluorescence in intact cells versus those lysed in

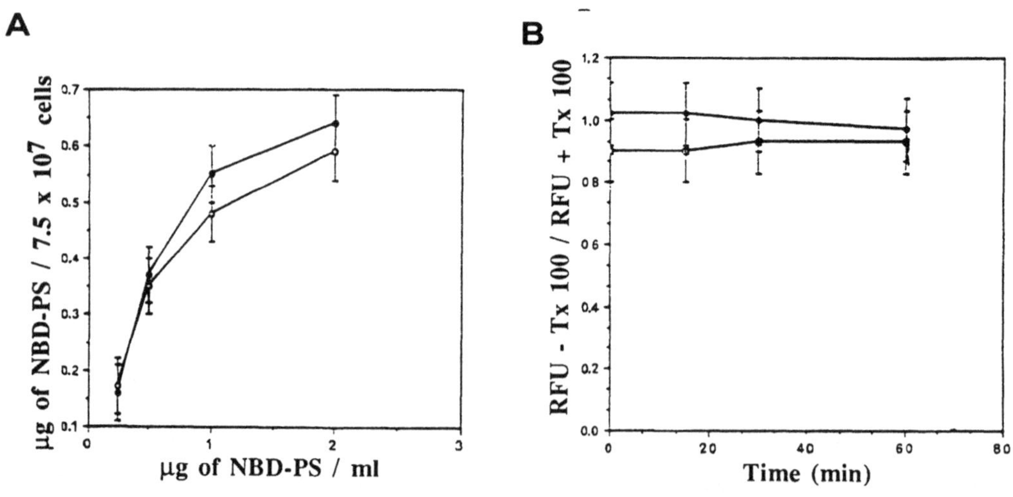

Figure 1. **A.** Transfer of C6NBD-PS to uninfected and infected erythrocyte membranes. 7.5 x 10^7 cells were incubated in 1 ml of RPMI 1640 at 0°C with varying amounts of C6NBD-PS for 15 min as described by Haldar et. al. 1989. The labelled cells were lysed in 1% Triton X-100 and the fluorescence was quantitated spectrophotometrically. (o-o-o) uninfected cells; (●-●-●) 69% ring and trophozoite infected cells. **B.** Quench curves for C6NBD-PS. Infected erythrocytes at 56% parasitemia and uninfected erythrocytes were labelled with C6NBD-PS as described in A and subsequently incubated at 10°C (Haldar et. al. 1989). Aliquots were removed at the indicated times and the cell associated fluorescence of each aliquot was determined in the presence and absence of 1%Triton X-100. The ratio of -Tx100/+Tx 100 was plotted for infected (●-●-●) and uninfected (o-o-o) cells.

internalisation of C6NBD-PC is not because of endocytosis but rapid protein mediated transbilayer lipid flip flop at the infected erythrocyte membrane, followed by monmer diffusion to the parasite. The evidence are the following: (a) rapid transport of C6NBD-PC from the erythrocyte membrane to the intracellular malaria parasite occurs at temperatures well below those required for endocytosis and at all stages of parasite growth; (b) bulk endocytosis of membrane lipid does not occur at the infected erythrocyte membrane; (c) L- -glycerophosphorylcholine inhibits internalisation of the label while the D- isomer does not, indicating a stereospecific protein mediated step in the translocation of C6NBD-PC from the erythrocyte membrane to the parasite; (d) transport of C6NBD-PC can be inhibited by chemical modification of cellular proteins; and (e) depleting cellular ATP inhibits internalisation of C6NBD-PC (Table I, Haldar et. al. 1989). In conclusion, our studies indicate that C6NBD-PC undergoes parasite induced transbilayer flip flop in the infected erythrocyte membrane possibly via plasmodial PC transporter.

Incubation	%label resistant to back extraction.
Uninfected red cells	
0°C, 10°C, 0 min	25.6 +/- 4.3
0°C, 30 min	28.2 +/- 5.2
10°C, 30min	26.8 +/- 5.6
+ azide, 10°C, 30 min	26.6 +/- 6.3
+ cyanide, 10°C, 30 min	27.9 +/- 7.3
Ring and Troph infected cells	
49% parasitemia	
0°C, 10°C, 0 min	26.2 +/- 6.2
0°C, 30 min	27.3 +/- 9.1
10°C, 30 min	53.2 +/- 10.8
+ azide, 10°C, 30 min	27.7 +/- 10.2
+ cyanide, 10°C, 30min	29.0 +/- 10.4
60% parasitemia	
0°C, 10°C, 0 min	26.1 +/- 6.8
0°C, 30 min	30.1 +/- 8.1
10°C, 30 min	67.4 +/- 12.4
+ azide, 10°C, 30 min	28.4 +/- 11.4
+ cyanide, 10°C, 30 min	31.6 +/- 11.1

Table I. Internalisation of C6NBD-PC by parasitised erythrocytes. Infected and uninfected cells were labelled with C6NBD-PC, incubated as described in methods at the indicated times and temperatures. For each batch of cells, i.e., uninfected, 49% parasitemia and 60% parasitemia, the labelling was repeated three times, in triplicate. Consequently the numbers reflect three independent labelling experiments and are an average of nine measurements (Haldar et. al. 1989).

Triton X-100 at different times during the incubation (Fig. 1B). The results indicate that the probe did not undergo self quenching and the observed distribution of fluorescence in both infected and uninfected cells was representative of the relative distribution of the probe in the cell.

The amount of C6NBD-PS internalised at 10°C was quantitated by back extracting the cells three times at 0°C with 2 mg/ml BSA (Haldar et. al. 1989). As shown in Fig 2C, the cell surface fluorescence of both infected and uninfected erythrocytes was considerably diminshed. The intracellular parasite associated fluorescence was resistant to back extraction. Equal amounts of label were resistant to back extraction in infected and normal erythrocytes, indicating that the same amount of probe was internalised in both cell types (Table II). Pretreating the cells with 2.5 mM N-ethylmaleimide (NEM) prior to labelling with C6NBD-PS, inhibit transmembrane movement of the probe in both infected and uninfected cells. The results indicate that that the steady state flip flop of PS catalysed by the host cell aminophospholipid translocator at 30 min at 10°C, is not altered by infection.

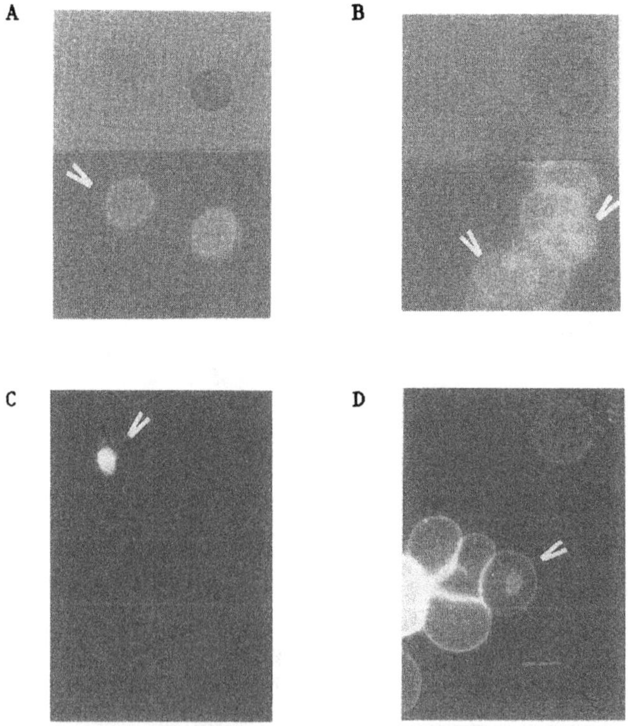

Figure 2. Fluorescence micrograph of uninfected and parasitised erythrocytes (indicated by the arrow) labelled with C6NBD-PS. A. Lower panel, labelled cells at 0°C; B. Lower panel, labelled cells incubated at 10°C for 30 min; C. Late rings incubated at 10°C for 30 min and back extracted; D. Non backextracted cells from C.

Internalisation of C6NBD-PE in parasitised erythrocytes.

When infected and uninfected erythrocytes were incubated with a second fluorescent aminophospholipid analogue 1-palmitoyl-2-(-4-nitrobenzo-2-oxa-1,3-diazole)-aminocaproyl phosphatidylethanolamine (C6NBD-PE), equal amounts of the probe were delivered to infected and uninfected cells and a cell surface flourescence was observed at 0°C (Fig. 3A; Fig. 4A). Internalisation of fluorescence was observed at 10°C as indicated in Fig. 4B, and the probe did not suffer self quenching during the incubation at 10°C (Fig. 3B). Pretreating the cells with 2.5 mM NEM inhibit internalisation of the probe in both normal and parasitised red cells. The amount of probe resistant to back extraction was the same in infected and uninfected cells as shown in Table II. In addition the steady state internalisation of C6NBD-PE was considerably lower than that of C6NBD-PS in both normal and parasitised erythrocytes. This was consistent with the amount of intracellular probe detected by microscopy. In Fig. 4, the cells had to be incubated for atleast 60 min at 10°C to observe prominent internalisation of C6NBD-PE.

Our data indicate that upto the late ring and early trophozoite stages, there is no measurable difference in the transmembrane flip flop of the fluorescent amino-phospholipids in infected and uninfected erythrocyte membranes. This strongly suggests that the red cell PS/PE transporter activity is unaltered by parasitism and provides the primary mechanism for amino-phospholipid flip flop across the infected erythrocyte membrane.

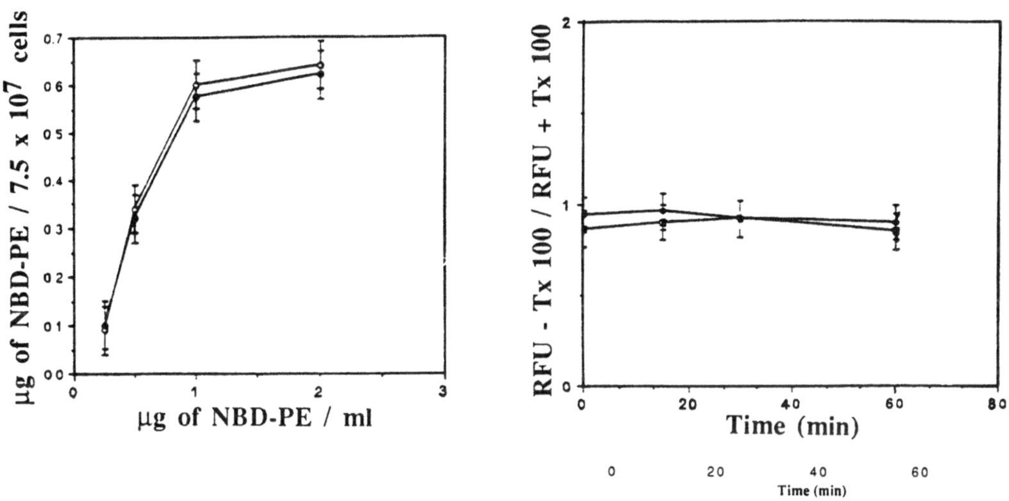

Figure 3. **A.** Transfer of C6NBD-PE to uninfected and infected erythrocyte membranes. Cells were labelled with C6NBD-PE and the fluorescence was quantitated under conditions identical to those described in Fig. 1A. (o-o-o) uninfected cells ; (●-●-●) 71% ring and trophozoite infected cells. **B.** Quench curves for C6NBD-PE. Samples were prepared as described in Fig. 1B. (●-●-●) infected cells; (o-o-o) uninfected cells.

A B C D

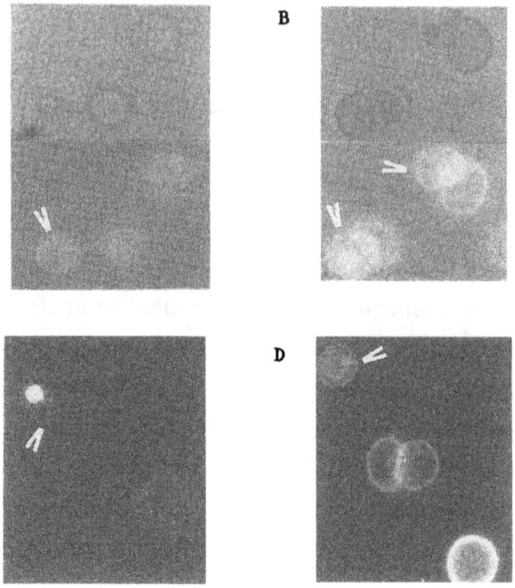

Figure 4. Fluorescence micrographs of uninfected and parasitised erythrocytes (indicated by the arrow) labelled with C6NBD-PE. A. Lower panel, labelled cells at 0°C; B. Lower panel, labelled cells incubated at 10°C for 30 min; C. Cells incubated at 10°C for 30 min and back extracted; D. Non back extracted cells from C.

Incubation (temperature andtime)	% PS resistant to back extraction	%PE resistant to back extraction
Uninfected red cells		
0°C, 10°C; 0 min	29.2 +/- 4.6	29.8 +/- 4.3
0°C; 30 min	30.6 +/- 4.9	31.3 +/- 3.9
10°C; 30 min	49.9 +/- 6.3	34.6 +/- 4.1
*2.5 mM Nem, 10°C; 30 min	28.9 +/- 4.8	28.9+/-3.8
67% ring and troph-infected cells		
0°C, 10°C; 0 min	32.0 +/- 5.2	30.1 +/- 4.9
0°C; 30 min	33.3 +/- 6.2	29.6 +/- 5.4
10°C; 30 min	51.6 +/- 5.8	36.7 +/- 5.2
*2.5 mN Nem, 10°C; 30 min	29.4 +/- 6.7	27.1 +/- 4.9

Table II. Internalisation of C6NBD-PS and C6NBD-PE by normal and parasitised erythrocytes. Infected and uninfected cells were labelled with C6NBD-PS or C6NBD-PE at 0°C, and incubated at the indicated times and temperatures. The labelling was repeated three times, and in triplicate with each batch of cells. Consequently the numbers reflect three independent labelling experiments and are an average of nine measurements. *Infected and uninfected cells were incubated 2.5 mM NEM for 15 min at 37°C, prior to labelling with the fluorescent probes.

Lack of constitutive endocytosis of the lipid bilayer at the infected erythrocyte membrane.

In order to investigate endocytotic uptake of lipids at the infected erythrocyte membrane, we investigated the intracellular transport of lipophillic marker 3,3-dihexadecylindocarbocyanine (diI), 5-(-4,6-dichlorotriazinyl)-aminofluorescein (DTAF) labelled cell surface glycoproteins, the pinocytotic marker lucifer yellow, and a red cell impermeant probe N-4-nitrobenzo-2-oxa-1,3-diazole dipalmitoyl phosphatidyl ethanolamine (N-NBD-PE), in parasitised erythrocytes. In previous studies (Haldar et. al. 1989) we described the uniform labelling of infected and uninfected erythrocyte membranes by diI at 0°C. Equal amounts of label were delivered to infected and uninfected cells at dye concentrations where there was no observed self quenching. When the cells were warmed to 37°C for upto 120 min there was no polarisation or internalisation of the cell surface label, indicating that constitutive endovesiculation of the lipid bilayer in general does not occur in the infected erythrocyte membrane, even at 37°C (Haldar et. al. 1989). Intergral membrane

Probe	Internalisation of fluorescence
Lucifer yellow	
Pretreatment	
Control	+
1 mM azide, 37°C	-
Lucifer yellow-dextran	
mw. 10, 000	-
40,000	-
70,000	-
N-NBD-L-α-PE	
Pretreatment	
Control	+
1 mM azide, 37°C	-
1 mM cyanide, 37°C	-
10 mM L-GPE, 37°C	+
10 mM ethanolamine, 37°C	+
10 mM L-GPC, 37°C	+
N-NBD-D-α-PE	+

Table III. Uptake of lucifer yellow and N-NBD-PE by parasitised erythrocytes. Ethanolic solutions of N-linked NBD-PE were introduced into parasitised cell suspensions in RPMI 1640 at 37°C in the presence of the indicated reagents. The reagents were present in subsequent washing and incubation steps. Cultures of infected erythrocytes were incubated with 1 mg/ml lucifer yellow or equivalent molar concentrations of lucifer yellow-dextran conjugates in RPMI 1640 at 37°C. Where indicated, 1 mM sodium azide was included in all steps of the incubations. Internalisation of fluorescent label was monitored in the microscope.

proteins of the infected cell surface tagged with a flourescent marker were not internalised in parasitised cells.

Seemingly contradictory to the lack of constitutive enodcytosis at the infected cells membrane was the energy dependent accumulation of the pinocytotic marker lucifer yellow (mw 457.25) by mature infected cells at 37°C. However when trophozoite and schizont infected erythrocytes were incubated with lucifer yellow dextran conjugates of molecular weights 10,000, 40,000 and 70,000, no internalisation of fluorescence was observed (Table III). The lack of internalisation of a 10,000 molecular weight complex indicates the absence of constitutive endocytotic internalisation and the presence of alternative energy dependent transport pathways for the uptake of lucifer yellow in the infected erythrocyte membrane. Our results indicate that constituve endocytosis of the erythrocyte bilayer does not occur in plasmodium infected red cells.

Uptake of N-NBD-PE by mature infected erythrocytes.

In earlier studies (Haldar et. al. 1989) we described the stage specific internalisation of N-NBD-PE from the surface of the red cell by the mature infected erythrocyte. Ethanolic solutions of N-NBD-PE were introduced into parasitised cell suspensions at 37°C. (This transfer was not observed at 0-2°C). The lipid was restricted to the cell surface in the presence of energy inhibitors. Ring stage parasites did not internalise the probe indicating that the mechanism for uptake of N-NBD-PE is induced at the trophozoite stage (27h) of the life cycle and is distinct from the transmenbrane flip flop of C6NBD-PC, C6NBD-PE and C6NBD-PS observed throughout the erythrocytic life cycle. In order to determine whether the internalisation of N-NBD-PE was a protein mediated lipid transfer event we investigated the effect of stereoisomers on the internalisation of the analogue. As shown in Table III, L-α-glycerophosphorylethanolamine L-α-glycerophosphorylcholine, and ethanolamine did not inhibit internalisation of the fluorescent lipid, suggesting that its transport is not catalysed via a stereospecific head group recognition process. Additionally when infected cells were labelled with N-NBD-D-o-glycerophosphatidylethanolamine, intracelluar accumulation of the fluorescent lipid was observed (Fig. 5). The internalisation of both D and L analogues of N-NBD-PE strongly argues against a stereospecific protein mediated lipid transfer mechanism. Energy dependent uptake of D-α and L-α isomers at 37°C would be suggestive of endocytotic uptake of the lipid from the host cell membrane. However since the infected red cell membrane is incapable of constitutive endocytosis, alternative mechanisms for N-NBD-PE transport must exist. Resonance energy transfer studies are under way to determine whether the dipalmitoyl N-NBD-PE lipid is directly inserted in the bilayer or present in vesicles clustered around the the infected cell surface. It is possible that vesicles or micelles of N-NBD-PE fuse with the cell surface membrane, delivering N-NBD-PE to the cytoplasmic face of the bilayer. Nonspecific lipid transfer proteins or possibly micelle formation at the cytoplasmic face may catalyse transport of the lipid to the parasite across the intraerythrocytic space. Studies investigating these mechanisms are currently underway.

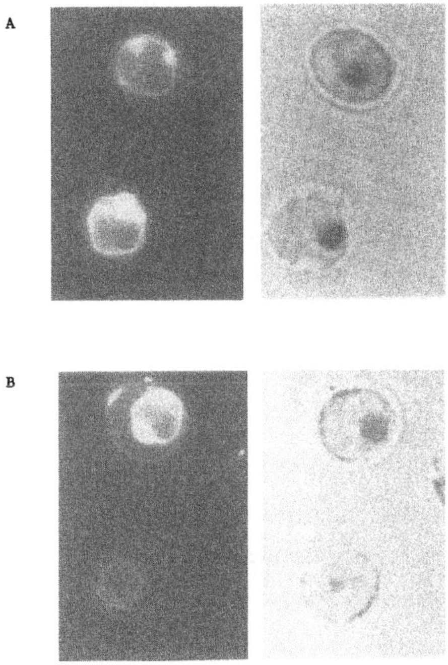

Figure 5. Fluorescence micrographs of parasitised erythrocytes incubated with N-NBD-PE;. Infected erythrocytes were labelled for 20 min in RPMI 1640 at 37°C in 1 mM sodium azide (Haldar et. al. 1989). A cellsurface fluorescence labelling pattern was observed with both the D and L analogues (not shown). The azide was removed and the labelled cells were subsequently incubated in RPMI 1640 at 37°C for 30 min and photographed. A. N-NBD-LαPE; B. N-NBD-D-αPE.

Conclusions.

Our studies and those of others indicate that in plasmodium infected erythrocytes there is active transport of phospholipids from the host cell membrane to the intracellular parasite. We have used fluorescent phospholipid probes to investigate mechanisms of lipid transport in malaria infected erythrocytes. These lipid analogues have been exploited to probe mechanisms maintaining phospholipid asymmetry across membranes. C6NBD-PC, C6NBD-PE and C6NBD-PS exhibit rapid monomer diffusion between membranes, but do not undergo spontaneous flip flop across lipid bilayers. It has been shown that, upon insertion into the plasma membrane of Chinese hamster fibroblasts, rapid flip flop of C6NBD-PE and C6NBD-PS occurs via a stereospecific protein mediated mechanism (Martin and Pagano, 1987; Sleight and Pagano, 1985). In contrast C6NBD-PC is not transported to the cytoplasmic face, but is taken up into the cell by endocytosis (Sleight and Pagano 1984), suggesting a phospholipid distribution similar to that of normal erythrocytes. Our data strongly support the presence of PC transporter in the infected erythrocyte membrane, either a plasmodium encoded protein or a modified host polypeptide. PC transporters have not been reported on cell surface membranes and may represent a unique feature of parasitism. Transbilayer movement of PS and PE across ring-infected erythrocyte membranes appears to be mediated by the endogenous red cell aminophospholipid translocator. At $10^{O}C$ there is no significant difference on steady state levels of PS and PE flip flop in normal and late ring-infected erythrocytes. These results contrast with the increased rates of radiolabelled PS and PE uptake observed in trophozoite/schizont-infected erythrocytes at $37^{O}C$ (Moll et. al. 1988). The reported increased uptake may reflect a rate limiting transfer of the phospholipids from the cytoplasmic face of the erythrocyte membrane to the parasite, which does not occur in uninfected red cells. The probes used in our studies are capable of rapid monomer diffusion across aqueous compartments and consequently their internalisation provides a measure of red cell transbilayer lipid movement alone. Alternativley, lipid flip flop at $10^{O}C$ may not provide a measure of translocation at $37^{O}C$. The stage specific internalisation of N-NBD-PE indicates that the lipid transport activities of the infected erythrocyte membrane are altered during intraerythrocytic parasite development. Late trophozoite/schizont infected erythrocytes are likely to have additional mechanisms which could account for increased radiolabelled PS and PE uptake. We were unable to quantitate increased accumulation of C6NBD-PS and C6NBD-PE in late trophozoites/early schizonts since back extraction of the cell surface fluorescent lipid induced significant cell lysis: probably due to the increased fragility of the mature infected red cell membrane.

The implications of altered lipid uptake on the phospholipid asymmetry of infected erythrocyte membranes remains unclear. Gupta and Mishra 1981 and Gupta et. al., 1982 indicated that the erythrocyte phospholipid organisation was altered in both infected and uninfected cells from monkeys infected with *Plasmodium knowlesi*, with PS exposed on the outer leaflet, and PC transported to the cytoplasmic face of the cell. A subsequent report (Joshi et. al. 1986) indicated that the observed changes in phospholipid asymmetry were limited to cells obtained from chronically infected splenectomised animals. Schwartz et. al. 1987 reported altered phospholipid organisation in *Plasmodium falciparum*-infected erythrocyte membranes. These results are contradictory to those of van der schaft et. al. 1987, which indicate that phospholipid organisation remains unaltered despite infection. Our results indicate that protein mediated PC translocation is restricted to infected erythrocyte membranes and does not occur in uninfected erythrocytes. Our cells are grown in in vitro culture and we have not investigated the effects of host factors like the spleen on cell surface phospholipid asymmetry.

The mechanisms of lipid transport across the intraerythrocytic space remain unclear. It is unlikely that there is a continuum of membrane extending from the red cell membrane to the parasitophorous vacuole membrane which mediates lipid transport. In the presence of a continuous membrane the flourescent lipid probes would not be restricted to labelling the erythrocyte membrane alone at 0^oC. The intraerythrocytic membrane compartment would also be labelled. DiI would be internalised. In the absence of constitutive endocytosis of the erythrocyte bilayer in infected cells, the results of Moll et al. 1988 and our data on the internalisation of N-NBD-PE strongly supports the presence of transfer/exchange proteins catalysing lipid transport in the intraerythrocytic space.

Plasmodia are incapable of chain elongation but engage in active phospholipid synthesis in the presence of fatty acids, glycerolipids and head group components. A recent report from Simoes et. al. 1989, indicates that the parasites re-model infected cell PC to contain16:0/18:2-phosphatidylcholine. Given this endogenous lipid biosynthetic activity, it remains unclear why the parasite accquires phospholipids from its host cell membrane. One explanation may be that it uses all available sources of lipid given the increased phospholipid content which is sustained through maturation. Alternatively the import of lipids from the host cell membrane may be necessary for the biogenesis of the vacuolar membrane.

References

Connor J and Schroit AJ (1988) Transbilayer movement of phosphatidylserine in erythrocytes: inhibition of transport and preferential labelling of a 31,000-dalton protein by sulfhydril reactive reagents. Biochemistry 27:848 -851

Daleke DL and Huestis WH (1985) Incorporation and translocation of amino-phospholipids in human erythrocytes. Biochemistry 24:5406-5416

Gupta CM and Mishra GC (1981) Transbilayerphospholipid asymmetry in Plasmodium knowlesi-infected host cell membrane Science (Wash. DC) 212:1047-1049

Gupta CM, Alam A, Mathur, PN and Dutta GP (1982) A new look at non parasitised red cells of malaria infected monkeys. Nature (Lond) 299:259-261

Haldar K, de Amorim AF and Cross GAM (1989) Transport of fluorescent phospholipid analogues from the erythrocyte membrane to the parasite in Plasmodium falciparum-infected cells. J. Cell Biol. 108: 2183-2192

Joshi P, Alam A, Chandra R Puri SK and Gupta CM (1986) Possible basis for membrane changes in non parasitised erythrocytes of malaria infected animals. Biochim. Biophys. Acta. 862:220-222.

Kates M (1975) Techniques of lipidology. *In* Laboratory Techniques in Biochemistry and Molecular Biology, editors Work TS and Work E vol 3: 558-565

Martin OC, and Pagano RE (1987) Transbilayer movement of fluorescent analogues of phosphatidylserine and phosphatidylethanolamine at the plasma membrane of cultured cells. J. Biol. Chem. 262: 5890-5898.

Moll G N, Vial HJ, Ancelin ML, Op den Kamp JAF, Roelofson B and van Deenen LL (1988) Phospholipid uptake by *Plasmodium knowlesi* infected erythrocytes. FEBS Lett. 232:341 - 346

Op den Kamp JAF (1979) Lipid asymmetry in membranes. Ann. Rev. Biochem. 48: 47-71

Schwartz RS, Olson JA, Ravento-Suarez C, Yee M, Heath RH, Lubin B and Nagel RL (1987) Altered plasma membrane phospholipid organisation in Plasmodium falciparum-infected erythrocytes. Blood 69:401-407

Simoes AP, Moll GN, Vial HJ, Roelofsen B and Op den Kamp JAF (1989) On the phosphatidylcholine molecular species composition in Plasmodium knowlesi infected erythrocyte membranes. Submitted for publication

Sleight RG and Pagano, RE (1984) Transport of fluorescent phosphatidylcholine analogue from the plasma membrane to the golgi apparatus. J. Cell Biol. 99:742-751

Sleight RG and Pagano RE (1985) Transbilayer movement of a fluorescent phosphatidylethanolamine analogue across the plasma membrane of cultutred mammalian cells. J. Biol. Chem. 26:1146-1154

Tilley L, Cribier S, Roelofsen B, Op den Kamp JAF and van Deenen LL (1986) ATP dependent translocation of aminophospholipids across the human erythrocyte membrane. FEBS Lett. 194:21 - 27

van der Schaft PH, Beaumelle B, Vial BH, Roelofsen B, Op den Kamp JJ and van Deenen LL (1987) Phospholipid organisation in monkey erythrocytes upon *Plasmodium knowlesi* infection. Biochim. Biophys. Acta. 901:1 - 14

Vial HJ, Thuet MJ and Phillipot JR (1982) Phospholipid biosynthesis in synchronous *Plasmodium falciparum* cultures. J. Parasitol. 68: 379 - 391

Zachowski A, Favre E, Cribier S, Herve P and Devaux P (1986) Outside-inside translocation of aminophospholipids in the human erythrocyte membrane is mediated by a specific enzyme. Biochemistry 25: 2585 -2590

Acknowledgements.

I would like to thank Dr. R. E. Pagano for his gift of C6NBD-PS. This work was partly supported by a grant from the United Nations Development Program/World Bank/World Health Organisation Special Program for Research and Training in Tropical Diseases (T16/18/M2/17J), a grant from the National Institutes of Infectious and Allergic Diseases (AI26670) and a grant from the MacArthur Foundation.

MEMBRANE PROPERTIES MODULATE THE ACTIVITY OF A YEAST PHOSPHATIDYLINOSITOL/ PHOSPHATIDYLCHOLINE TRANSFER PROTEIN

G. Szolderits, A. Hermetter, F. Paltauf and G. Daum
Institut für Biochemie und Lebensmittelchemie
Technische Universität Graz
Schlögelgasse 9/III, A-8010 Graz
Austria

Summary

The substrate specificity and the influence of membrane properties of donor and acceptor vesicles (lipid composition, charge effects, fluidity) on the activity of a 2800-fold enriched phosphatidylinositol / phosphatidylcholine transfer protein (PI/PC-TP) from yeast was tested. A fluorescence assay employing pyrene-labeled phospholipids was used for transfer measurements *in vitro*. The rate of transfer catalyzed by this protein is markedly higher for phosphatidylinositol than for phosphatidylcholine. Although yeast cells contain neither alkylacyl glycerophospholipids nor plasmalogens (alkenylacyl glycerophospholipids) these subclasses of phospholipids are readily transferred by the yeast PI/PC-TP; the transfer rate decreases in the order alkylacyl glycerophospholipid > plasmalogen > diacyl glycerophospholipid. A moderate but significant net transfer of phosphatidylcholine is observed when acceptor membranes consist only of non-transferable phospholipids, e.g. phosphatidylethanolamine.

The yeast PI/PC-TP forms complexes with phosphatidylinositol and phosphatidylcholine, respectively, the affinity for phosphatidylinositol being considerably higher. This result correlates with the preferential transfer of phosphatidylinositol as mentioned above.

Phosphatidylcholine transfer catalyzed by the yeast PI/PC-TP is inhibited by negatively charged phospholipids. Differences in the accessibility of the charged groups of lipids to the transfer protein might account for the different inhibitory effects, which occur in the order phosphatidylserine > phosphatidylglycerol > phosphatidylinositol >

NATO ASI Series, Vol. H 40
Dynamics and Biogenesis of Membranes
Edited by J.A.F. Op den Kamp

cardiolipin > phosphatidic acid. Mitochondrial membranes containing high quantities of negatively charged phospholipids do not show the same inhibitory effect as vesicles prepared from phospholipids extracted from these membranes. This observation points to a compensation of the negative charge of phospholipids by membrane proteins. Ergosterol, which decreases the membrane fluidity, reduces the rate of phosphatidylcholine transfer. The notion that membrane fluidity modulates the transfer rate, is also supported by experiments using acceptor vesicles consisting of dipalmitoyl phosphatidylcholine: the rate of phosphatidylcholine transfer was significantly reduced when assays were carried out below the phase transition temperature of the acceptor vesicle phospholipid.

Introduction

Phospholipid transfer proteins are ubiquitous proteins which have been isolated from mammalian and plant cells as well as from prokaryotic and eukaryotic microorganisms (for recent reviews see Helmkamp, 1986; Daum and Paltauf, 1989). They catalyze the transport of phospholipid molecules between artificial and/or biological membranes *in vitro*; the role of these proteins *in vivo* is still unclear. They are thought to be involved in membrane biogenesis and in the maintenance of membrane lipid composition and asymmetry. Vice versa the lipid composition of membranes modulates the activity of phospholipid transfer proteins (Helmkamp, 1986; Somerharju et al., 1983).

The cytosol of yeast cells contains proteins with transfer activities for several phospholipids (Cobon et al., 1976; Daum and Paltauf, 1984). A phosphatidylinositol / phosphatidylcholine specific transfer protein has been purified to homogeneity (Daum and Paltauf, 1984). A yeast phospholipid transfer protein with somewhat different properties has been described by Bozzato and Tinker (1987).

Methods

The phosphatidylinositol / phosphatidylcholine transfer protein (PI/PC-TP) of the yeast *Saccharomyces cerevisiae* D273-10B was isolated as described by Daum and Paltauf (1984) with the following modifications: hydroxyapatite chromatography and chromatofocusing were omitted, and hydrophobic interaction chromatography using Phenyl Sepharose was employed as the last step of purification (Szolderits et al., in preparation).

Fluorescence assays for phospholipid transfer and binding measurements were carried out using the following fluorescent phospholipids as substrates: 1 -palmitoyl- 2 -pyrenedecanoyl-*sn*-glycero- 3 -phosphocholine, 1-O-(1'-alkenyl)-2-pyrenedecanoyl-*sn*-glycero-3-phosphocholine, 1-O-hexadecyl-2-pyrenedecanoyl-*sn*-glycero-3-phosphocholine, 1-stearoyl-2-pyrenedecanoyl-*sn*-glycero-3-phosphocholine and 1-stearoyl-2-pyrenedecanoyl-*sn*-glycero-3-phosphoinositol.

Donor vesicles were prepared by injection of an ethanolic solution of the pyrene-labeled phospholipid (100 μM) into 10 mM TrisHCl, pH 7.1, 0.02 % NaN_3 with stirring at 37°C (Kremer et al., 1977). The buffer was routinely filtered (Sartorius membrane filters, 0.45 μm) before use. The final donor lipid concentration was 0.1 μM. For comparative studies of phosphatidylcholine and phosphatidylinositol transfer phosphatidylcholine donor vesicles contained 10 % 1-stearoyl-2-pyrenedecanoyl-glycerophosphocholine, 70 % egg phosphatidylcholine, 10 % phosphatidylinositol and 10 % N-trinitrophenyl phosphatidylethanolamine as an internal quencher (Somerharju et al., 1987); phosphatidylinositol donor vesicles consisted of 10 % 1-stearoyl-2-pyrenedecanoyl-glycerophosphoinositol, 80 % egg phosphatidylcholine and 10 % N-trinitrophenyl phosphatidylethanolamine. This vesicle lipid composition allowed transfer studies with a standardized vesicle phospholipid composition at a low surface charge.

Acceptor vesicles were prepared by mixing egg phosphatidylcholine with the appropriate amount of other lipids. The mixtures were dried under a stream of nitrogen and dissolved either in DMSO/ethanol (1:3, v/v) or in ethanol to a final concentration of 5 mM. This solution was injected into 10 mM TrisHCl, pH 7.1, 0.02 % NaN_3 under stirring at 37°C. Aliquots were used in the transfer assay to give a final acceptor lipid concentration of 5 μM. In cases where the donor lipid concentration was increased, e.g.

during comparative measurements of phosphatidylcholine and phosphatidylinositol transfer, the amount of acceptor vesicles was increased correspondingly to maintain the 50-fold excess of acceptor vesicle lipid. Mitochondrial membranes used as acceptor in some experiments were isolated by the method of Daum et al. (1982).

The fluorescence assay for phospholipid transfer measurements *in vitro* was similar to that described by Somerharju et al. (1987) and van Paridon et al. (1988a, 1988b). The excitation wavelength was 342 nm (2 mm slit), and the emission wavelength 380 nm (10 nm slit). In a standard assay 50 pmol of pyrene-labeled phospholipid were incubated with a 50-fold excess of acceptor membrane vesicles in a total volume of 0.5 ml 10 mM TrisHCl, pH 7.1, 0.02 % NaN_3 at $40^{\circ}C$. The spontaneous transfer of fluorescently labeled phospholipids to unlabeled acceptor membranes was followed continously by measuring the increase of pyrene monomer fluorescence intensity on a Shimadzu RF-540 spectrofluorimeter. Protein catalyzed transfer was calculated by subtracting spontaneous transfer from transfer observed in the presence of 5-100 ng PI/PC-TP.

Complex formation of fluorescently labeled phospholipids with the PI/PC-TP at $30^{\circ}C$ was measured essentially as described by Somerharju et al. (1987) and van Paridon et al. (1988a). Phospholipid vesicles containing the fluorescent phospholipid and 10 % N-trinitrophenyl phosphatidylethanolamine were prepared as described above. The final concentration of labeled phospholipids was 0.1 nmol/ml; the total assay volume was 0.5 ml. Portions of PI/PC-TP were added, and the resulting increase of pyrene monomer fluorescence intensity was recorded. Parameter for fluorescence measurements were the same as described above for transfer assays. Binding affinities between the transfer protein and phospholipids were calculated from the initial slope of plots of fluorescence intensities (arbitrary units) versus increasing amounts of the PI/PC-TP.

Results and discussion

The phospholipid transfer protein used for the experiments described here was 2800-fold enriched from yeast cytosol. This protein has an apparent molecular weight of 37000 as judged by SDS-PAGE, and an isoelectric point of 5.2 as determined by isoelectric focusing. It catalyzes specifically the transfer of phosphatidylinositol and phosphatidylcholine. Using fluorescently labeled phospholipids the rates of phosphatidylinositol and phosphatidylcholine transfer were compared. To standardize assay conditions the lipid compositions of donor and acceptor vesicle, respectively, were kept constant. Donor vesicles contained 10 % phosphatidylinositol and 80 % phosphatidylcholine (including the fluorescently labeled substrates, respectively) as well as 10 % N-trinitrophenyl phosphatidylethanolamine as an internal quencher. Acceptor vesicles consisted of 10 % phosphatidylinositol and 90 % phosphatidylcholine and were used in a 50-fold excess. Under these conditions the rate of phosphatidylinositol transfer was 19 times higher than that of phosphatidylcholine. Similar results have been reported by van Paridon et al. (1987) for the bovine brain phosphatidylinositol transfer protein.

It is a standing question whether or not phospholipid transfer proteins catalyze lipid exchange or net transfer (Helmkamp, 1986). It has been shown that the phosphatidylinositol transfer protein from bovine brain is able to catalyze net transfer of phosphatidylinositol and/or phosphatidylcholine to membranes, which initially did not contain the respective phospholipid. We addressed this problem in an experiment where acceptor membranes consisted only of the non-transferable phospholipid stearoyloleoyl glycero phosphoethanolamine. Under these conditions a moderate, but significant net transfer of phosphatidylcholine occurred, the rate being approximately 10 % of the control where stearoyl oleoyl glycerophosphocholine was used as the acceptor lipid. In this respect the yeast PI/PC-TP is similar to the bovine liver phosphatidylcholine transfer protein (Berkhout et al., 1984). The bovine brain phosphatidylinositol transfer protein, on the other hand, does not catalyze phosphatidylcholine transfer to vesicles consisting only of non-transferable phospholipids (phosphatidylethanolamine and phosphatidic acid) (Kasper and Helmkamp, 1981).

Experiments using the bovine liver phosphatidylcholine transfer protein (Kamp et al., 1973) showed that plasmalogens were preferentially transferred as compared to the corresponding diacyl glycerophospholipid species (G. Szolderits, unpublished observation). This result prompted us to test the transfer of alkenylacyl and alkylacyl glycerophosphocholine catalyzed by the yeast PI/PC-TP, mainly because yeast cells, in contrast to mammalian cells, do not contain ether phospholipids. As can be seen from Table I alkenylacyl and alkylacyl glycerophosphocholine as components of donor vesicles are transferred to egg phosphatidylcholine acceptor vesicles at an approximately 3-fold higher rate than the diacyl species. The transfer rates of the individual subclasses of donor phospholipids are not significantly altered by the presence of alkylacyl or alkenylacyl glycerophosphocholine in acceptor vesicles (data not shown). In contrast to data obtained with the yeast PI/PC-TP the non-specific lipid transfer protein from bovine liver (Crain and Zilversmit, 1980) has a minor preference for the transfer of diacyl glycerophosphocholine as compared to alkylacyl or alkenylacyl glycerophosphocholine.

Table I

Transfer of different subclasses of choline glycerophospholipids catalyzed by the yeast PI/PC-TP or by the non-specific lipid transfer protein from bovine liver

	Relative transfer rates (%)	
	Yeast	Bovine liver
Donor lipid	PI/PC-TP	nsLTP
Diacyl-GPC	100	100
Alkenylacyl-GPC	250	93
Alkylacyl-GPC	295	85

GPC: glycerophosphocholine
The transfer rates of fluorescent diacyl glycerophosphocholine to egg phosphatidylcholine acceptor vesicles were set 100 %.

One possible explanation for the differences in transfer rates would be differences in the affinity of the transfer protein for distinct phospholipids. Complex formation of the transfer protein with single phospholipid molecules is thought to be the initial step in the transfer process. Data presented in Fig. 1 support this notion only in part. The binding affinity is approximately 12-15 times higher for phosphatidylinositol than for phosphatidylcholine with the same acyl chain composition. Binding affinities of alkylacyl and alkenylacyl glycerophosphocholine are only 15 % increased over the binding affinity of diacyl phosphatidylcholine. Therefore the 2,5 to 3-fold higher rates of alkylacyl and alkenylacyl glycerophosphocholine transfer as compared to diacyl glycerophosphocholine transfer cannot be attributed to protein binding affinities of the two former subclasses (see Table I). From these results it is evident that binding affinities and transfer rates are not necessarily correlated. Similar observations have been reported before by Somerharju et al. (1987) and van Paridon et al. (1988a) who studied mammalian phospholipid transfer proteins. Transfer rates might also depend

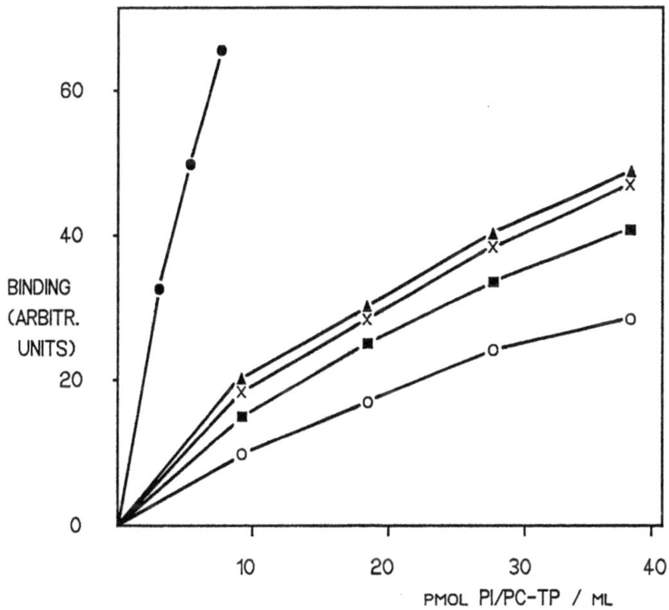

Fig. 1

Complex formation of fluorescent phospholipids with the PI/PC-TP from yeast

●——● Stearoyl-pyrenedecanoyl-glycerophosphoinositol
o——o Stearoyl-pyrenedecanoyl-glycerophosphocholine
■——■ Palmitoyl-pyrenedecanoyl-glycerophosphocholine
x——x Hexadecyl-pyrenedecanoyl-glycerophosphocholine
▲——▲ 1'-Alkenyl-pyrenedecanoyl-glycerophosphocholine

on kinetic parameters such as activation energy, or properties of the membrane surface or interface.

Effects of membrane properties on rates of phospholipid transfer were tested by manipulating the membrane phospholipid composition and by comparing model membranes with protein- containing biological membranes. Negatively charged phospholipids exert a strong inhibitory effect on phospholipid transfer in the order phosphatidylserine > phosphatidylglycerol > phosphatidylinositol > cardiolipin > phosphatidic acid (Table II); the zwitterionic phospholipids phosphatidylethanolamine and sphingomyelin had no effect. Inhibition of phosphatidylcholine transfer by phosphatidylinositol can partly be attributed to competition between the two phospholipids for the substrate binding site of the protein. The mechanism of inhibition by the other negatively charged phospholipids is not clear; it might be similar to that observed with mammalian phospholipid transfer proteins. In this case inhibition by negatively charged

Table II

Modulation of the rate of protein catalyzed phosphatidylcholine transfer by acceptor membrane lipid composition

Lipid added to acceptor vesicles	% of lipid in acceptor vesicles leading to a 50 % inhibition of transfer rate
Phosphatidylinositol	10
Phosphatidylserine	7
Phosphatidylglycerol	8
Cardiolipin	16
Phosphatidic acid	50
Phosphatidylethanolamine	n.i.0-50 %*
Sphingomyelin	n.i.0-50 %*
Ergosterol	> 50

*No significant inhibition of transfer between 0 and 50 %.
The transfer assay using fluorescently labeled phosphatidylcholine is described in the Methods section. Values are expressed as mol % of the respective phospholipids leading to a 50 % inhibition of the phosphatidylcholine transfer rate when present in egg phosphatidylcholine acceptor vesicles.

phospholipids has been attributed to a positively charged membrane interaction site on the protein (van Paridon et al., 1988b). Why one phospholipid inhibits more effectively than another in spite of both bearing the same net negative charge is not completely understood; accessibility of charged groups at the membrane surface might play a role.

As was to be expected from the above results, a combination of negatively charged and zwitterionic lipids, resembling the phospholipid composition of a yeast membrane, was completely inhibitory to phosphatidylcholine transfer (Table III). This result is at variance with data presented by Bozzato et al. (1987) who found maximum rates of phosphatidycholine transfer employing the same lipid mixture as acceptor membranes. The reason for this discrepancy might be different preparations of transfer proteins or different assay systems. When phospholipids extracted from yeast mitochondria were used for the preparation of acceptor membranes phosphatidylcholine transfer was essentially zero (see Table III). However, mitochondrial membranes did effectively serve as acceptors. Clearly membrane proteins mask the effect of negatively charged phospholipids, either by neutralizing the charge or by burying it at the membrane surface. This modulating effect of membrane proteins might play a role in the regulation of phospholipid transfer *in vivo*.

Table III

Phosphatidylcholine transfer to complex acceptor membranes

Acceptor membrane components	Relative transfer rate (%)
Egg phosphatidylcholine	100
PC:PE:PI:PS (4:3:2:1)	0
Yeast mitochondria	71*
Lipid extract from yeast mitochondria	0*

*Yeast mitochondria contain PC:PE:PI:PS:PG:CL:ergosterol at an approximate molar ratio of 4:3:1:0.5:0.6:0.6:1.2.

Another parameter which affects phospholipid transfer rates is membrane fluidity. Incorporation of increasing amounts of ergosterol into acceptor phospholipid vesicles decreases phosphatidylcholine transfer rates; with 1:1 molar mixtures of phosphatidylcholine and ergosterol a 40 % reduction was observed (see Table II). When dipalmitoyl phosphatidylcholine formed the acceptor vesicles, the transfer rates of phosphatidylcholine were distinctly lower at temperatures below the phase transition temperature of dipalmitoyl phosphatidylcholine (Fig. 2). The temperature dependence was much less pronounced with egg phosphatidylcholine as this phospholipid was in the liquid crystalline state at the temperatures tested.

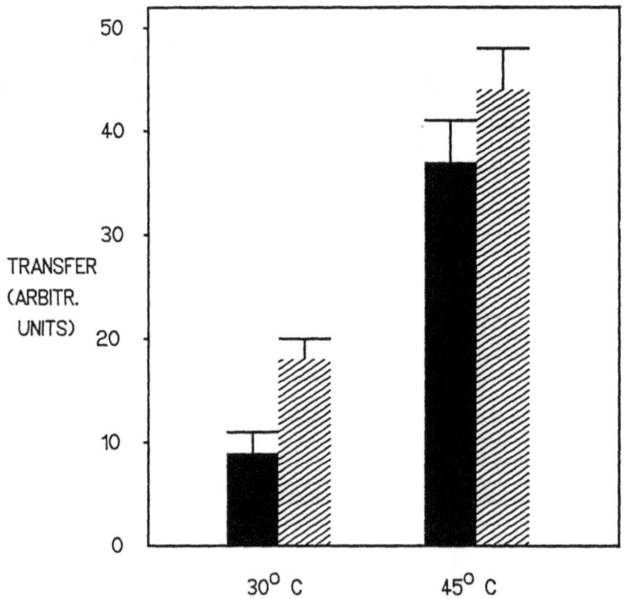

Fig.2

Transfer of fluorescent phosphatidylcholine to acceptor membrane vesicles with different fatty acyl chain unsaturation
■ Dipalmitoyl glycerophosphocholine
▨ Egg phosphatidylcholine

Acknowledgements

The authors gratefully acknowledge the precious gift of two fluorescent phospholipids by P. Somerharju, Helsinki. This work was financially supported by the Fonds zur Förderung der wissenschaftlichen Forschung in Österreich (project 5746 to A.H., and project 6958 to G.D.).

References

Berkhout TA, van den Bergh C, Mos H, de Kruijff B, Wirtz KWA (1984) Regulation of the activity of phosphatidylcholine transfer protein by vesicle phosphatidic acid and membrane curvature: A fluorescence study using 2-parinaroyl phosphatidylcholine. Biochemistry 23:6894-6900

Bozzato RP, Tinker DO (1987) Purification and properties of two phospholipid transfer proteins from yeast. Biochem Cell Biol 65:195-202

Bozzato RP, Woolley D, Tinker DO (1987) Catalytic properties of the yeast phospholipid transfer protein. Biochem Cell Biol 65:203-210

Cobon GS, Crowfoot PD, Murphy M, Linnane AW (1976) Exchange of phospholipids between mitochondria and microsomes in vitro is stimulated by yeast cytosol. Biochim Biophys Acta 441:255-259

Crain RC, Zilversmit DB (1980) Two nonspecific phospholipid exchange proteins from beef liver. 1. Purification and characterization. Biochemistry 19:1433-1439

Daum G, Böhni PC, Schatz G (1982) Import of proteins into mitochondria. Cytochrome b_2 and cytochrome c peroxidase are located in the intermembrane space of yeast mitochondria. J Biol Chem 257:13028-13033

Daum G, Paltauf F (1984) Phospholipid transfer in yeast. Isolation and partial characterization of a phospholipid transfer protein from yeast cytosol. Biochim Biophys Acta 794:385-391

Helmkamp GM Jr (1986) Phospholipid transfer proteins: mechanism of action. J Bioenerg Biomembr 18:71-91

Kamp HH, Wirtz KWA, van Deenen LLM (1973) Some properties of phosphatidylcholine exchange protein purified from beef liver. Biochim Biophys Acta 318:313-325

Kasper AM, Helmkamp GM Jr (1981) Intermembrane phospholipid fluxes catalyzed by bovine brain phospholipid exchange protein. Biochim Biophys Acta 664:22-32

Kremer JMH, v.d.Esker MWJ, Pathmamanoharan C, Wiersema PH (1977) Vesicles of variable diameter prepared by a modified injection method. Biochemistry 16:3932-3935

Paltauf F, Daum G (1989) Phospholipid transfer in microorganisms. In: Hilderson H (ed) Subcellular Biochemistry, vol 16 Plenum Publishing Corporation, New York London, in press

Somerharju PJ, van Loon D, Wirtz KWA (1987) Determination of the acyl chain specificity of the bovine liver phosphatidylcholine transfer protein. Application of pyrene-labeled phosphatidylcholine species. Biochemistry 26:7193-7199

Somerharju P, van Paridon P, Wirtz KWA (1983) Phosphatidylinositol transfer protein from bovine brain. Substrate specificity and membrane binding properties. Biochim Biophys Acta 731:186-195

Van Paridon PA, Gadella TWJ Jr, Somerharju PJ, Wirtz KWA (1987) On the relationship between the dual specificity of the bovine brain phosphatidylinositol transfer protein and membrane phosphatidylinositol levels. Biochim Biophys Acta 903:68-77

Van Paridon PA, Gadella TWJ Jr, Somerharju PJ, Wirtz KWA (1988a) Properties of the binding sites for the *sn*-1 and *sn*-2 acyl chains on the phosphatidylinositol transfer protein from bovine brain. Biochemistry 27:6208-6214

Van Paridon PA, Gadella TWJ Jr, Wirtz KWA (1988b) The effect of polyphosphoinositides and phosphatidic acid on the phosphatidylinositol transfer protein from bovine brain: a kinetic study. Biochim Biophys Acta 943:76-86

PROTEIN GLYCOSYLATION: OLIGOSACCHARYL TRANSFERASE AND A NOVEL RECOGNITION PROTEIN

Robert Noiva, Howard A. Kaplan, M. Geetha-Habib, and William J. Lennarz
Department of Biochemistry and Molecular Biology
The University of Texas M. D. Anderson Cancer Center
1515 Holcombe Boulevard
Houston, Texas 77030
USA

INTRODUCTION

The amount of information about the pathway of secretory and plasma membrane glycoprotein synthesis has increased dramatically during the past decade. It is clear that not only the initial process of translation/glycosylation, but also the subsequent steps of polypeptide and oligosaccharide chain processing occur in the endomembrane system consisting of the rough endoplasmic reticulum, the Golgi complex, and vesicles derived from these subcellular organelles. Although there has been good progress in acquiring information about the overall pathway of N-linked glycoprotein synthesis in this endomembrane system, detailed characterization of the properties of the enzymes has been restricted to those involved in processing reactions in the Golgi complex. Indeed, a number of these enzymes have been isolated in pure form and some have been cloned and sequenced.

In contrast, the enzymes assembled within the rough endoplasmic reticulum that catalyze reactions involving the dolichylphosphate-linked intermediates have been refractory to purification to homogeneity by the classical techniques of detergent solubilization followed by ion-exchange chromatography, gel filtration, etc. As a consequence, none of these enzymes, including the one that is the focus of this chapter, oligosaccharyl transferase, has been characterized in great detail. Presumably the refractory properties of these enzymes are reflections of the extraordinary hydrophobicity of one of the substrates, dolichyl phosphate, which has a very apolar domain about 75-100 Å in length. Given these facts, it is likely that these enzymatic reactions occur either within or at the interface of the hydrophobic environment of the phospholipid bilayer of the endoplasmic reticulum. If so, it is not surprising that conventional manipulations during purification that require disruption of the bilayer lead to a loss of enzymatic activity.

NATO ASI Series, Vol. H 40
Dynamics and Biogenesis of Membranes
Edited by J.A.F. Op den Kamp

RATIONALE

We have concentrated our efforts on oligosaccharyl transferase because it plays a pivotal role in the synthesis of N-linked glycoproteins, being central in the pathway, i.e. at the end of the dolichylphosphate-linked oligosaccharide assembly sequence and prior to the processing reactions. This enzyme catalyzes the *en bloc* transfer of the oligosaccharide chain from dolichyl-PP-$GlcNAc_2Man_9Glc_3$ to the β-amido nitrogen atom of asparagine residues within -Asn-X-Ser/Thr- sequences in polypeptide chains (reviewed in Waechter & Lennarz, 1976; Kornfeld & Kornfeld, 1980; Struck & Lennarz, 1980; Hubbard & Ivatt, 1981). As shown in Figure 1, in this reaction the oligosaccharide chain of oligosaccharyl-PP-Dol, the end product of the assembly pathway in the rough endoplasmic reticulum, is transferred to the carboxyamido side chain of an asparagine residue in a polypeptide. Additional reasons why this enzyme is of particular interest are: 1) In most cases the reaction it catalyzes is cotranslational; that is, it occurs while the polypeptide chain is growing. 2) The substrates are unusual, one being a molecule consisting of a hydrophilic oligosaccharide chain attached to an extremely hydrophobic lipid and the other being the side chain of an amino acid residue in a growing polypeptide chain. 3) The reaction apparently involves a carboxyamido group as a nucleophile and Dol-PP as a leaving group. 4) Only specific Asn-X-Ser/Thr sites in a polypeptide are glycosylated. Given the above considerations, a detailed understanding of the properties and specificity of oligosaccharyl transferase is important if we are to understand glycoprotein synthesis.

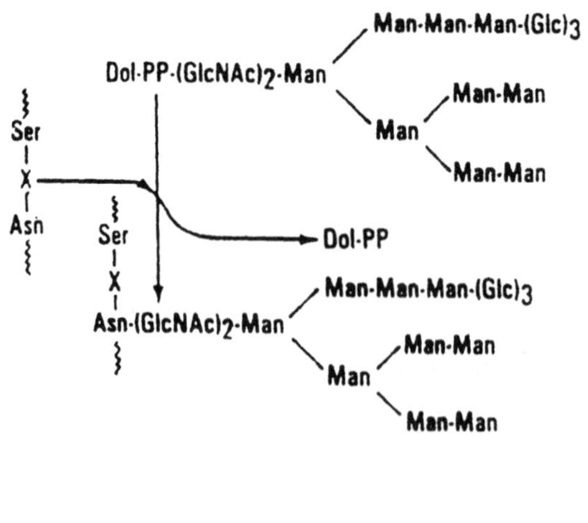

FIGURE 1 - Reaction catalyzed by oligosaccharyl transferase.

SUBSTRATES FOR OLIGOSACCHARYL TRANSFERASE

The earliest studies relevant to the peptide recognition site were those showing that the oligosaccharide and polypeptide chains are linked via an N-glycosidic bond to an asparaginyl residue (reviewed in Struck & Lennarz, 1980). Examination of the then very limited number of instances in which the amino acid sequence adjacent to the Asn bearing the carbohydrate chain was known led Marshall (1974) to propose that the Asn must be part of the sequence -Asn-X-Ser- or -Asn-X-Thr-, where X was any one of the 20 amino acids. Moreover, the observation that many such tripeptide sequences in proteins are not glycosylated led to the conclusion that the primary structure of this tripeptide was necessary, but not sufficient, for glycosylation. Subsequent surveys of known amino acid sequences have fully confirmed these generalizations.

In view of the great increase in the number of proteins that has been sequenced, an earlier survey (Struck & Lennarz, 1980) of the distribution of the 20 amino acids in the X position has been updated (Kaplan et al., 1987). Approximately 2000 -Asn-X-Ser/Thr- tripeptides were identified, of which about half (1000) were found to be -Asn-X-Ser-. Of these, 4.2% (85 sites) were glycosylated. This percentage is misleadingly low, however, because cytoplasmic proteins, which would not be expected to be glycosylated, were included among the 1000. Of interest was the finding that, although there were equal numbers of Asn-X-Ser and Asn-X-Thr tripeptides, -Asn-X-Thr- was almost three times more likely to be glycosylated than -Asn-X-Ser-. Thus 11.3% (227 sites) of the Thr-containing tripeptides were glycosylated. Earlier possible exclusions of several amino acids from the X position (Struck & Lennarz, 1980) were proved incorrect by the additional data of the more recent survey and the only conclusion that can be drawn is that Cys, Trp and Pro in the X position are very rare in glycosylated -Asn-X-Ser/Thr- sites. Overall, 7.5% of the -Asn-X-Ser/Thr- tripeptides in sequenced proteins are known with certainty to be glycosylated. The true percentage, however, is much higher in membrane or secretory proteins because (a) in many other proteins that have been sequenced oligosaccharide chains are known to be present, but have not yet been assigned to a specific site, and (b) as discussed above, it is unrealistic to include in this tally the cytoplasmic proteins, because they do not have an opportunity to interact with oligosaccharyl transferase during their synthesis.

Further insight into the factors controlling glycosylation was initially provided by the finding that unfolded proteins could serve as substrates for oligosaccharyl transferase (Pless & Lennarz, 1977; Kronquist & Lennarz, 1978). Subsequently, it became clear that peptides generated from such proteins (Struck et al., 1978) and, in fact, synthetic tripeptides (Hart et al., 1979) serve as substrates, providing that the C- and N-termini are blocked. Earlier studies on such peptides from a number of laboratories have been reviewed previously (Struck & Lennarz, 1980), but several more recent studies have increased our understanding of the glycosylation process. Bause (1983) demonstrated that a tripeptide containing the epoxy analog of Thr, epoxyethylglycine, as the third residue was an irreversible inhibitor of the enzyme, probably because it became covalently linked to the protein in the vicinity of the active site. With respect to the other amino acid that obviously is crucial, studies on modification of the Asn residue by Welply et al. (1983) revealed that peptides containing altered Asn residues were neither substrates nor inhibitors. Studies with β-fluoroasparagine, which is toxic to cells (Stern et al., 1982), suggests that the toxic effect is the result of inhibition of glycosylation (Hortin et al, 1983). Consistent with these findings, it was shown (Rathod et al., 1986) that synthetic peptides containing *threo*-β-fluoroasparagine are extremely poor substrates for *in vitro* glycosylation. Studies by Welply et al. (1983) showed that increased hydrophobicity of the Asn-X-Ser(Thr) tripeptide by use of hydrophobic blocking groups decreased the apparent K_m of the peptide. Finally, it was found that such acceptor tripeptides, when added to an *in vitro* system for cotranslational glycosylation of the nascent chain of VSV G protein or ovalbumin, inhibited the glycosylation of these polypeptides, presumably by saturating the -Asn-X-Ser(Thr)- recognition site of oligosaccharyl transferase (Lau et al., 1983).

Much less information is available about the specificity of oligosaccharyl transferase for the other substrate, oligosaccharyl-PP-dolichol. Although the results of some earlier studies are difficult to interpret, the overall conclusion is that the complete oligosaccharide-lipid, containing three glucosyl units, is a more effective substrate than shorter oligosaccharides (reviewed in Waechter & Lennarz, 1976; Hubbard & Ivatt, 1981). However, shorter oligosaccharide-lipids, including N,N'-diacetylchitobiosyl-lipid, can be transferred to protein acceptors in oviduct (Chen & Lennarz, 1977) and yeast (Sharma et al., 1983) membrane preparations. In fact, in yeast membranes both $GlcNAc_2$-PP-Dol and $Glc_3Man_9GlcNAc_2$-PP-Dol exhibit the same K_m (Sharma et al., 1983). Furthermore, *in vivo* studies using a mutant that lacks mannosylphosphoryldolichol synthase indicate that transfer of a shorter oligosaccharide-lipid containing Glc_3, but lacking the Man residues that arise from Man-P-Dol, does occur (Kornfeld et al., 1979).

Genetic evidence concerning yeast mutants supports the idea that the glucosyl residues on oligosaccharyl-PP-Dol can enhance its efficiency as a donor (Huftaker & Robbins, 1983; Ballou et al., 1986). Indeed a Glc-P-Dol synthase defective mutant exhibited reduced efficiency of glycosylation of the major yeast glycoprotein, invertase, presumably because glucose residues were absent from the oligosaccharide chain (Ballou et al., 1986). Thus, based on both the *in vivo* and *in vitro* studies, oligosaccharyl transferase exhibits a rather broad specificity with respect to the saccharide-lipid. At present, however, little is known about how the presence of the "correct" oligosaccharide-lipid substrate affects recognition of the polypeptide substrate. Studies with oligosaccharyl transferase in yeast membranes by Sharma et al. (1983) indicate that the structure of the oligosaccharide-lipid affects the apparent K_m of the enzyme for the peptide substrate. Thus, the K_m for the peptide in the presence of the "natural" substrate, $Glc_3Man_9GlcNAc_2$-PP-Dol, is 10-fold lower than the K_m observed when the "unnatural" substrate, $GlcNAc_2$-PP-Dol, is added.

CHEMICAL MECHANISM OF THE TRANSFER REACTION

As noted in the Introduction, the reaction catalyzed by oligosaccharyl transferase is unusual because it involves the reaction between the β-amido nitrogen atom of the Asn residue, apparently acting as a nucleophile, and oligosaccharyl-PP-Dol. Normally a β-amido nitrogen atom of the type found in an Asn residue would lack nucleophilic properties. A model proposed by Bause (Bause, 1984; Bause & Legler, 1981) to explain how this moiety could acquire nucleophilicity is supported by the above-mentioned finding that a tripeptide containing epoxyethylglycine in place of Ser or Thr is an irreversible inhibitor of oligosaccharyltransferase. In this model, the oxygen atom of the hydroxyl group of Ser or Thr, by virtue of interaction with a basic residue in the protein, serves as an electron donor to one of the hydrogen atoms on the amido nitrogen. This, in turn, increases the nucleophilicity of the nitrogen, which reacts at C-1 of the oligosaccharide donor, releasing dolichylpyrophosphate. The observation that the epoxide derivative inhibits the enzyme can be explained if one assumes that the epoxide mimics the function of the hydroxyl group in Ser or Thr, but in this process a basic residue in the enzyme is alkylated, thereby inactivating the enzyme.

TOPOLOGICAL ASPECTS OF PROTEIN GLYCOSYLATION

A series of *in vivo* studies (reviewed in Hanover & Lennarz, 1981) demonstrated that the glycosylation of newly translated protein occurs in the rough endoplasmic reticulum and that, in most cases studied, it is a cotranslational process. Several lines of evidence suggest that oligosaccharyl transferase is situated in the membrane in such a way that oligosaccharide transfer to the growing nascent polypeptide chain occurs at the luminal rather than the cytoplasmic face.

1. Completed oligosaccharyl-lipid is not accessible to concanavalin A added to microsomes (Snider & Rogers, 1984). Similarly, newly synthesized glycoproteins are not degraded by trypsin added to microsomes (Katz et al., 1977; Lingappa et al., 1978; Rothman & Lodish, 1975).

2. Free oligosaccharide, released from oligosaccharyl-PP-Dol by either chemical or enzymatic hydrolysis, remains entrapped in the lumen of the rough endoplasmic reticulum (Hanover & Lennarz, 1982).

3. After attachment of Asn 293 to the growing ovalbumin nascent chain, at least 32 more amino acid residues (a number sufficient for the polypeptide chain to reach the luminal face) are added before the carbohydrate is attached (Glabe et al., 1980).

4. When small acceptor peptides (which freely permeate the microsome) are glycosylated most of the resulting glycopeptide is trapped within the lumen (Welply et al., 1983).

PROPERTIES OF THE OLIGOSACCHARYL TRANSFERASE OF HIGHER EUKARYOTES

Given the central position of oligosaccharyl transferase between the assembly and processing stages of glycoprotein synthesis, it is important to have detailed knowledge of its properties. Oligosaccharyl transferase from several sources has been solubilized with detergent (Sharma et al., 1983; Das & Heath, 1980; Ronin, 1980; Aubert et al., 1982). In two cases, the solubilized enzyme was partially purified from hen oviduct by affinity chromatography on resins containing covalently attached peptide acceptor substrates (Das & Heath, 1980; Aubert et al., 1982). Unfortunately, purification of enzymatically active oligosaccharyl transferase to homogeneity has not been accomplished using these and other conventional techniques, probably because of the extreme lability of the enzyme after it has been solubilized. In this context it is important to cite a recent study (Chalifour & Spiro, 1988) that, following earlier observations (Frank & Bouchilloux, 1984), suggests that specific phospholipids may be important for activity.

In contrast to the enzyme *per se*, considerable knowledge has been accumulated about the peptide acceptor substrate. In microsome preparations from at least two tissues, simple tripeptides containing Asn-X-Thr/Ser sequences, blocked at both the amino and carboxy termini, were shown to be acceptors for N-glycosylation (Hart et al., 1979; Ronin et al., 1981). An unmodified asparagine side chain in position 1 (Welply et al., 1983) and threonine, serine, or perhaps cysteine in position 3 (Bause & Legler, 1981) are general requirements for acceptor activity. Of particular significance were the findings that the acceptor activity of an Asn-X-Thr/Ser tripeptide could be increased up to 10-fold by the presence of hydrophobic groups at the amino terminus (Welply et al., 1983), and that any amino acid except proline was permitted in the X position (Bause & Legler, 1981).

DEVELOPMENT OF PHOTOAFFINITY PROBES FOR A COMPONENT OF OLIGOSACCHARYL TRANSFERASE

We hoped to utilize the information obtained on peptide specificity to develop labeled photoaffinity derivatives of peptide substrates. Such derivatives would then be used to identify and isolate labeled oligosaccharyl transferase. Following identification of the polypeptide labeled by this technique, the plan was to isolate it (not necessarily in an enzymatically active state), prepare antibody, and proceed with the usual approaches that antibodies afford for structural studies (see below).

Given the latitude permissible in the X position of Asn-X-Ser/Thr and the desirability of having a hydrophobic peptide, the first generation photoprobe prepared had the structure shown in Figure 2, where the N-terminus is acetylated and the C-terminus is amidated. Initially it was established that this peptide, N^{α}-[^{3}H]-Ac-Asn Lys(N^{ϵ}-p-azidobenzoyl)-Thr-NH_2, in the presence of hen oviduct microsomes containing endogenous oligosaccharyl pyrophosphoryl dolichol, was converted to a glycopeptide (Welply et al., 1985). Moreover, ^{3}H-peptide addition to microsomes followed by UV irradiation resulted in covalent linkage to a protein component, with a concomitant 10-15% inhibition of the oligosaccharyl transferase activity.

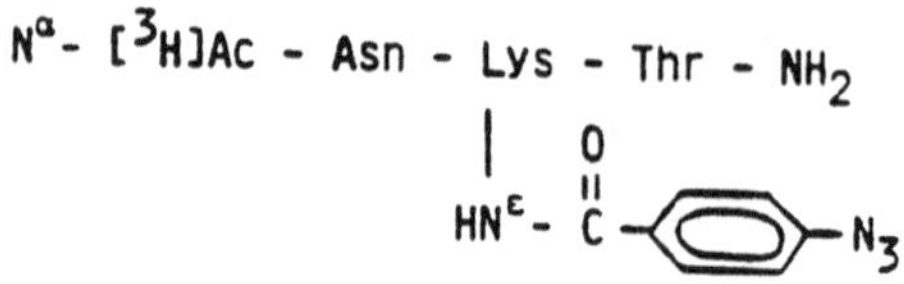

FIGURE 2 - Structure of first generation photoaffinity probe.

As determined by SDS/PAGE, the molecular mass of the photolabeled putative oligosaccharyl transferase from hen oviduct microsomes was 60 kDa. Radiolabeling of the 60-kDa polypeptide was completely dependent upon photolysis of the azido peptide in the presence of the hen oviduct endoplasmic reticulum preparation and Mn^{2+}, which is required for enzymatic activity. Labeling of the enzyme was not inhibited in the presence of a 10-fold excess of nonacceptor peptides, unacetylated Asn-Lys(N^{ϵ}-p-azidobenzoyl)-Thr-NH_2, or unacetylated Asn-Leu-Thr-NH_2, whereas it was completely abolished by the presence of a 10-fold excess of the competing acceptor peptide, N^{α}-Bz-Asn-Leu-Thr-NH_2. Thermal inactivation of oligosaccharyl tranferase activity was achieved by heating endoplasmic reticulum preparations to 60°C. This loss of enzyme activity at 60°C was paralleled by a decrease in radiolabeling of the 60-kDa polypeptide, whereas temperatures of 50°C and lower had no effect on either process.

To increase the signal-to-noise ratio in such labeling studies a second generation probe was synthesized (Welply et al., 1986). In this probe, the structure of which is shown in Figure 3, the C-terminus is still amidated but the hydrophobicity was increased by use of a more hydrophobic moiety at the N-terminus. In addition, the specific activity was enhanced at least 500-fold by use of ^{125}I instead of ^{3}H. As shown in Figure 4, this probe could be used in assays of binding and the overall glycosylation process.

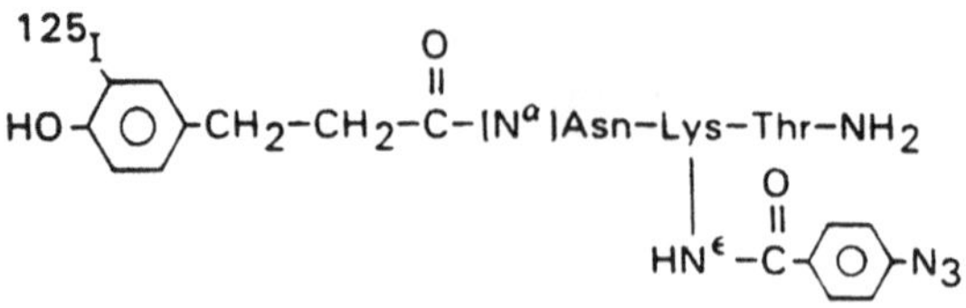

FIGURE 3 - Structure of second generation photoaffinity probe.

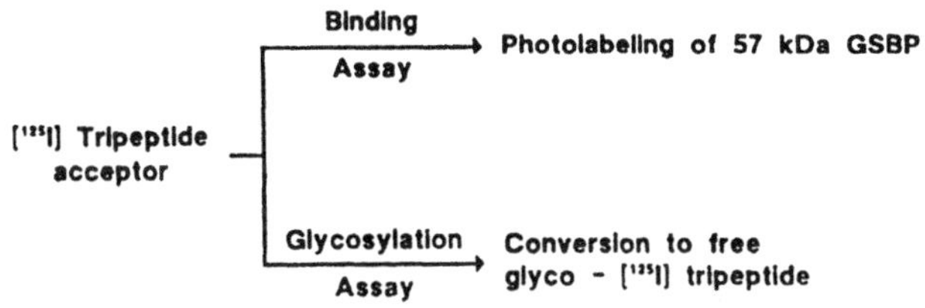

FIGURE 4 - Use of labeled probe to measure either binding to GSBP or glycosylation.

Studies carried out with this new probe revealed that all procedures that would be expected to perturb the lipid bilayer resulted in a dramatic loss of oligosaccharyl tranferase activity. Treatment of sealed microsomal membrane vesicles with phospholipase A resulted in nearly complete enzyme inactivation; in contrast, phospholipase C or D had moderate or little effect, respectively. Taken together, these results suggested that the hydrophobic environment of the membrane was required for oligosaccharyl transferase activity. Trypsin treatment of intact vesicles diminished enzyme activity by nearly 70%, but it had no effect on binding of the ^{125}I-labeled photoaffinity probe nor did it reduce the molecular weight of the ^{125}I-labeled protein. These results led us to propose two models (Figure 5). In one (A) the domain of oligosaccharyl transferase that recognizes glycosylation sites (GSB) was suggested to be luminally disposed; a second domain (OS) bound and transferred the carbohydrate donor, oligosaccharyl-PP-Dol, and was inactivated by external trypsin. In the second model (B), two separate subunits were proposed: the lumen subunit (GSBP) recognized the peptide acceptor site and the membrane subunit (OSP) recognized and transferred the oligosaccharide chain of oligosaccharyl-PP-Dol. These two alternative models were consistent with the data available at that time.

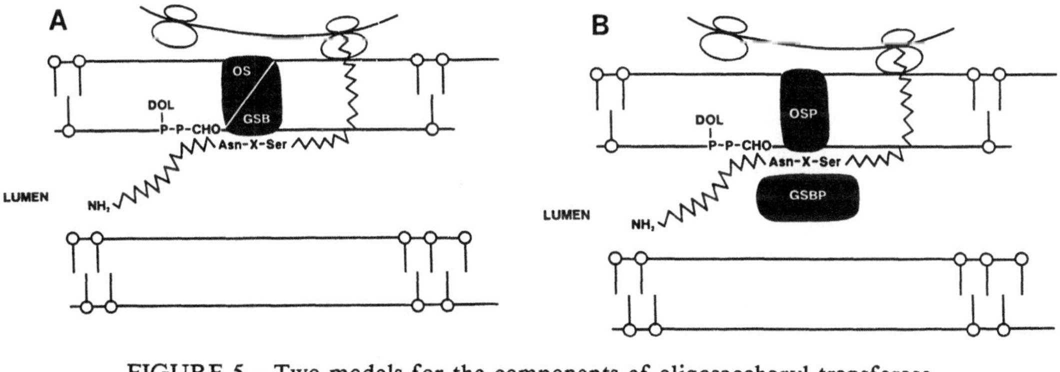

FIGURE 5 - Two models for the components of oligosaccharyl transferase.

Two subsequent studies cast light on the nature and the molecular weight of the polypeptide that became photolabeled in these two models. In the first of these studies (Kaplan et al., 1988), it was established that the ^{125}I-tripeptide probe actually bound to a 57-kDa protein of the endoplasmic reticulum that was subsequently glycosylated at the newly introduced glycosylation site and thereby converted to the previously reported 60-kDa form. The 57-kDa protein, isolated from microsomes by two-dimensional gel electrophoresis, was used as immunogen to prepare polyclonal antisera. The specificity of the antibody was established on the basis of its ability to 1) recognize the 57-kDa protein by immunoblotting and 2) immunoprecipitate the photolabeled protein. The antibody also recognized photolabeled protein from different tissues and organisms. The 57-kDa protein isolated by immunoprecipitation retained its ability to interact with the photoaffinity probe, but was inactive in catalyzing glycosylation of peptides. This finding supported the idea that the 57-kDa protein is the component of oligosaccharyl transferase that recognizes the glycosylation site in polypeptides. The demonstration of the existence of this component (named glycosylation site binding protein, GSBP), coupled with the earlier experiment on lack of trypsin sensitivity of this polypeptide when it is in the ER, clearly supported model B.

The second study (M. Geetha-Habib et al., 1988) focused on the component that binds the glycosylation site, i.e. GSBP. A surprisingly close relationship was revealed between two seemingly unrelated aspects of the endoplasmic reticulum-polypeptide glycosylation and resident proteins in the lumen of this subcellular organelle. With respect to glycosylation, the experiments with saponin revealed that depletion of GSBP from the endoplasmic reticulum resulted in loss of ability of the membrane to catalyze oligosaccharide transfer to protein, although the released GSBP still recognized the glycosylation site because it could be specifically labeled with the photoaffinity probe. This observation, suggesting a requirement for GSBP in the overall glycosylation process, is in agreement with earlier thermal inactivation studies (Welply et al., 1983) showing that loss of ability to photolabel GSBP was accompanied by loss of oligosaccharylation. Further correlative support for this idea comes from studies on muscle cell differentiation and *in vivo* studies on rabbit cervix development (see below). All of these findings are consistent with model B. In this model, luminal GSBP recognizes glycosylation sites in nascent polypeptides and thereby facilitates interaction of these sites with the subunit of the transferase that carries out the oligosaccharide transfer step, OSP. However, further studies involving reconstitution of glycosylation activity in microsomes that have been depleted of GSBP will be necessary to test this model.

The second interrelated aspect of this work concerns luminal proteins of the endoplasmic reticulum. While this work was in progress, it became apparent that the ER contains a resident group of luminal proteins that are retained in this compartment by virtue of a tetrapeptide (-KDEL) sequence at their C termini (Koch, 1987). The luminal distribution of GSBP led us to focus on these proteins. Three of them, the β-subunit of prolyl hydroxylase, protein disulfide isomerase (PDI), and thyroid hormone binding protein, have in common a molecular mass of 55-60 kDa, and recent cloning and sequencing have led to the conclusion that all three have high sequence similarity, if not identity (Pihlajaniemi et al., 1987; Cheng et al., 1987; Koivu et al., 1987). Moreover, biochemical studies on two of the three, the β-subunit of prolyl hydroxylase and PDI from either rat liver or chick embryos, have shown that these proteins have the same peptide maps and immunoreactivity (Boado et al., 1988). In addition, the β-subunit of prolyl hydroxylase has been shown to possess PDI activity (Boado et al., 1988). Recently a fourth protein, triiodothyronine monodeiodinase (Grant et al., 1986), has been added to this list of apparently identical proteins.

The similarity in subcellular distribution and molecular weight of GSBP and that of the three other proteins, and the cross-hybridization of the GSBP cDNA clones with a PDI cDNA clone, suggested that GSBP is related, if not identical, to these other proteins. Conclusive proof for the high sequence similarity of GSBP and these other three luminal proteins was obtained by sequence analysis of the 2.4 kbp GSBP cDNA. This analysis revealed that the degree of similarity in the coding region between chicken GSBP and rat PDI is 94% if conservative substitutions are allowed (Geetha-Habib et al., 1988). Excluding the signal sequence, only 52 amino acid residues are different between these two proteins. Moreover, the proposed active sites of PDI (Pihlajaniemi et al., 1987), involving two Trp-Cys-Gly-His-Cys-Lys sequences, are also conserved in GSBP.

The cDNA-derived amino acid sequence of the chicken GSBP is also very similar to those of the human β-subunit of prolyl hydroxylase and human thyroid hormone binding protein and triiodothyronine monodeiodinase (Geetha-Habib et al., 1988; Parkloneu et al., 1988). The coding region similarity between chicken GSBP and these three proteins exceed 90% if conservative substitutions are allowed. The finding of a deduced carboxy-terminal -KDEL sequence in GSBP is consistent with its location in the lumen of the ER (Geetha-Habib et al., 1988). The validity of these interspecies comparisons is underscored by the earlier cited study with rabbit anti-hen oviduct GSBP, which suggested that at least some portion of the primary structure of GSBP is conserved in a wide variety of higher eukaryotes (Kaplan et al., 1987)

Although it remains to be unequivocally established that in a given species GSBP and the other luminal proteins are one and the same molecule, our observations that protein disulfide isomerase can be labeled with the photoaffinity probe (unpublished data) and that GSBP can be immunoprecipitated with antibody to thyroid hormone binding protein strongly support the idea of identity (Geetha-Habib et al., 1988). If, in fact, they are identical, it raises the very interesting question of how this single multifunctional protein functions in co- and/or post-translational modifications of polypeptides that are synthesized in the endoplasmic reticulum. Except for the thyroid hormone-related proteins, the other three activities involve recognition of a polypeptide at a specific site prior to a covalent modification. That is, recognition of an "appropriate" prolyl, cystinyl, or asparaginyl residue must occur prior to hydroxylation, reduction, or glycosylation, respectively.

PHYLOGENETIC AND REGULATORY ASPECTS OF GSBP

Even before the elucidation of GSBP as the component that was labeled by the photoaffinity probe, phylogenetic studies (Kaplan et al., 1987) using photoaffinity labeling indicated that the 57-kDa species was present in all higher eukaryotes examined, including *Drosophila melanogaster*. When antibody to GSBP became available and was used to immunoprecipitate photolabeled proteins from a variety of sources, the same conclusion was reached (Kaplan et al., 1988), i.e. in all higher eukaryotes, including *Drosophila*, a 57-kDa labeled protein was precipitated with rabbit antibody to chicken oviduct GSBP.

These findings, indicating widespread distribution and apparent high homology of GSBP in higher eukaryotes, are relevant to findings on the relationship between the level of this component of the glycosylation system and the overall glycosylation process. This relationship was examined by separately measuring the amount of GSBP that can be photolabeled and the rate of conversion of the labeled peptide to glycopeptide in microsomes as outlined earlier. One of the two biological systems studied was a nonfusing myoblast cell line, BC_3H1 (Grant et al., 1986), which was chosen because transfer of proliferating undifferentiated myoblast cells to mitogen-depleted medium results in withdrawal from the cell cycle and induction of a variety of muscle-specific gene products including cell surface glycoproteins (Olson et al., 1983). As noted above, using an acceptor peptide the overall glycosylation activity was measured by formation of [^{3}H]glycopeptide and the level of GSBP was measured on the basis of binding of the ^{125}I-labeled photoreactive acceptor tripeptide. Both of these independent assay methods revealed a marked increase in the oligosaccharyl transferase system when differentiation was induced by serum depletion. Moreover, mitogenic stimulation of differentiated cells, thereby inducing de-differentiation, resulted in a return of the oligosaccharyl transferase components to near basal levels. With the availability of a cDNA probe for GSBP this system has also been studied with respect to changes in the level of GSBP mRNA. Similarly, changes in the level of PDI activity, which would be expected to be identical to those of GSBP, were measured. The results of these measurements, along with the earlier findings, all indicate coordinate changes in GSBP/PDI levels and peptide glycosylation activity, as expected if GSBP is a required component in the overall glycosylation process and if GSBP and PDI are the same molecule (unpublished data).

The second biological system examined was the developing rabbit cervix (Chilton et al., 1988). N-linked glycoprotein synthesis was studied during sexual maturation and following ovariectomy of mature rabbits. As described above, GSBP levels were measured using the photoprobe and overall glycosylation activity was measured by assessing the conversion of the peptide to glycopeptide. The results of these two independent assay methods revealed an estrogen-dependent, 10- to 15-fold increase in both GSBP and peptide glycosylation activity during cervical differentiation. Consistent with these developmental effects, ovariectomy of adult rabbits reduced titers of serum estradiol and reduced glycosylation activity by 75%. Similar declines were observed when estrous animals were made pseudopregnant. The treatment of ovariectomized animals with estradiol resulted in the restoration of the level of GSBP and the glycosylation activity to estrous control values. Collectively, these results provide the first definitive evidence that hormones can regulate the level and activity of components involved in the attachment of N-linked oligosaccharide chains to proteins in the endocervix.

STUDIES ON YEAST OLIGOSACCHARYL TRANSFERASE

Studies with the photoprobe have revealed that GSBP in a diversity of higher eukaryotes, including *Drosophila*, has an apparent mass of 57 kDa. Studies using the photoaffinity probe with yeast microsomes failed to reveal the presence of a 57-kDa protein. Rather, an 80-kDa microsomal protein was labeled (Kaplan et al., 1987). Given the relative lack of information on the specificity of the peptide glycosylation in this system, controls were performed to document the specific nature of the labeling by the photoprobe. The results of experiments to test the specificity of photolabeling shown in Figure 6 indicate that unlabeled acceptor peptide (A) blocks labeling and glycosylation, whereas a peptide in which the hydroxy amino acid in position 3 has been replaced by Ala does not (B). These results strongly support the idea that the 80-kDa protein that is labeled specifically recognizes the glycosylation site sequence.

The finding of an 80-kDa rather than 57-kDa protein coupled with the observation that anti-GSBP did not immunoprecipitate the photolabeled 80-kDa yeast protein suggested that this organism might have an oligosaccharyl transferase system distinct from that found in higher eukaryotes. Recent experiments support this idea. As shown in Figure 7 the yeast enzyme can be solubilized in active form with detergent and in the presence of exogenous oligosaccharyl pyrophosphoryl dolichol shows high activity. This observation contrasts with the instability of the solubilized mammalian oligosaccharyl transferase. A further difference was that, unlike GSBP, the 80-kDa photolabeled component was not released from microsomes by Na_2CO_3 treatment or by sonication (unpublished observations), suggesting that it is an integral membrane protein.

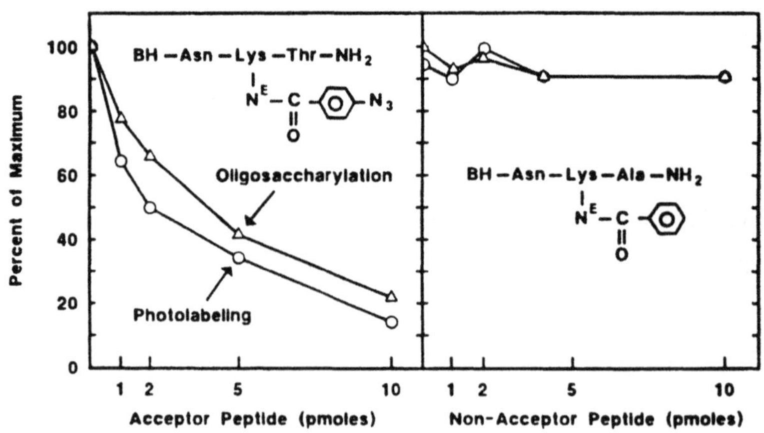

FIGURE 6 - Specificity of labeling of yeast oligosaccharyl transferase.

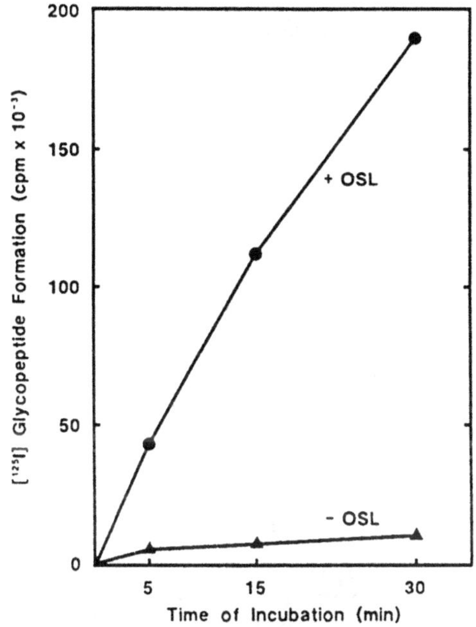

FIGURE 7 - Activity of solubilized yeast oligosaccharyl transferase.

Finally, using yeast microsomes it was found that trypsin treatment caused a parallel loss in oligosaccharylation and the ability to photolabel the 80-kDa protein. Similar parallel losses were caused by EDTA or heat treatment.

Based on these observations the current working model for yeast oligosaccharyl transferase is one in which a soluble luminal subunit, i.e. GSBP, is not present. Instead, as postulated earlier in Figure 5, Model A, both the oligosaccharyl-PP-Dol and the -Asn-X-Ser/Thr- recognition components, OS and GSB, respectively are proposed to reside in the same polypeptide chain. Recently, sufficient quantities of the 80-kDa protein have been isolated to prepare an antibody. The antibody binds to an 80-kDa protein in yeast microsomes upon Western blot analysis but does not react with oviduct microsomes. Conversely, anti-GSBP antibody reacts with oviduct but not with yeast microsomes. In addition, the yeast antibody is capable of precipitating an 80-kDa photolabeled protein. Currently this antibody is being used to screen a yeast genomic library. Hopefully in the near future it will be possible to clone and then sequence yeast oligosaccharyl transferase.

ACKNOWLEDGMENTS

This work was supported by NIH grant GM33185 to WJL. Dr. Lennarz, a Robert A. Welch Professor of Chemistry, gratefully acknowledges the Robert A. Welch Foundation.

REFERENCES

Aubert, J.P., Chiroutre, M., Kerckaert, J.P., Helbecque, N. and Loucheux-Lefebvre, M.H. (1982). Biochem. Biophys. Res. Commun. **104**, 1550-1559.
Ballou, L., Gopal, P., Krummel, B., Tammi, M. and Ballou, C. (1986). Proc. Natl. Acad. Sci USA **83**, 3081-3085.
Bause, E. (1983). Biochem. J. **209**, 323-330.
Bause, E. (1984). Biochem. Soc. Trans. **12**, 514-517.
Bause, E. and Legler, G. (1981). Biochem. J. **195**, 639-644.
Boado, R.J., Campbell, D.A. and Chopra, I.J. (1988). Biochem. Biophys. Res. Comm. **155**, 1297-1304.
Chalifour, R.J. and Spiro, R.C. (1988). J. Biol. Chem. **263**, 15673-15680.
Chen, W.W. and Lennarz, W.J. (1977). J. Biol. Chem. **252**, 3473-3479.
Cheng, S., Gong, Q., Parkison, C., Robinson, E.A., Appella, E., Merlino, G.T. and Pastan, I. (1987) J. Biol. Chem. **262**, 11221-11227.
Chilton, B.S., Kaplan, H.A. and Lennarz, W.J. (1988). Endocrinology **123**, 1237-1244
Das, R. and Heath, E.C. (1980). Proc. Natl. Acad. Sci. USA **77**, 3811-3815.
Edman, J.C., Ellis, L., Blacher, R.W., Roth, R.A. and Rutter, W.J. (1985). Nature **317**, 267-270.
Franc, J.-L. and Bouchilloux, S. (1984). Biochim. Biophys. Acta **800**, 166-170.
Geetha-Habib, M., Noiva, R., Kaplan, H.A. and Lennarz, W.J. (1988). Cell **54**, 1053-1060.
Glabe, C.G., Hanover, J.A. and Lennarz, W.J. (1980). J. Biol. Chem. **255**, 9236-9242.
Grant, S.R., Welply, J.K., Olson, E.N. and Lennarz, W.J. (1986). Arch. Biochem. Biophys. **248**, 424-428.
Hanover, J.A. and Lennarz, W.J. (1981). Arch. Biochem. Biophys. **211**, 1-19.
Hanover, J.A. and Lennarz, W.J. (1982). J. Biol. Chem. **257**, 2787-2794.
Hart, G.W., Brew, K., Grant, G.A., Bradshaw, R.A. and Lennarz, W.J. (1979). J. Biol. Chem. **254**, 9747-9753.
Hortin, G., Stern, A.M., Miller, B., Abeles, R.H. and Boime, I. (1983). J. Biol. Chem. **258**, 4047-4050.
Hubbard, S.C. and Ivatt, R.J. (1981). Ann. Rev. Biochem. **50**, 555-583.
Huftaker, T.C. and Robbins, P.W. (1983). Proc. Natl. Acad. Sci. USA **80**, 7466-7470.
Hummel, J.P. and Dreyer, W.J. (1962). Biochim. Biophys. Acta **63**, 530-537.
Kaplan, H.A., Naider, F. and Lennarz, W.J. (1988). J. Biol. Chem. **263**, 7814-7820.
Kaplan, H.A., Welply, J.K. and Lennarz, W.J. (1987). Biochim. Biophys. Acta **906**, 161-173.
Katz, F.N., Rothman, J.E., Lingappa, U.R., Blobel, G. and Lodish, H.F. (1977). Proc. Natl. Acad. Sci. USA **74**, 3278-3282.
Koch, G.L.E. (1987). J. Cell Sci. **87**, 491-492.
Koivu, J., Myllyla, R., Helaakoski, T., Pihlajaniemi, T., Tasanen, K. and Kivirikka, K.I. (1987). J. Biol. Chem. **262**, 6447-6449.
Kornfeld, R. and Kornfeld, S. (1980). The Biochemistry of Glycoproteins and Proteoglycans (Lennarz, W.J., ed.). pp. 1-34, Plenum Press, New York.
Kornfeld, S., Gregory, W. and Chapman, A. (1979). J. Biol. Chem. **254**, 11649-11654.
Kronquist, K.E. and Lennarz, W.J. (1978). J. Supramol. Struct. **8**, 51-65.
Lau, J.T.Y., Welply, J.K., Shenbagamurthi, P., Naider, F. and Lennarz, W.J. (1983). J. Biol. Chem. **258**, 15255-15260.
Lingappa, V.R., Lingappa, J.R., Prasad, R., Ebner, K.E. and Blobel, G. (1978). Proc. Natl. Acad. Sci. USA **75**, 2338-2341.
Marshall, R.D. (1974). Biochem. Soc. Symp. **40**, 17-26.
Olson, E.N., Glaser, L., Merlie, J.P., Sebane, R. and Lindstrom, J. (1983). J. Biol. Chem. **258**, 13946-13953.
Parkloneu, T., Kivirikko, K.I. and Pihlajaniemi, T. (1988). Biochem. J. **256**, 1005-1011.
Pihlajaniemi, T., Helaakoski, T., Tasanen, K., Myllyla, R., Huhtala, M.-L., Koivu, J. and Kivirikko, K.I. (1987). EMBO J. **6**, 643-649.

Pless, D.D. and Lennarz, W.J. (1977). Proc. Natl. Acad. Sci. USA **74**, 134-138.
Rathod, P.K., Tashjian, A.H., Jr. and Abeles, R.H. (1986). J. Biol. Chem. **261**, 6461-6469.
Ronin, C. (1980). FEBS Lett. **113**, 340-344.
Ronin, C., Granier, C., Caseti, C., Bouchilloux, S. and Rietschoten, J.V. (1981). Eur. J. Biochem. **118**, 159-164.
Rothman, J.E. and Lodish, H.F. (1975). Nature (London) **269**, 775-780.
Sharma, C.B., Lehle, L. and Tanner, W. (1983). Eur. J. Biochem. **116**, 101-108.
Snider, M.D. and Rogers, O.C. (1984). Cell **36**, 753-761.
Stern, A.M., Foxman, B.M., Tashjian, A.H., Jr. and Abeles, R.H. (1982). J. Med. Chem. **25**, 544-550.
Struck, D.K. and Lennarz, W.J. (1980). The Biochemistry of Glycoproteins and Proteoglycans (Lennarz, W.J., ed.), pp. 35-84, Plenum Press, New York.
Struck, D.K., Lennarz, W.J. and Brew, K. (1978). J. Biol. Chem. **253**, 5786-5794.
Waechter, C.J. and Lennarz, W.J. (1976). Ann. Rev. Biochem. **45**, 95-112.
Welply, J.K., Kaplan, H.A., Shenbagamurthi, P., Naider, F. and Lennarz, W.J. (1986). Arch. Biochem. Biophys. **246**, 808-819.
Welply, J.K., Shenbagamurthi, P., Lennarz, W.J. and Naider, F. (1983). J. Biol. Chem. **258**, 11856-11863.
Welply, J.K., Shenbagamurthi, P., Naider, F., Park, H.R. and Lennarz, W.J. (1985). J. Biol. Chem. **260**, 6459-6465.

MUTANTS: A FRUITFUL APPROACH TO THE UNDERSTANDING OF ASPARAGINE-LINKED GLYCOSYLATION IN EUKARYOTIC CELLS

James Stoll, Anne Rosenwald, and Sharon S. Krag
Department of Biochemistry
School of Hygiene and Public Health
The Johns Hopkins University
615 North Wolfe Street
Baltimore, Maryland
USA

Proteins which have glycans attached on one to six asparagine residues (N-linked glycoproteins) are important constituents of eukaryotic cells. They occur as soluble or membrane-associated components either inside or outside of cells. Secreted proteins such as thyroglobulin and immunoglobulins and intravesicular proteins such as the lysosomal glycosidases are N-linked glycoproteins. Many membrane-associated proteins found on the cell-surface are this class of glycoproteins - insulin receptor, EGF receptor, LDL receptor, and transferrin receptor, as are enzymes associated with many cellular membranes such as Na^+/K^+ ATPase, galactosyl transferase, sialyl transferase and HMG CoA reductase. Fibronectin and laminin, proteins found at the cell surface, also contain N-linked glycans.

The role of the glycan portion of these molecules varies from protein to protein. Carbohydrate moieties can influence the protein's solubility, structure, turnover, and compartmentalization. While there is little evidence that the glycan plays a direct role in enzymatic function, sugars are known to be involved in cell-cell interactions, cell-virus interactions, and receptor-ligand interactions (Rademacher, et al 1988 and Krag, 1985).

NATO ASI Series, Vol. H 40
Dynamics and Biogenesis of Membranes
Edited by J.A.F. Op den Kamp

The oligosaccharides attached to asparagine residues have diverse structures, with either a complex composition (N-acetylglucosamine, mannose, galactose, sialic acid, and fucose) or a simple, high mannose composition (with only N-acetylglucosamine and mannose) (Kornfeld and Kornfeld, 1985). One protein can have oligosaccharides attached to both asparagine residues and serine/threonine residues (O-linked glycans) (Kornfeld and Kornfeld, 1980). While a specific asparagine residue in a protein has either high mannose or complex oligosaccharides attached, a protein can have both types of glycans (Hsieh, et al, 1983).

Despite their diversity in structure, all oligosaccharides found attached to asparagine residues originate from a common biosynthetic pathway (Hubbard and Ivatt, 1981). This pathway is complex, involving at least fifty reactions in three different subcellular compartments. The reactions in this pathway can be grouped into the following four major steps: (1) synthesis of the carrier lipid; (2) assembly of the oligosaccharide-lipid intermediate; (3) transfer of the oligosaccharide unit to an asparagine residue on the protein; and (4) modification of the oligosaccharide unit on the protein.

The carrier lipid involved in these reactions is dolichyl phosphate, a long-chain polyisoprenoid lipid containing 17 to 20 isoprene units (Figure 1). It is synthesized from acetate through mevalonate (Schroepfer, 1981), using enzymes found in the cytoplasm and the endoplasmic reticulum. Intermediates in the synthesis include hydroxymethylglutaryl CoA, isopentenyl pyrophosphate, farnesyl pyrophosphate, and polyprenyl pyrophosphate. The precise end product of the *de novo* synthesis which is catalyzed by prenyl transferase (Poulter and Rilling, 1974; Adair, et al 1984) is unclear (Ekstrom, et al 1984; Keller, 1986; Wellner and Lucas, 1979; Adair and Keller, 1982; Daleo, et al, 1977; Wong and Lennarz, 1982). As shown in Figure 1, many interconversions of dolichyl derivatives are known, catalyzed by polyprenol reductase, dolichol kinase, dolichol acyl transferase, dolichyl esterase, dolichyl phosphate phosphatatase, and dolichyl pyrophosphate phosphatase (Allen et al, 1978; Appelkvist, et al, 1981; Belcopitow and Boscoboinik, 1982; Kato, et al, 1980; Wedgewood and Strominger, 1980; Burton, et al, 1981; Rip, et al, 1981; Radominska-Pyrek, et al, 1979; Scher and Waechter, 1981).

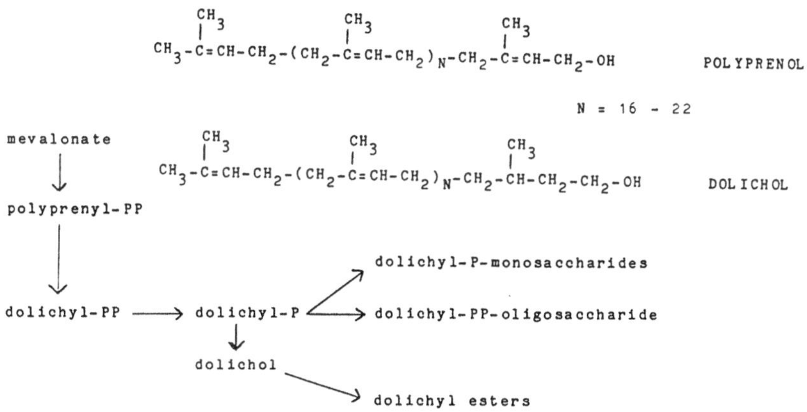

Figure 1. Lipid carrier dolichol and its precursor polyprenol and the various dolichyl derivatives

The second step in the biosynthesis of these glycans is the assembly of an oligosaccharide on the dolichyl phosphate carrier lipid; the oligosaccharide contains N-acetylglucosamine (GlcNAc), mannose (Man), and glucose. The sugar transferases presumably occur in the rough endoplasmic reticulum and face either the cytoplasmic or the lumenal face (Hirschberg and Snider, 1987). As depicted in Figure 2, sugars are donated by either sugar nucleotide carriers or lipid carriers depending on the transferase. During its biosynthesis, the saccharide moiety attached to the dolichyl phosphate must be transported from the cytoplasmic face of the endoplasmic reticular membrane to the lumenal face, although the point(s) at which this translocation occurs is still unclear (Snider and Rogers, 1984; Hanover and Lennarz, 1982).

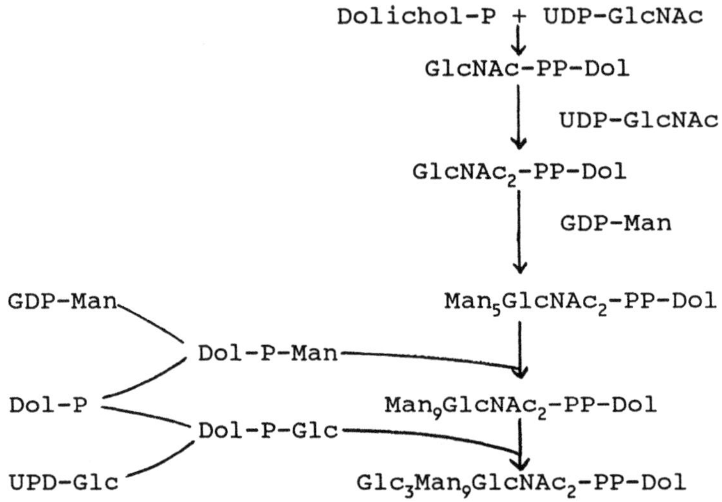

Figure 2. Summary of steps in the assembly of the oligosaccharide-lipid intermediate in glycoprotein biosynthesis

The third step in the biosynthesis of N-linked glycans is the transfer of the oligosaccharide as a unit to proteins which have the sequence Asn-X-Ser/Thr (Marshall, 1974). In many cases the transfer occurs as the protein is being translated and translocated into the lumen of the rough endoplasmic reticulum. Dr. Lennarz will be presenting the current understanding of this enzymatic step in his presentation.

Finally, the oligosaccharide which has been transferred from the lipid to the protein, in most cases $Glc_3Man_9GlcNAc_2$, is restructured to its mature form while on the protein by a combination of membrane-associated glycosidases and glycosyl transferases, located in the rough endoplasmic reticulum and Golgi (Kornfeld and Kornfeld, 1985; Kornfeld, 1982).

In order to understand this complex biosynthetic pathway, its regulation, and the function of the glycose moiety on proteins, a combination of genetic and biochemical approaches must be used. Inhibitors such as tunicamycin, deoxynojirimycin, and swainsonine are important tools in this analysis (Schwarz and Datema, 1982; Elbein, 1987). The approach of purification of the individual enzymes and substrates and the in vitro reconstitution of the pathway will be aided by large amounts of protein available by utilizing current recombinant DNA technologies.

A third approach has been to isolate and characterize glycosylation mutants. Characterization of mutants can lead to an understanding of the effects of an alteration in a specific part of the pathway. Mutants are often the source of substrates for reconstitution studies. Finally, mutants can be used as tools in the isolation of the genes for enzymes in the pathway in order to obtain large amounts of the enzymes for reconstitution studies.

Both selection and screening procedures, with or without enrichment techniques, have been used to isolate glycosylation mutants (Figure 3). Many of these mutants have been isolated in Chinese hamster ovary (CHO) cells (Stanley, 1984; Krag, 1985) because they are relatively easy to grow, have a stable karyotype in culture, and have an extensive functional hemizygosity of their genome (Adair and Siciliano, 1986). There has been a recent report of one apparently diploid locus (Kingsley and Krieger, 1984). Recently many glycosylation mutants have been isolated in yeast which are most amenable to genetic analysis (Huffaker and Robbins, 1982; Runge et al, 1984).

Some selection procedures use lectins, proteins that bind specific sugar residues on the surface of cells and are cytotoxic when taken up by cells (Stanley, 1984). For example, studies on Lec 1 CHO cells, selected either by resistance to phytohemagglutinin (Stanley, et al, 1975) or ricin (Gottlieb, et al, 1975), were important in elucidating the sequence of reactions during processing of newly-glycosylated proteins. Lec 1 cells lack GlcNAc transferase I activity and accumulate the endo-beta-N-acetylglucosaminidase H-sensitive $Man_5GlcNAc_2$ structure on protein (Kobata, 1978; Tarentino, et al, 1978). The accumulation of this species led to the conclusion that the processing pathway is ordered such that GlcNAc must be transferred from the sugar nucleotide to $Man_5GlcNAc_2$ prior to the removal of the two mannosyl residues on the alpha 1,6 branch by mannosidase II to generate $GlcNAc_1Man_3GlcNAc_2$ structures on proteins (Kornfeld and Kornfeld, 1985).

Isolation of Glycosylation Mutants

I. Selections based on defective cell surface molecules
   - Lectins
   - Man 6-P Receptor
   - LDL Receptor

II. Screens
   - Low intracellular lysosomal activities
   - Secretion
   - Low mannose incorporation into protein

Figure 3. Strategies for the isolation of glycosylation mutants

Other selection procedures also exploit the function or structure of particular cell-surface molecules. For example, the CHO cell B4-2-1 was isolated by a protocol (Figure 4, Robbins, et al, 1981) that selected for cells with surface mannose 6-phosphate receptors having reduced binding to lysosomal enzymes; these receptors recognize proteins bearing a mannose 6-phosphate moiety. Mutagenized CHO cells were incubated with a conjugate of Man_5P-ricin in the presence of lactose. Cells with a functional mannose 6-phosphate receptor on the cell surface would bind to the conjugate, internalize it, and be killed by the action of the toxin ricin. The presence of lactose during the incubation prevented the toxin from binding to the cells via the ricin molecule. Among those cells which survived this selection procedure would be cells unable to produce a functional receptor because of alterations in the glycosylation of the receptor. A secondary screen was performed by incubating colonies on replica discs with the conjugate, again in the presence of lactose (Robbins, et al, 1981). Colonies in which the cells incorporated [^{35}S]methionine into protein under these conditions were chosen for further study. It was determined that one of these clones, B4-2-1, was a glycosylation mutant (Stoll et al, 1982).

Defect: Mannosylphosphoryldolichol Synthase

mutagenized Chinese Hamster Ovary (CHO) cells

SELECTION ↓ incubation with Man_5P-ricin in presence of lactose

survivors

SCREEN ↓ during incubation of colonies on replicas with Man_5P-ricin in the presence of lactose, measure incorporation of [^{35}S]methionine into protein

B4-2-1

Figure 4. Isolation of CHO mutant B4-2-1.

B4-2-1 cells were found to lack mannosylphosphoryldolichol synthase activity (Stoll, et al 1982). Lymphoma cells with the identical enzymatic defect (Chapman, et al, 1980) survived an immunoselection procedure because they did not have Thy antigen on their surface (Hyman and Trowbridge, 1977). Studies using these two cell lines provided the evidence as to which mannosyl donor (GDP-mannose or mannosylphosphoryldolichol) was used for each of the mannoses on the oligosaccharide-lipid intermediates (see Figure 2, Stoll, et al, 1982; Chapman, et al, 1980; Rearick, et al, 1981). In cells lacking mannosylphosphoryldolichol, no $Glc_3Man_9GlcNAc_2$-P-P-lipid was found, rather the major mannose-labelled oligosaccharide intermediate was $Man_5GlcNAc_2$-P-P-lipid; this oligosaccharide is resistant to endo-beta-N-acetylglucosaminindase H. This structure had previously been identified as a minor species in CHO cells (Li and Kornfeld, 1979). Thus, the five mannose residues present in $Man_5GlcNAc_2$-P-P-lipid are donated by GDP-mannose while the additional four mannoses present in $Glc_3Man_9GlcNAc_2$-P-P-lipid are donated by mannosylphosphoryldolichol.

Another example of a selection procedure based on the functioning of a cell-surface receptor which was used to isolate a glycosylation mutant was that described by Kreiger, et al, 1985 (Figure 5).

Defect: UDP-Gal/UDP-GalNAc 4-Epimerase

mutagenized CHO cells

SELECTION ↓ incubation with LDL reconstituted with 25-hydroxy cholesteryl oleate

survivors

SCREEN ↓ incubation with LDL reconstituted with fluorescent cholesteryl oleate

mutants

Figure 5. Isolation of a CHO mutant lacking UDP-Gal/UDP-GalNAc 4-Epimerase

In this selection, mutagenized CHO cells were incubated with LDL reconstituted with a toxic compound 25-hydroxy cholesteryl oleate. Survivors of this selection presumably had an LDL receptor incapable of binding or internalizing the reconstituted LDL. Survivors were then screened after incubation with LDL reconstituted with fluorescent cholesteryl oleate for those cells unable to bind or internalize LDL. One of the mutants isolated by this procedure was characterized to be deficient in the enzyme UDP-Gal/UDP-GalNAc 4-epimerase, an important enzyme in the synthesis of sugar nucleotide donors involved in the synthesis of both O- and N-linked oligosaccharides (Kingsley, et al, 1986)

In addition to selection procedures, direct screening methods (Figure 3) have been used successfully to isolate glycosylation mutants. For example, a procedure was developed to screen a mutagenized population of CHO cells with low intracellular levels of two lysosomal enzymes (Hall, et al, 1986). The colonies were evaluated for two enzymes simultaneously in order to eliminate those reduced in only one activity due to a defect in that particular protein.

The rationale for using lysosomal enzyme activity for this screen was two-fold. First, oligosaccharide sidechains of lysosomal hydrolases in CHO cells are primarily of the high mannose type (Krag and Robbins, 1982); the oligosaccharide transferred to these proteins requires minimal processing to achieve its final structure. Therefore, alteration in an early glycosylation step should directly affect the final oligosaccharide structure of lysosomal hydrolases; alterations in later steps, i.e. addition of terminal sugars and extensive removal of mannose residues, should not. Second, two glycosylation mutants were known to have a marked reduction in intracellular alpha-L-iduronidase activity (Robbins and Myerowitz, 1981, Krag and Robbins, 1982). One of these mutants also had low intracellular levels of beta-glucuronidase activity. Therefore, cells were screened for lowered intracellular levels of alpha-L-iduronidase and beta-glucuronidase activities. Among the mutants isolated was CHB 11-1-3, which seems to have reduced levels of translocation of oligosaccharide-lipid intermediates across the membrane of the endoplasmic reticulum (Hall, et al, 1986).

Another screen which has resulted in the isolation of glycosylation mutants was that developed by Novick and Schekman (1979) to isolate yeast cells defective in secretion (Figure 6). Mutagenized yeast cells were first separated on a

Ludox density gradient in order to enrich the population for dense cells. These cells were then screened for those cells which were temperature sensitive (ts) for growth at 37°. The ts cells were then screened for those which did not secrete invertase and acid phosphatase. Among these mutants in secretion was sec53, which was defective in phosphomannomutase, an enzyme involved in the synthesis of GDP-mannose. Interestingly, this same mutant was also isolated in yeast by a screen of mutagenized cells for the lack of incorporation of tritiated mannose into macromolecules (Figure 6). In this case, the population was first enriched for mutants by using a tritium suicide procedure (Huffaker and Robbins, 1982).

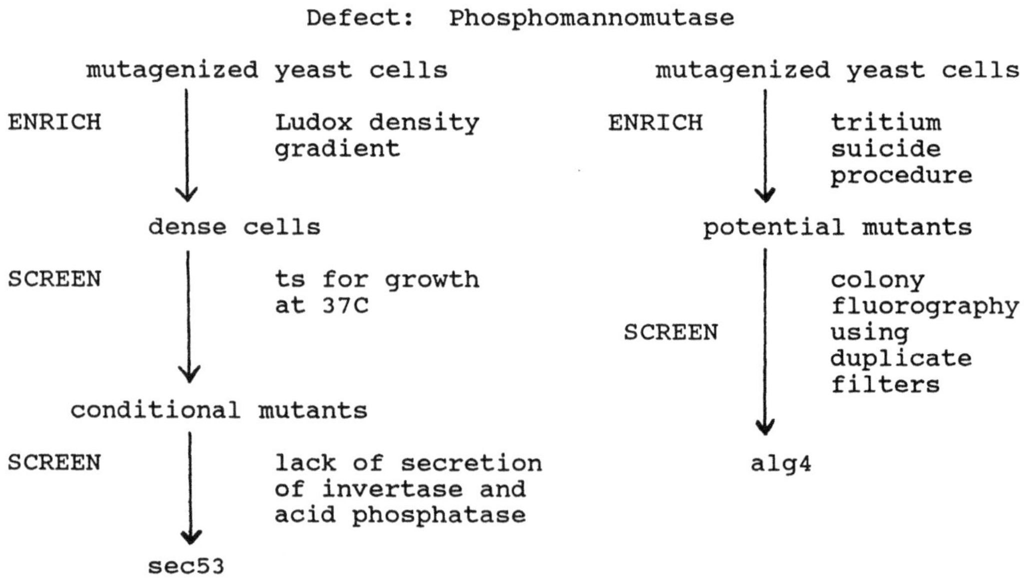

Figure 6. Isolation of yeast mutants defective in phosphomannomutase using different screening procedures.

Finally, a mutant defective in dolichol synthesis was isolated from mutagenized B4-2-1 cells by a screen <u>in situ</u> for reduced tritiated mannose incorporation into macromolecules

(Figure 7). Because defects in the biosynthesis of N-linked glycoproteins may be lethal, a temperature shift step was included in the isolation procedure. No prior enrichment or selective procedures were used.

Defect: Polyprenol Reductase

mutagenized B4-2-1 CHO cells

SCREEN ↓ colony fluorography measuring mannose incorporation into protein

F2A8

Figure 7. Isolation of F2A8, a CHO mutant deficient in dolichol biosynthesis.

One of the mutants, F2A8, isolated by this procedure was chosen for detailed biochemical characterization (Stoll and Krag, 1988; Figure 8).

1. F2A8 cells incorporate 15% the amount of labeled mannose into oligosaccharide-lipid and protein as does parental cells.
2. F2A8 has the same oligosaccharide on oligosaccharide-lipid and on protein as does parental cells.
3. F2A8 synthesizes altered forms of glycoprotein G of VSV
4. VSV virus produced by F2A8 cells is ten-fold less infectious than virus produced by parental cells.

Figure 8. Phenotype of CHO glycosylation mutant, F2A8

Incorporation of radioactivity into mannosylsaccharide-lipids and total mannosylglycopeptides was reduced seven-fold compared to parental cells. F2A8 cells labelled comparable amounts of GDP-mannose, and synthesized the same oligosaccharides found attached to lipid and protein as did parental cells, but in reduced amounts. The viral glycoprotein G of vesicular stomatitis virus synthesized in F2A8 cells had a mobility on SDS-PAGE intermediate between normally glycosylated and unglycosylated G. The VSV virus produced by F2A8 cells was ten-fold less infectious than virus produced by parental cells.

The biochemical analysis (Stoll, et al, 1988) completed thus far on F2A8 is summarized in Figure 9.

1. F2A8 cells have lower amounts of endogenous lipid phosphate available for _in vitro_ glycosylation reactions.

2. F2A8 cells accumulate neutral unsaturated polyprenol.

3. F2A8 cells have normal amounts of total phosphorylated polyisoprenol lipids but less lipid phosphate and less oligosaccharide-lipid than parental cells.

4. Polyprenols in F2A8 cells are the same chain lengths as dolichols in parental cells.

5. F2A8 cells use unsaturated polyprenol not dolichol in their glycosylation reactions.

6. The lower glycosylation activity of F2A8 appears to be due to the limited activity of unsaturated polyprenol derivatives.

Figure 9. Biochemical characterization of F2A8.

In vitro assays using membrane preparations showed that F2A8 had parental levels of glucosylphosphoryldolichol synthase activity and UDP-GlcNAc:dolichyl phosphate:GlcNAc-phosphotransferase activity (Figure 2) when the enzymatic determinations were done in the presence of exogenous substrate, dolichyl phosphate. F2A8 had no mannosylphosphoryldolichol synthase activity since it was isolated from B4-2-1 which lacks this enzymatic activity (see above). However, in assays relying on endogenous lipid substrate, six-fold less glucosylphosphoryldolichol synthase activity was detected in membranes of F2A8 compared to membranes of parental cells. F2A8 appeared to have reduced amounts of dolichyl phosphate available for its glycosylation reactions.

The amounts and identities of polyisoprenoid lipids were determined in F2A8, B4-2-1, and wild-type CHO cells after incubation for twenty-four hours with [2-^{3}H]mevalonate. The neutral lipids incorporating the most labelled mevalonate, ubiquinone 45 (coenzyme Q9), cholesterol, and cholesteryl esters, were labelled equally in all three cells types. In wild-type and B4-2-1 cells, mevalonate incorporation was ten-fold higher into the glycosylated, phosphorylated derivatives of dolichol than into free dolichol or dolichyl esters (see Figure 1). These anionic lipids were soluble in both chloroform and choloroform:methanol:water mixtures. In contrast, F2A8 cells accumulated label in neutral polyisoprenol lipids, so that the ratio of neutral lipids to anionic lipids was 1:1 rather than 1:10 as in parental and wild-type cells. In wild-type and B4-2-1 cells, the polyisoprenoid lipid found as free alcohol and as phosphorylated, glycosylated forms was shown by hplc using silica columns to be primarily dolichol, a polyisoprenol that has the terminal isoprene units saturated (Figure 1). In F2A8, the polyisoprenoid lipid found primarily as the free alcohol and as phosphorylated, glycosylated forms was unsaturated polyprenol. The distribution of chain lengths found in the labelled polyisoprenoid alcohol of F2A8, B4-2-1, and wild-type was similar, as determined by hplc analysis using a reverse phase column, with the predominant chain length being 19 isoprene units.

In summary, F2A8 appears to be defective in dolichol biosynthesis, in particular in the enzyme polyprenol reductase which converts polyprenyl derivatives to dolichol derivatives (Figure 10). Studies are underway to test this hypothesis directly _in vitro_.

The effect of this enzymatic defect is an alteration in the steady-state levels of the various polyisoprenyl derivatives in F2A8 compared to B4-2-1. Specifically, there is a reduction in polyprenyl phosphate, an increase in glucosylphosphoryl polyprenol, a reduction in oligosaccharyl polyprenol, and an increase in neutral polyprenols. There was no change in the amount of total polyisoprenoids in F2A8 compared to B4-2-1. Thus, it appears that different enzymes in the oligosaccharide lipid biosynthetic pathway can use polyprenol derivatives effectively (for example, glucosylphosphoryldolichol synthase), while others cannot (for example, UDP-GlcNAc: dolichyl phosphate: GlcNAc-phosphotransferase). F2A8 glycosylates proteins at 15% the level of its parent, B4-2-1. Therefore, these results combined with our studies on the phenotype of this mutant indicated that the unsaturated polyprenyl derivatives do not function as well as dolichyl derivatives in cellular glycosylation reactions.

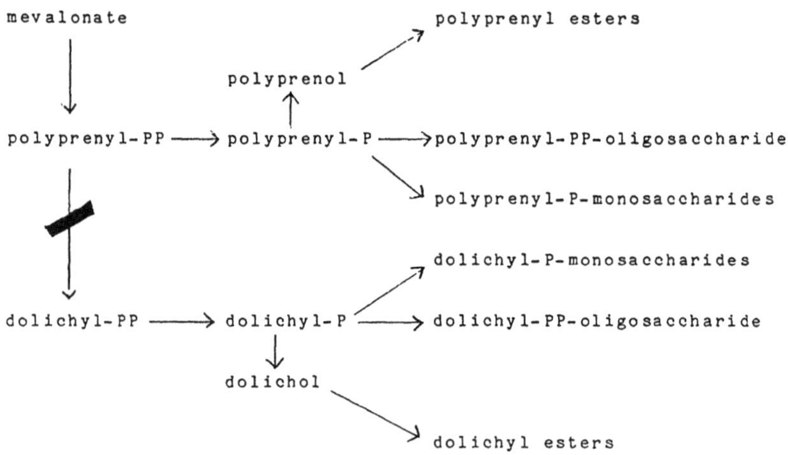

Figure 10. F2A8 appears to be defective in polyprenol reductase

ACKNOWLEDGEMENT:

This work was funded through grant RO1-CA20421 to SS Krag from the National Cancer Institute, DHHS, USA.

REFERENCES:

Adair GM and Siciliano MJ (1986) Somatic Cell and Mol. Genetics 12:111-119
Adair WL Jr and Keller RK (1982) J. Biol. Chem. 257:8990-8996
Adair WL Jr, Cafmeyer N and Keller RK (1984) J. Biol. Chem. 259:4441-4446
Allen CM Jr, Kalin JR, Sack J and Verizzo D (1978) Biochemistry 17:5020-5026
Appelkvist EL, Chojnacki T and Dallner G (1981) Biosci. Rep. 1, 619:626
Belcopitow E and Boscobionik D (1982) Eur. J. Biochem. 125:167-173
Burton WA, Scher MG and Waechter CJ (1981) Arch. Biochem. Biophys. 208:409-417
Chapman A, Fujimoto K and Kornfeld S (1980) J. Biol. Chem. 255:4441-4446
Daleo GR, Hopp HE, Romero PA and Pont Lezica R (1977) FEBS Lett. 81:411-414
Ekstrom RJ, Chojnacki T and Dallner G (1984) J. Biol. Chem. 259:10460-10468
Elbein AD (1987) Ann. Rev. Biochem. 56:497-534
Hall CW, Robbins AR and Krag SS (1986) Mol. Cell. Biochem. 72:35-45
Hanover JA and Lennarz WJ (1982) J. Biol. Chem. 257:2787-2794
Hirschberg CB and Snider MD (1987) Ann. Rev. of Biochem. 56:63-87
Hsieh P, Rosner MR and Robbins PW (1983) J. Biol. Chem. 258: 2555-2561
Hubbard SC and Ivatt RJ (1981) Ann. Rev. Biochem. 50:555-583
Huffaker TC and Robbins PW (1982) J. Biol. Chem. 257:3203-3210
Hyman R and Trowbridge I (1977) Cold Spring Harbor Symp. Quant. Biol. 41:407-415
Kato S, Tsuji M, Nakaniski Y and Suzuki S (1980) Biochem. Biophys. Res. Commun. 95:770-776
Keller RK (1986) J. Biol. Chem. 261:12053-12059
Kingsley DM, Kozarsby KF, Hobbie, L and Krieger M (1986) Cell 44:749-759.
Kingsley DM and Krieger M (1984) Proc. Natl. Acad. Sci. USA 81:5454-5458
Kobata A (1978) Methods in Enzymol. 50:567-573
Kornfeld S (1982) The Glycoconjugates (MI Horowitz ed.) Academic Press, New York, pp. 3-23
Kornfeld R and Kornfeld S (1980) In The Biochemistry of Glycoproteins and Proteoglycans, ed. W.J. Lennarz, Plenum Press, New York, pp.1-34
Kornfeld R and Kornfeld S (1985) Ann. Rev. Biochem 54:631-664
Kreiger M, Brown MS and Goldstein JL (1981) J. Mol. Biol. 150:167-184
Krag S (1985) Current Topics in Membranes and Transport 24:181-249
Krag S and Robbins AR (1982) J. Biol. Chem. 257:8424-8431
Li E and Kornfeld S (1979) J. Biol. Chem. 254:2754-2758
Marshall RD (1974) Biochem. Soc. Symp. 40:17-26
Novick P and Schekman R (1979) PNAS, 76:1858-1862

Poulter CD and Rilling HC (1974) In MTP International Review of Science: Biochemistry of Lipids. ed. by TW Goodwin, Series One, Vol. 4, Butterworth, London, pp. 162-224
Rademacher TW, Parekh RB and Divek RA (1988) Ann. Rev. Biochem. 57:785-838
Radominska-Pyrek A, Chojnacki T and Zulczyk W (1979) Acta Biochemica Polonica 26:125-134
Rearick JL, Fujimoto K and Kornfeld S (1981) J Biol. Chem. 256: 3762-3769
Rip JW, Rupar CA, Chaundhary N and Carroll KK (1981) J. Biol. Chem. 256:1929-1934
Robbins AR and Myerowitz R (1981) J. Biol. Chem. 256:10623-10627
Robbins AR, Myerowitz R, Youle RJ, Murray, GJ and Neville, DM Jr. (1981) J. Biol. Chem 256:10618-10622
Runge KW, Huffaker TC and Robbins PW (1984) J. Biol. Chem. 259:412-417
Scher MG and Waechter CJ (1981) Biochem. Biophys. Res. Commun. 99:675-681
Schroepher GJ Jr (1981) Ann. Rev. Biochem. 50:585-621
Schwarz RT and Datema R (1982) Adv. Carbohydr. Chem. Biochem. 40:287-379
Snider MD and Rogers OC (1984) Cell 36:753-761
Stanley P (1984) Ann. Rev. Genetics 18:525-552
Stanley P, Narasimhan S, Siminovitch L and Schachter H (1975) Proc. Natl. Acad. Sci. USA. 72:3323-3327
Stoll J and Krag SS (1988) J. Biol. Chem. 263:10766-10773
Stoll J, Robbins AR and Krag SS (1982) Proc. Natl. Acad. Sci. USA. 79:2296-2300
Stoll J, Rosenwald AG and Krag SS (1988) J. Biol. Chem. 263:10774-10782
Tarentino AL, Trimble RB and Maley F (1978) Methods in Enzymol. 50:574-579
Wedgewood JF and Strominger JL (1980) J. Biol. Chem. 255:1120-1123
Wellner RB and Lucas JJ (1979) FEBS Lett. 104:379-383
Wong RK and Lennarz WJ (1982) J. Biol. Chem. 257:6619-6624

FUNCTIONAL AND STRUCTURAL ASPECTS OF GRAMICIDIN-LIPID INTERACTIONS

J.A. Killian[1], H. Tournois[1] and B. De Kruijff[1,2]
[1]Centre for Biomembranes and Lipid Enzymology and [2]Institute of Molecular Biology and Medical Biotechnology, University of Utrecht, Padualaan 8, 3584 CH Utrecht, The Netherlands

INTRODUCTION

Lipids in isolated form show polymorphism. Depending on the type of lipid and experimental conditions, different macromolecular aggregate structures are formed. For membrane lipids hydrated in excess water, under physiological conditions of ionic strength, pH and temperature, three major liquid-crystalline phases are observed. These are: the lamellar phase, the inverted (type II) hexagonal phase (Fig. 1), and the normal (type I) micellar phase. The shape-structure concept of lipid polymorphism as originally proposed by Israelachvili et al. (1976) and adapted by Cullis and De Kruijff (1979) explains remarkably well most of the current data on lipid polymorphism (Fig. 2). If the surface area in the head group region matches that of the hydrocarbon region (such as in phosphatidylcholines) the molecule has an overall cylindrical shape and prefers to organize in extended bilayers. When the head group area exceeds the hydrocarbon area, as for instance in case of detergent-like lipids such as lysophosphatidylcholines, the molecule on average is cone shaped and prefers an organization with a convex (type I) surface curvature such as found in micelles or in the related normal (type I) hexagonal phase occurring for these lipids at low water content. Phosphatidylethanolamines are key representatives of the opposite case of an excess hydrocarbon area. This results in the concave (type II) surface curvature as found in the H_{II} phase. Head group hydration and inter- and intramolecular interactions are important determinants for the shape of the lipid molecules in the shape-structure concept. The strictly regulated (Goldfine et al., 1987; Wieslander et al., 1981) presence of large amounts of type II lipids (lipids which in hydration prefer to organize in type II structures) and their abundance in virtually all biomembranes suggests important structural and functional roles for these types of lipids.

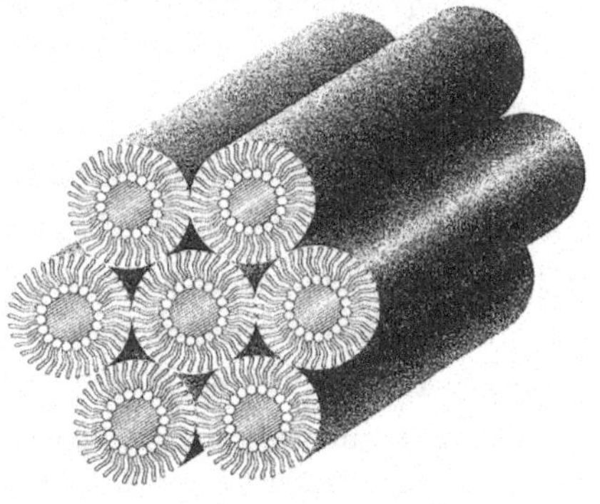

Fig. 1. The inverted hexagonal phase (H_{II} phase). The phase is built up of a large number of tubes, which are hexagonally organized and in which the lipid molecules surround with their head group a narrow aqueous channel.

NATO ASI Series, Vol. H 40
Dynamics and Biogenesis of Membranes
Edited by J.A.F. Op den Kamp

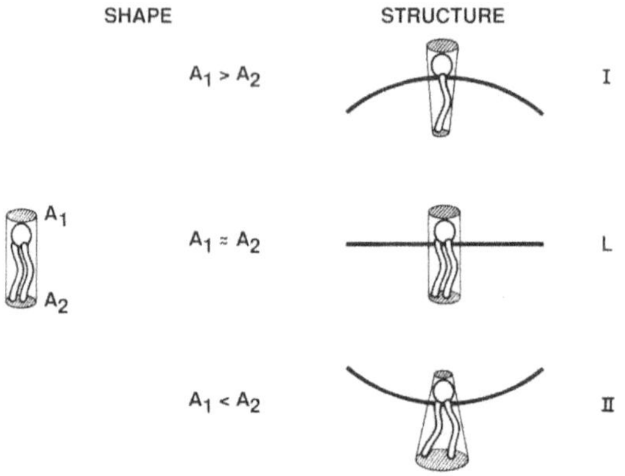

Fig. 2. Shape-structure concept.

The hexagonal H_{II} phase as such only occurs in biological membranes under pathological conditions, which can be readily understood by considering the harmful effect these large structures will have on the barrier function of the membrane. Numerous studies on model and biological membranes (for reviews see De Kruijff, 1987, and De Kruijff et al., 1985), indicated which roles the type II lipids might play in the structure and function of membranes. The major ones are shown in Fig. 3, which illustrates that the cone-shaped lipids can: (i) form contact points between membranes, (ii) induce local high membrane curvature, (iii) allow close juxtaposition of bilayers because of their low head group hydration, (iv) cause packing defects (shaded areas) which can be stabilized by hydrophobic molecules such as polyprenols, (v) form stable interbilayer connections, (vi) act as vehicles for transbilayer transport of lipids (flip-flop) and facilitate membrane translocation of polar components as proteins or Ca^{2+}, and (vii) correctly integrate proteins into membranes.

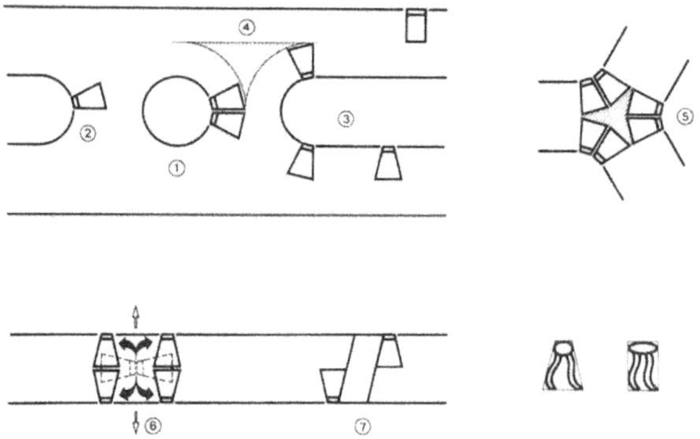

Fig. 3. Type II lipids and the structure and function of biomembranes. See text for details. Bottom right: cone-shaped (type II) and cylindrical shaped bilayer phase preferring lipid.

Coupled with the realization that all biological membranes undergo essential processes such as membrane fusion and endo- and exocytosis, during which the lipid bilayers must be locally and transiently disrupted, it can be anticipated from a functional point of view that mechanisms must exist in biomembranes for regulation of local and transient lipid organization. Lipid-protein interactions are obvious candidates to be involved in such regulatory mechanisms. That these interactions in model system can greatly influence macroscopic lipid organization in a highly variable way depending on the type of protein, lipid and environmental condition has been well documented (for reviews see De Kruijff et al., 1985; Killian & De Kruijff, 1986).

To understand the way in which lipid structure is changed by the interaction with proteins at the molecular level, detailed studies are required on more simple, well-characterized polypeptides of which melittin (reviewed in Batenburg & De Kruijff, 1988) and gramicidin are good examples. It is the aim of this article to describe our current insights in the pronounced lipid structure-modulating ability of gramicidin and to indicate its relationships with other known functional abilities of biological membranes such as fusion and lipid flip-flop.

The linear gramicidins are produced by *Bacillus brevis* strain ATCC8185. Like other peptide antibiotics they are synthesized by multi-enzyme complexes (Katz & Demain, 1977; Lipmann, 1980). Gramicidin A (Fig. 4) is with 80% the dominant species in the natural mixture. The less abundant gramicidins B (5%) and C (15%) differ from gramicidin A by the presence of a Phe or Tyr residue at position 11, respectively. The gramicidins are uncharged, very hydrophobic pentadecapeptides with the N- and C-terminus blocked by a formyl and ethanolamine group, respectively. During biosynthesis a conversion takes place of L-Val and L-Leu to their D-conformers (Akashi & Kurahashi, 1977), resulting in a unique alternation of D- and L-amino acids in the sequence, which provides the peptide with a remarkable ability to organize in many different mono- or oligomeric ß-helical conformations as shown in a number of different organic solvents (Veatch et al., 1974; Sychev et al., 1980). In biological and model membranes gramicidin can form cation-selective transmembrane channels. In the channel configuration the peptide is present as an N to N hydrogen-bonded single-stranded $\beta^{6.3}$ helical dimer (Urry, 1985; Urry et al., 1983; Andersen, 1984). The unique clustering of bulky tryptophans towards the C-terminus will be shown to be of special importance in the studies to be described.

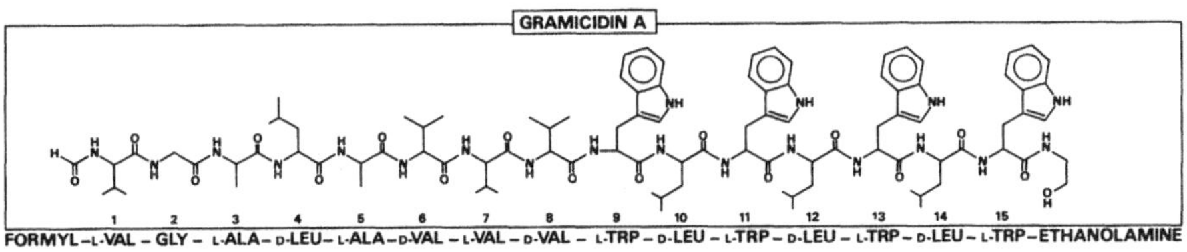

Fig. 4.

RESULTS

Gramicidin induces bilayer structures in mixtures with lyso-PC

1-Palmitoyl-lysophosphatidylcholine (lyso-PC), as a typical type I lipid, has strong detergent properties: it disrupts the bilayer organization in membranes and solubilizes membrane proteins by forming mixed micelles. Therefore, the observation that gramicidin induces bilayer structures in mixtures with lyso-PC was an unexpected result (Killian et al., 1983; Pasquali-Ronchetti et al., 1983). Figure 5 illustrates this by ^{31}P and ^{2}H NMR for samples prepared by hydration of a mixed film. ^{31}P and ^{2}H NMR methods can distinguish different lipid organizations because the lipids experience characteristic motions in the various phases and therefore give rise to different line shapes (Cullis et al., 1985). In short, when phospholipids undergo isotropic motion which is rapid on the NMR time scale ($\pm$ 10^{-5} sec), the chemical shift anisotropy (CSA) in case of ^{31}P or the nuclear quadrupolar interaction $\Delta\upsilon q$ (in case of ^{2}H in ^{2}H-labelled lipids) are fully averaged resulting in a narrow isotropic line shape. In Fig. 5 this is illustrated for [12-$^2H_2$]-16:0-lysophosphatidylcholine (lyso-PC), which in water forms micelles allowing rapid isotropic motion of the lipids by micelle tumbling and by diffusion of the lipids in the micelle surface. In case of organization in extended bilayers, the CSA and $\Delta\upsilon q$ are only partially averaged by rapid local motions and fast axial rotation. This results in case of ^{31}P NMR in an asymmetrical line shape with a high field peak and a low field shoulder separated by $\Delta\sigma$, the residual CSA, the value of which depends on the type of lipid and local order of the phosphate segment. The deuterium spectrum in this case is characterized by two peaks, separated by $\Delta\upsilon q$, the quadrupolar splitting, and two shoulders and greatly resembles a 'double' ^{31}P NMR spectrum symmetrical around the zero chemical shift position. $\Delta\upsilon q$ is a sensitive measure of the local order at the site of the deuteron (Seelig, 1977). Figure 5 shows that both at the level of the head group and the acyl chain region NMR spectra typical for a bilayer organization are observed for the lyso-PC/gramicidin (4:1, molar) mixed system.

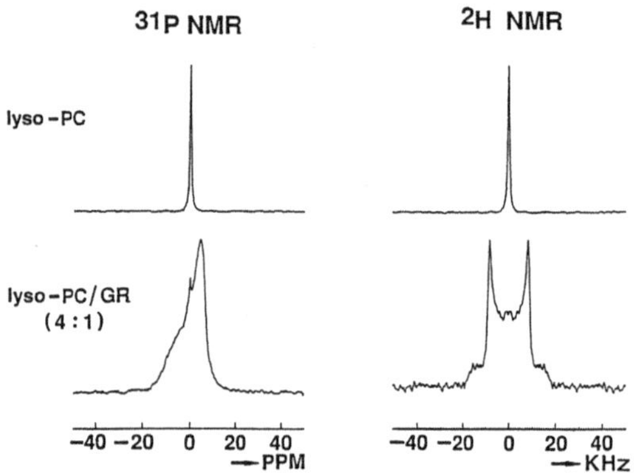

Fig. 5. Effect of gramicidin incorporation of the 121.4 MHz ^{31}P and 46.1 MHz ^2H NMR spectra of aqueous 1-[12-$^2H_2$]-palmitoyl-*sn*-glycero-3-phosphocholine dispersions. For experimental details see Killian et al. (1986a).

The stoichiometry of the peptide-lipid complex in the lamellar phase appeared to depend upon the conformation of the peptide. Circular dichroism (CD) studies showed that the conformation of gramicidin in a lipid environment is determined by the method of sample preparation (Tournois et al., 1987; Killian et al., 1988; Killian & Urry, 1988) and is further influenced by prolonged incubations at elevated temperatures (Killian et al., 1988; Killian & Urry, 1988; Masotti et al., 1980; Shungu et al., 1986). When the peptide is introduced via trifluorethanol (TFE) or dimethylsulphoxide (DMSO), in which solvents it is believed to be present as a $\beta^{6.3}$ helix (Glickson et al., 1972; Urry et al., 1972; Hawkes et al., 1987), it directly incorporates in this conformation as evidenced from the characteristic CD pattern (Urry, 1985). Introduction from ethanol, however, in which gramicidin is dominantly present as an anti-parallel double helix (Veatch et al., 1974; Veatch & Blout, 1974) results in a very different CD pattern. Figure 6 (left panel) illustrates this for samples of 16:0 lyso-PC and gramicidin (8:1, molar ratio) prepared by hydration of a mixed film that was obtained by co-dissolving peptide and lipid in either TFE or ethanol followed by removal of the solvent under reduced pressure. The spectrum obtained upon introduction of the peptide from a solution in TFE is characteristic for the $\beta^{6.3}$ helical conformation, which is the preferred conformation of the peptide in this solvent. Assuming that, also when introduced from ethanol, gramicidin initially retains its solvent conformation, this suggests that the corresponding CD pattern reflects an anti-parallel double helical conformation. Upon incubation of the samples at higher temperatures no changes in CD patterns were observed when the sample was introduced via TFE, but a slow conformational change in a direction towards the $\beta^{6.3}$ helix did occur when ethanol was used (Fig. 6, right panel), indicating that the $\beta^{6.3}$ helix is the thermodynamically most stable and preferred conformation in a lipid environment. Similar solvent dependent differences in conformation of gramicidin occurred when the peptide was introduced as an organic solution to lipid dispersions (Tournois et al., 1987).

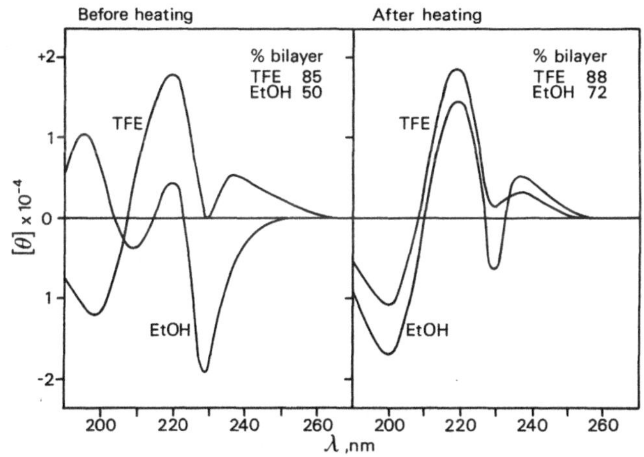

Fig. 6. Solvent dependence (left panel) and effect of heat incubation (20 h at 68°C; right panel) on CD patterns of gramicidin in 16:0-lyso-PC (1:8, molar ratio). Peptide/lipid dispersions were sonicated and diluted to the desired peptide concentration prior to CD measurements. Heat incubation was carried out on unsonicated samples. The inserts show the percentage of bilayer formation induced by gramicidin in these same systems, as calculated from 40 MHz ^{31}P NMR spectra of unsonicated samples. For details see Killian & Urry (1988).

From the extent of bilayer formation (Fig. 6, insert) it could be calculated that the stoichiometry of the peptide-lipid complex is significantly higher when gramicidin is in the $\beta^{6.3}$ helical configuration (± 7/1) than when present in a non-channel configuration (± 4/1). These differences in stoichiometry could result either directly from differences in conformation or indirectly from differences in aggregational behavior. Aggregation has been proposed to play a role in channel functioning (Spisni et al., 1983; Stark et al., 1986) and also is believed to occur in dispersions of gramicidin and LPC, based on ^{2}H NMR studies on acyl chain labelled lyso-PCs (Killian et al., 1986a) and on freeze-fracture electron microscopy studies (Spisni et al., 1983). Recently, gramicidin aggregation could be directly measured in those systems by ^{2}H NMR on ^{2}H labelled gramicidin (Macdonald & Seelig, 1988).

The most simple explanation for bilayer formation by gramicidin in the lyso-PC system is to assume that gramicidin is cone-shaped and orients parallel to the acyl chains with the smaller end of the cone at the water-lipid interface. Due to the complementarity of shapes, a lamellar complex can be formed analogous to mixtures of type I and type II lipids (Madden & Cullis, 1982). In the $\beta^{6.3}$ helix gramicidin is indeed cone-shaped due to the unidirectional location of the bulky tryptophans towards the C-terminus. An orientation of the N-terminus of gramicidin in the $\beta^{6.3}$ helix towards the aqueous phase was calculated to be the energetically most favorable orientation of gramicidin at the air-water interface (Brasseur et al., 1986). The prediction from this model (Fig. 7) is that gramicidin can induce a type II lipid organization for lamellar type of systems. This is indeed the case.

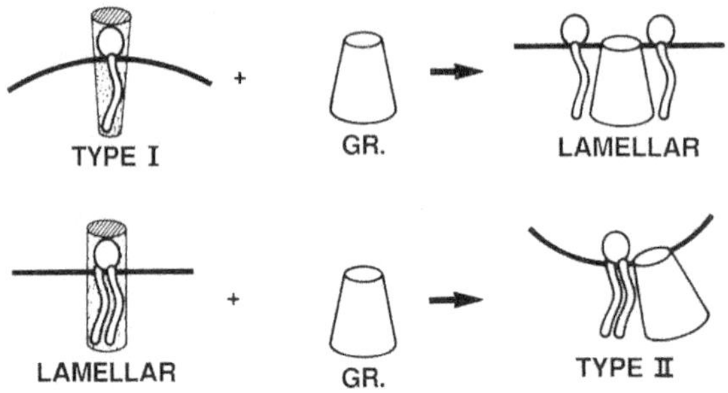

Fig. 7. Model to explain the effect of gramicidin (GR, schematically shown as $\beta^{6.3}$ monomer) on organization of type I and lamellar lipid systems.

Gramicidin induces hexagonal H_{II} phase formation in bilayer organization preferring (model) membrane systems

Since the original observation by Van Echteld et al. (1982) that gramicidin induces the hexagonal H_{II} phase in phosphatidylcholine systems, when the acyl chain length exceeds 16 carbon atoms much information on the gramicidin-induced H_{II} phase formation in membrane systems has accumulated. Because the thickness of the hydrophobic part of a 16:0/16:0-PC (DPPC) bilayer in the liquid-crystalline phase (26 Å; Lewis & Engelman, 1983) is equal to the length of the gramicidin channel (26 Å; Urry et al., 1979), it was proposed that a mismatch in length is required for gramicidin-induced H_{II} phase formation with the bilayer thickness exceeding the length of the gramicidin dimer (Van Echteld et al., 1982). This could be demonstrated using length analogs of gramicidin with a Leu-Ala-Leu-Ala sequence inserted between Ala[5] and Val[6] (length of dimer 32 Å). If a mismatch is required these peptides are predicted to loose their ability to induce H_{II} phase formation in $18{:}1_c/18{:}1_c$-PC (DOPC) model membranes, but not in $22{:}1_c/22{:}1_c$-PC model membranes (thickness resp. 27 and 34 Å; Lewis & Engelman, 1983). This was investigated using ^{31}P NMR. In the hexagonal H_{II} phase the lipids can undergo rapid lateral diffusion around the axes of the tubes. In the absence of changes in local order upon transition from a bilayer to the H_{II} phase this results in a reversed line shape with a low field peak and a high field shoulder and a two-fold reduced $\Delta\sigma$. The ^{31}P NMR spectra in Fig. 8 therefore show that con-Leu-Ala-Leu-Ala-gramicidin induces H_{II} phase formation for part of the lipids in $22{:}1_c/22{:}1_c$-PC model membranes but not in DOPC, demonstrating that indeed such a mismatch is required. This was confirmed by small-angle X-ray diffraction measurements (Killian et al., 1989).

Another factor that appeared to be of importance for gramicidin-induced H_{II} phase formation is peptide aggregation. Gradient centrifugation experiments revealed that H_{II} formation is accompanied by phase separation and peptide aggregation, such that the H_{II} phase which is highly enriched in gramicidin can be isolated in a pure form (Fig. 9).

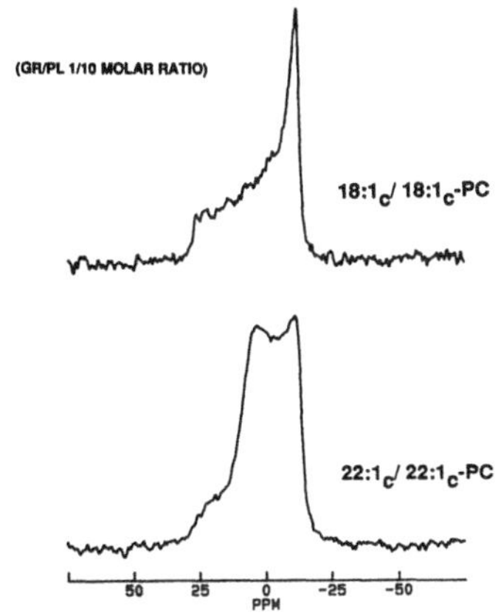

Fig. 8. 121.4 MHz ^{31}P NMR spectra of aqueous dispersions of con-Leu-Ala-Leu-Ala-gramicidin with $18:1_c/18:1_c$-PC and $22:1_c/22:1_c$-PC. For details see Killian et al. (1989).

H_{II} phase promotion was observed in different diacyl phospholipid model systems (Van Echteld et al., 1982; Killian et al., 1986b) and in erythrocyte membranes (Tournois et al., 1987). Particularly relevant for the mechanism of H_{II} phase formation are the observations that also in a variety of different acidic phospholipids H_{II} phase formation is induced (Killian et al., 1986b).

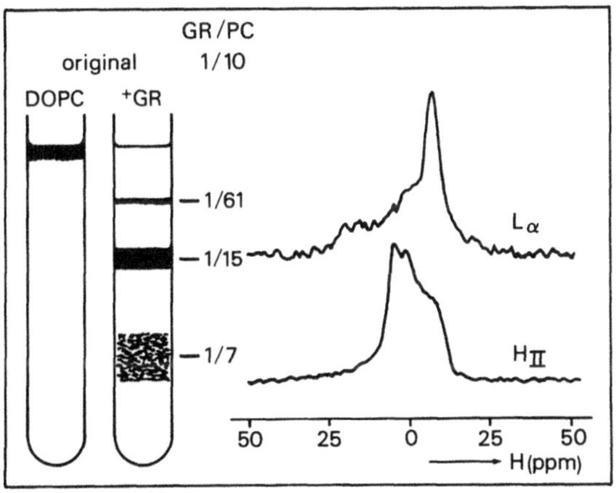

Fig. 9. Density gradient centrifugation of aqueous dispersions of DOPC and DOPC-gramicidin (1:10, molar). The samples were spun in a linear gradient of 3-20% sucrose and the fractions banding at a density of 1.0284 and 1.0408 g/ml, which were shown to have a DOPC/gramicidin molar ratio of 1:15 and 1:7 were analyzed by ^{31}P NMR spectroscopy. For further details on the procedures and the nature of the band with a gramicidin/DOPC molar ratio of 1:61 the reader is referred to Killian et al. (1987).

The energetically unfavorable concentration of negative charges in the narrow aqueous channel within the H_{II} tubes is apparently overcome by the highly favored location of gramicidin in the H_{II} phase. There is much experimental evidence that gramicidin aggregation in the membrane is essential for H_{II} phase formation (Killian et al., 1987; Killian & De Kruijff, 1985a,b; Chupin et al., 1987). One line of evidence is illustrated in Fig. 10. Gramicidin incorporation reduces the energy content (ΔH) of the gel -> liquid-crystalline phase transition in bilayers of DPPC in a biphasic way (Killian & De Kruijff, 1985; Chapman et al., 1977). At low concentrations gramicidin causes a nearly linear decrease in ΔH. Around a molar ratio of 1 gramicidin to 15 DPPC molecules this effect levels off most likely as a result of peptide aggregation (Killian & De Kruijff, 1985; Chapman et al., 1977), preventing a further increase in the extent of gramicidin-DPPC interactions. Interestingly, in the DOPC system at around the same gramicidin concentration H_{II} phase formation is induced (Fig. 10). From a combination of SAXS, ^{31}P and ^{2}H NMR studies on chain-labelled phospholipids the unique character of the gramicidin-induced H_{II} phase could be unambiguously demonstrated (Chupin et al., 1987).

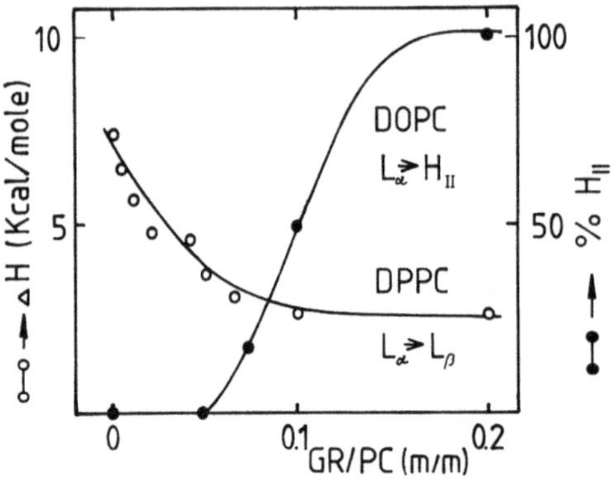

Fig. 10. Effect of gramicidin incorporation on the ΔH of the gel -> liquid-crystalline phase transition in DPPC dispersions and the extent of H_{II} phase formation in DOPC dispersions. For experimental details see Chupin et al. (1987), and Killian & De Kruijff (1985).

In case of an H_{II} phase formed by pure type II lipids as $18{:}1_c/18{:}1_c$-phosphatidylethanolamine (DOPE), an increase in temperature will result in a decrease in chain order and effective lipid length (Fig. 11). This is experimentally observed as a decrease in $\Delta\upsilon q$ in the ^{2}H NMR spectrum of [11,11-$^{2}H_2$]-DOPE and a decrease in H_{II} tube diameter (d) derived from the long spacing of the H_{II} phase (Fig. 12, top). In contrast, in the gramicidin-induced H_{II} phase in the [11,11-$^{2}H_2$]-DOPC system a temperature-independent $\Delta\upsilon q$ and tube diameter is observed. This strongly suggests that gramicidin as a rigid molecule forms the structural backbone of the H_{II} phase.

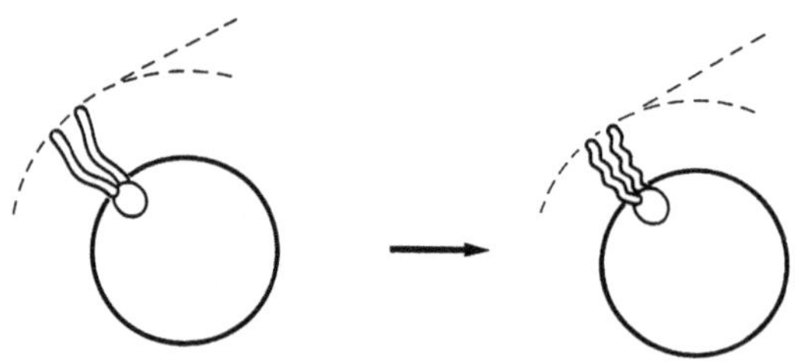

Fig. 11. Temperature increase results in decreased acyl chain order and effective lipid length in H_{II} tubes of a pure lipid system.

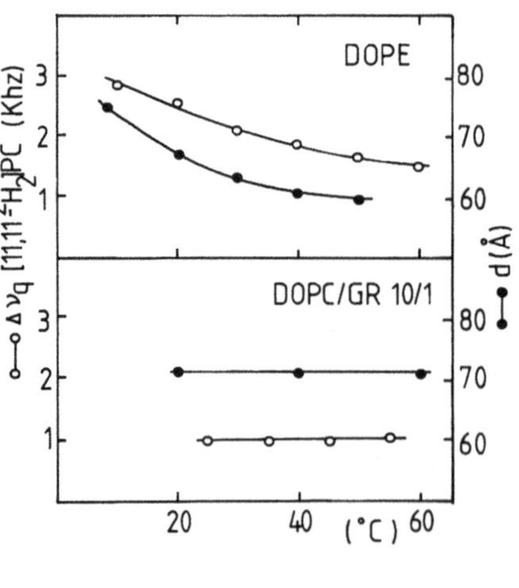

Fig. 12. Effect of temperature on the quadrupolar splitting ($\Delta\upsilon q$) and tube diameter (d) of the H_{II} phase in [11,-11-$^{2}H_2$]-DOPE and [11,11-$^{2}H_2$]-DOPC/-gramicidin (10:1, molar). For details see Chupin et al. (1987) and Killian & De Kruijff (1985).

Structural requirements of gramicidin-induced changes in lipid polymorphism

Different gramicidin analogs were tested for their ability to induce bilayer formation in lyso-PC systems (Killian et al., 1986b; Aranda et al., 1987), and H_{II} phase promotion in DOPC (Killian et al., 1985,1986b,1987) and $18{:}1_t/18{:}1_t$-phosphatidylethanolamine (DEPE) systems (Killian et al., 1986b; Aranda et al., 1987). Surprisingly, major changes at the C- and N-terminus, i.e. introduction of either negatively or positively charged groups had virtually no effect on the H_{II} phase-inducing capacity of the peptide (Killian et al., 1986b). In strong contrast large effects were observed upon modification of the tryptophans.

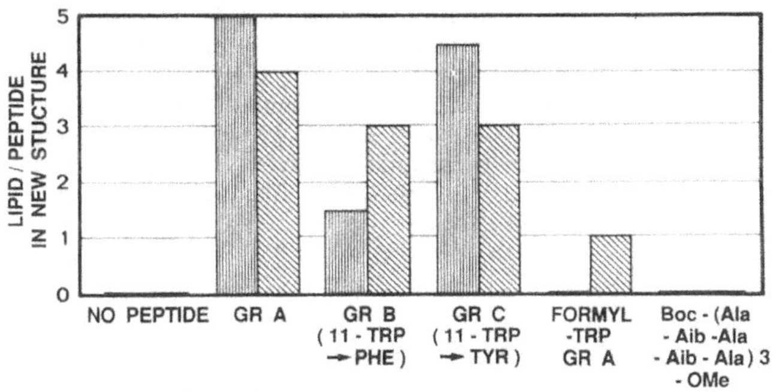

Fig. 13. Effect of different gramicidin analogs on H_{II} phase formation in DOPC (vertical lines) and bilayer formation with lyso-PC (slanted lines). The efficiency of lipid structure modulation is given as the number of lipid molecules induced per peptide into either the H_{II} or bilayer organization.

Figure 13 shows that formylation of all four tryptophans (formyl-Trp-gramicidin A') resulted in complete loss of H_{II} phase induction (Killian et al., 1985) and greatly reduced the bilayer forming ability in the lyso-PC system (Aranda et al., 1987). But even much more subtle changes such as Trp^{11} -> Phe^{11} or Trp^{9} -> Phe^{9} substitutions (not shown) greatly decreased the efficiency of the peptide to change lipid organization. Interestingly, the Trp^{11} -> Tyr^{11} substitution was much less effective than the Trp^{11} -> Phe^{11} substitution (Killian et al., 1987). The uniqueness of gramicidin in these effects is further demonstrated by the complete inability of the α-helical (Jung et al., 1981,1983) BOC-(Ala-Aib-Ala-Aib-Ala)$_3$-OMe pentadecapeptide to change lipid organization despite its incorporation into the model membrane (Aranda et al., 1987). Not only the chemical structure but also the conformation of gramicidin is an essential determinant in these effects as was already discussed for the lyso-PC/gramicidin system. Introduction of gramicidin from trifluoroethanol (TFE) or dimethylsulfoxide into DOPC liposomes immediately results in H_{II} phase formation (Fig. 14; Tournois et al., 1987a). In contrast, addition from ethanol initially does not lead to changes in lipid structure (Fig. 14), but after 24 h incubation at 35°C, H_{II} phase formation did slowly occur also in these samples (Tournois et al., 1987). The effects on lipid structure could be directly related to differences in conformation of gramicidin, as discussed for the LPC/gramicidin system and as shown in Fig. 14. From these results it is clear that only in the $\beta^{6.3}$ conformation gramicidin can induce the H_{II} phase in DOPC model membranes.

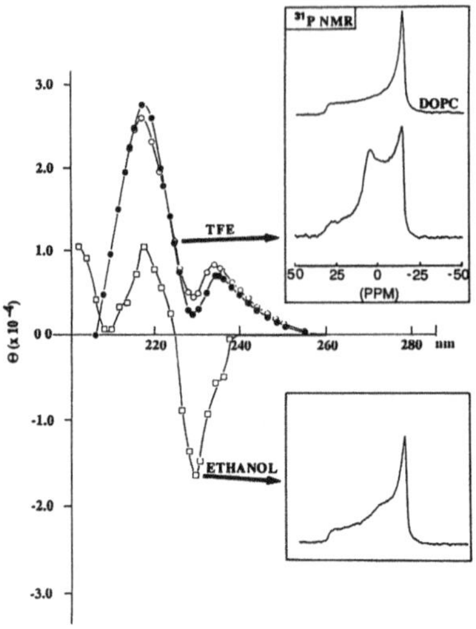

Fig. 14. Circular dichroism spectra of gramicidin added to DOPC vesicles in a 1:25 peptide to lipid molar ratio from a solution of trifluoroethanol (o), dimethylsulfoxide (o) and ethanol ( ). The inserts show the 121.4 MHz ^{31}P NMR spectra obtained from aqueous dispersions of DOPC and DOPC to which gramicidin was added in a 1:10 peptide to lipid molar ratio from a solution of trifluoroethanol or ethanol. For details see Tournois et al. (1987a).

Gramicidin-induced lipid flip-flop

So far, the described results emphasized the structural consequences of gramicidin-lipid interactions. However, gramicidin incorporation also affects lipid dynamics, not only within the lipid molecule (Chupin et al., 1987; Rice & Oldfield, 1979), but also of the entire lipid molecule. For instance, the rate of lateral diffusion of lipids in the bilayer (Tank et al., 1982) as well as around the H_{II} tubes is decreased by the presence of high concentrations of gramicidin (Tournois et al., 1987b; Gasset et al., 1988). The erythrocyte membrane was chosen as a test system to investigate the consequences of gramicidin incorporation for the transversal diffusion (flip) of the lipids. Normally, transbilayer movement of di- and monoacyl phospholipids is very slow in this membrane as can also be concluded from the very low rate constant for the lyso-PC flip in this membrane (Fig. 15; Classen et al., 1987). Evidently, incorporation of gramicidin greatly increases the rate of 16:0-lyso-PC flip when it is inserted under non-lytic conditions from a DMSO solution in the membrane. The first significant increase in 16:0 lyso-PC flip occurs already at a molar ratio of 1 gramicidin per 2000 phospholipids. At gramicidin concentrations exceeding 10 μmol/l (corresponding to a gramicidin/phospholipid molar ratio of 1:100) reliable determination of flip is impossible due to occurring hemolysis. Flip enhancement is accompanied by an increase in the aspecific permeability of intermediate sized solutes (Classen et al., 1987). These effects are mechanistically related to H_{II} phase formation in the red cell membrane, which occurs at higher peptide concentrations. The insert in Fig. 15 shows the presence of a 'hexagonal H_{II}' component in the ^{31}P NMR spectrum of erythrocyte ghosts to which gramicidin was added. Addition of tryptophan formylated gramicidin (NFG) does not result in H_{II} phase formation despite its membrane incorporation and also does not increase the lyso-PC flip or aspecific permeability in the red cell membrane (Classen et al., 1987). The gramicidin-induced flip is not restricted to 16:0-lyso-PC but also occurs for other lyso-PC's and acyl carnitines (Classen et al., 1987). Apparently, flip proceeds via defects in membrane lipid organization caused by the peptide. Such defects, which can be considered to be precursor states towards H_{II} phase formation might also be responsible for the gramicidin-induced enhanced susceptibility of the outer layer membrane phospholipids in the red cell membrane towards phospholipases (Classen et al., 1987). Using the solvent approach described above in Fig. 14, it could be established that the $\beta^{6.3}$ conformation of gramicidin is required for enhancement of flip and aspecific permeability (Tournois et al., 1988). The channels formed by gramicidin (N to N $\beta^{6.3}$ dimer) in the red cell membrane as such are not responsible for flip enhancement as they can be detected at peptide concentrations which are much lower than those giving rise to flip enhancement (Classen et al., 1987). The very large difference in gramicidin concentration required to induce cation selective channels and to change lipid organization is illustrated by the observation that one gramicidin molecule per 500,000 lipids is sufficient to cause rapid release of all Na^+ enclosed in large unilamellar DOPC vesicles in 5 min (Tournois et al., 1987a), whereas gramicidin induces the H_{II} phase in DOPC dispersions at a molar ratio of around 1 gramicidin per 15 lipids (Chupin et al., 1987).

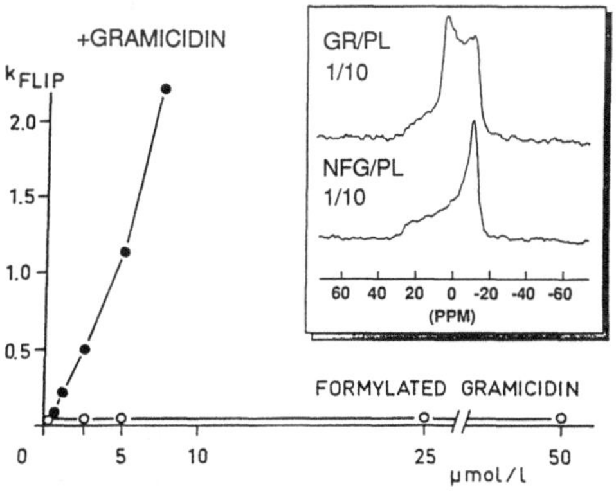

Fig. 15. Effect of gramicidin (GR) and tryptophan formylated gramicidin (NFG) on the rate constant of 16:0-lyso-PC flip (k flip) in human red cell membranes. The insert shows the 121.4 MHz ^{31}P NMR spectra of erythrocyte ghosts to which gramicidin or tryptophan formylated gramicidin were added in a 1:10 molar ratio (Tournois et al., 1987b). The lyso-PC flip was measured via the bovine serum extraction method (Classen et al., 1987). For further details see this reference.

Gramicidin-induced membrane fusion

As outlined in the introduction section, lipid/protein interactions are likely candidates to regulate local and transient lipid organization in biological membranes. One obvious example of a membrane process during which such regulation must occur is fusion of biomembranes. Type II lipids are ideally suited to be involved in such processes, because (1) they allow close apposition of adjacent bilayers, because of their low head group hydration, and (2) they can readily form inverted structures that are attractive intermediates in the fusion process (De Kruijff et al., 1985; Siegel, 1987). Moreover, during the last years an increasing amount of evidence is presented that factors that have a tendency to promote organization of lipids in type II structures such as the H_{II} phase, also promote membrane fusion (De Kruijff et al., 1985; Batenburg et al., 1987; Ellens et al., 1986; Valtersson et al., 1985; Hope & Cullis, 1981).

The ability of gramicidin to induce H_{II} phase formation in DOPC model membranes can be modulated by either varying the conformation, using different solvents to introduce the peptide, or by changing the primary structure by modification of the Trp residues. Therefore it is an excellent model to establish the relationship between peptide structure and indication of non-bilayer structures on one hand and the involvement of these structures in the fusion process on the other.

Membrane fusion can be monitored conveniently either by determination of the extent of lipid mixing between different vesicle populations or by mixing of aqueous contents of separate vesicles. In the experiments described here the extent of lipid mixing was determined by incorporating a fluorescent lipid probe (pyrene-PC) in unilamellar DOPC vesicles. This assay is based on the characteristic ability of pyrene groups to form excimers, a process that is strongly concentration-dependent (Doody et al., 1980). To monitor lipid mixing, large unilamellar DOPC vesicles containing 10 mol% pyrene-PC were mixed (1:20, mol/mol) with unlabelled, non-pyrene PC-containing vesicles. The ratio of excimer (E) and monomer (M) fluorescence intensities serves as a measure for the surface concentration of pyrene PC and decreases when lipid mixing between labelled and unlabelled vesicles occurs. For further details see Tournois et al. (1989). Figure 16 (left panel) shows that incorporation of gramicidin added from DMSO induces lipid mixing in a concentration-dependent way. Lipid mixing is markedly reduced when either gramicidin is added from ethanol (not shown) or when NFG is used.

Mixing of vesicle contents was measured by the fluorescein isothiocyanate (FITC)-dextran assay, adapted from Stutzin (1986). In this assay, aqueous content mixing is monitored as dequenching of the fluorescence of FITC dextran (FD-10; MW 9400) enclosed in self-quenching concentrations (20 mM) in DOPC vesicles, due to dilution of vesicle contents in a 20-fold excess of a non-fluorescent dextran-containing acceptor vesicles. Corrections for an increase of fluorescence due to leakage of FITC-dextran into the medium were made by measuring changes in the fluorescence intensity in the absence of acceptor vesicles. Figure 16 (right panel) shows that gramicidin, added from DMSO induces mixing of aqueous contents between different vesicle populations. In contrast, addition of tryptophan-formylated gramicidin, or addition of gramicidin from an ethanolic solution (not shown) does not lead to a significant extent of contents mixing. Thus, gramicidin is able to induce fusion of DOPC vesicles when added from DMSO in a molar ratio of peptide to lipid of 1:100 or higher. The experiments using NFG and different solvent conformations further demonstrate that there is a close parallel between the ability of gramicidin to induce H_{II} formation and its ability to induce vesicle fusion.

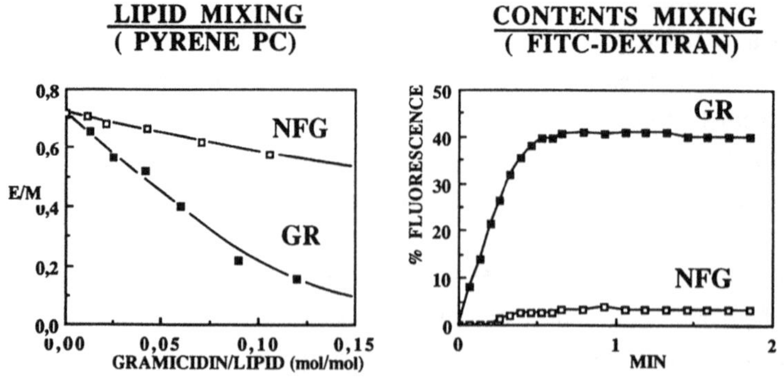

Fig. 16. Gramicidin-induced lipid mixing (left panel) and contents mixing (right panel) of large unilamellar DOPC vesicles (150 μM). Gramicidin (GR) or its formylated analog (NFG) were added from a 0.5 mM DMSO solution. Lipid mixing was measured at various peptide to lipid ratios 1 min after addition of the peptide. The time course of contents mixing was measured after adding peptide in a molar ratio of 0.07 as dequenching of the FITC-dextran fluorescence.

SUMMARIZING DISCUSSION

The results described in this paper demonstrate the unique and potent ability of gramicidin to influence lipid organization and dynamics in membranes. We will discuss now the architecture of the gramicidin-induced hexagonal H_{II} phase and then try to give a possible link between the mechanism of H_{II} phase formation and gramicidin-induced flip-flop and fusion.

In Fig. 17 a model of one tube of the gramicidin-induced H_{II} phase is presented which accommodates all experimental data. In this model cone-shaped gramicidin molecules present in aggregates form the structural backbone of the tube. The gramicidin molecules are pictured as $\beta^{6.3}$ monomers, oriented perpendicular to the axis of the tube and with the N-terminus pointing towards the aqueous channel present within the tube. The H_{II} phase is always very rich in gramicidin. For the H_{II} phase induced in DOPC at least one gramicidin is present per 7 DOPC, but a much higher amount of gramicidin can be accommodated in the tubes (Killian et al., 1987). The lipid molecules fill the remainder of the space in the tubes. Water molecules can rapidly exchange between the channel in the gramicidin molecule and the H_{II} tubes (Killian & De Kruijff, 1985). Besides for the arguments given in the Results section, experimental evidence in favor of this model is further provided in references (Killian & De Kruijff, 1985a,b; Chupin et al., 1987). Of this latter we would like to emphasize the virtually membrane system independent tube diameter (Killian et al., 1986b; Brasseur et al., 1987), the reduction in rate of lateral diffusion of lipids about the tubes in the gramicidin-induced H_{II} phase (Tournois et al., 1987b; Gasset et al., 1988), and the favored mode of self-association of gramicidin in aggregates which have a curvature matching that of the tubes of the H_{II} phase (Brasseur et al., 1987). Gramicidin can thus be considered to be the first member of a new class of intramembrane polypeptides which together with lipids can induce tubular membrane structures, similar for instance to the tight junctions (Kachar & Reese, 1982) which separate the basolateral and the apical domain of the plasma membrane of polarized cells.

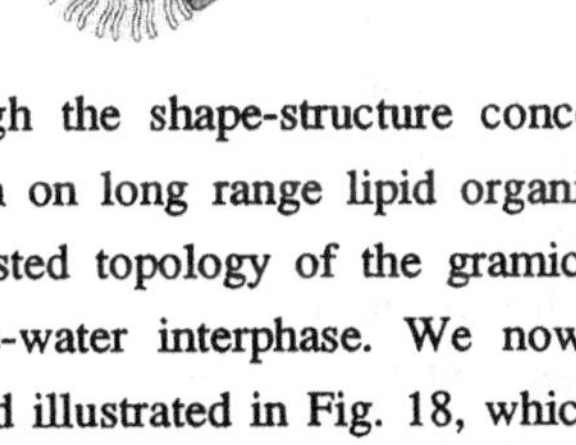

Fig. 17. Model of the gramicidin-induced H_{II} phase.

Although the shape-structure concept depicted in Fig. 6 explains very well the effect of gramicidin on long range lipid organization it is immediately clear that it seems to contradict the suggested topology of the gramicidin channel in which the C-terminus is localized at the membrane-water interphase. We now present a theory, adapted from Killian & De Kruijff (1988) and illustrated in Fig. 18, which links the gramicidin channel structure to the gramicidin-induced H_{II} phase. When gramicidin is added from ethanol it initially incorporates as a double-stranded anti-parallel dimer, such as also found for the crystal structure of gramicidin (Wallace & Ravikumar, 1988; Langs, 1988). In these dimers maximal inter- and intramolecular hydrogen bonding occurs. Because of the even distribution of the 8 tryptophans over the surface, the dimer is more or less cylindrical and thus fits nicely in a PC bilayer. In time the double-stranded dimer converts to the N to N single-stranded $\beta^{6.3}$ dimer (the channel), being the energetically most favorable conformation in a lipid bilayer. Upon introduction of gramicidin from TFE or DMSO it directly incorporates in the $\beta^{6.3}$ helical conformation and may form channels by hydrogen bonding of the N-termini.

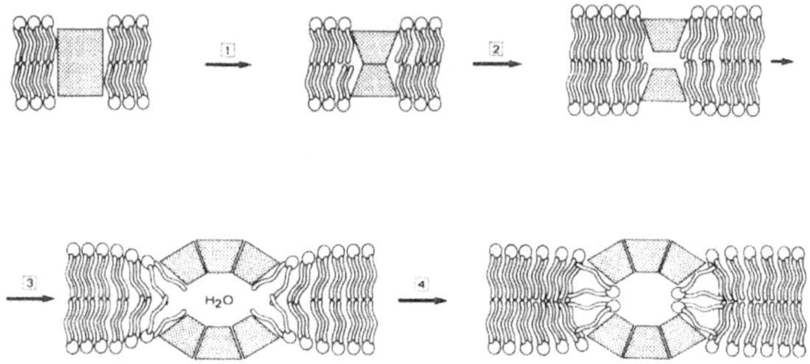

Fig. 18. Schematic representation of the mechanism of H_{II} phase formation upon addition of gramicidin. When added from ethanol, gramicidin is present as an anti-parallel double helix, which is depicted with an overall cylindrical shape. In time the conformation of the peptide converts to the channel conformation in which it is directly present when added from TFE or DMSO, consisting of two cone-shaped single-stranded $\beta^{6.3}$ helices (step 1). When the bilayer thickness exceeds the length of the gramicidin dimer, the dimer dissociates and water fills the space between monomers (step 2). The monomers undergo lateral self-association (step 3), whereby the specific aggregate structure causes a strong decrease in order of the adjacent lipids which in addition, because of their unfavorable topology, flip (step 4). Further intra- and interbilayer aggregation and fusion processes then lead to H_{II} phase formation. All drawings are side views. For further details and experimental support of this model see Killian & De Kruijff (1988).

When there is a mismatch in length, as for instance in case of the longer chained DOPC, it is proposed that there is a shift in the equilibrium between monomers and dimers, such that a larger amount of monomers is present. This has been proposed on basis of the observed increase in rates of dissociation of channels upon increasing the lipid chain length in black lipid membrane experiments (Kolb & Bamberg, 1977; Elliott et al., 1983) and is consistent with CD measurements showing that the conformation is independent of acyl chain length and does not discern between monomers and dimers (Killian et al., 1988). It is possible that such dissociation results from interactions between the lipid carbonyl groups and the Trp or ethanolamine of gramicidin (Meulendijks et al., 1988; Wallace, 1986). By lateral clustering of monomers in the bilayer the energetically preferred curved gramicidin aggregates are formed (Brasseur et al., 1987). Dimer dissociation and monomer clustering could be interactive processes. As a result of the geometry of the gramicidin cluster, pockets of water are induced in the center of the bilayer. In addition, the lipid molecules immediately adjacent to the gramicidin aggregate become highly disordered. Such disordered lipids were identified by different NMR techniques (Chupin et al., 1987). The packing constraints and unfavorable topology of those boundary lipids can be overcome by phospholipid reorientation. The gramicidin-induced lipid flip-flop in the erythrocyte membrane could very well be the result of such a mechanism. The structure of the aggregate after lipid reorientation is basically similar to the organization in the resulting H_{II}

phase, which could be formed by further growth of these precursor structures within the bilayers and by gramicidin-induced fusion with adjacent bilayers.

The tryptophans of gramicidin play an essential role in the lipid structure-modulating activity of the peptide. They contribute largely to the cone shape of the molecule. Furthermore, stacking interactions between Trp^9 and Trp^{15} could stabilize the $ß^{6.3}$ structure (Spisni et al., 1983; Killian et al., 1987; Prasad et al., 1983; Boni et al., 1986). In addition, they appear to play an important role in the formation of gramicidin aggregates in the bilayer (Killian et al., 1987). Energy calculations (Brasseur et al., 1987) showed that the favored aggregate structure of gramicidin is to a large extent the result of the close proximity of the tryptophans of adjacent molecules.

From experiments with Trp-modified gramicidin analogs and using different organic solvents to introduce the peptide it is clear that a close parallel exists between gramicidin-induced H_{II} phase formation, flip-flop and membrane fusion. The results of the present study on gramicidin-lipid interactions as well as a number of other recent studies unambiguously show that factors that induce the formation of type II structures, such as the H_{II} phase, are the same factors that at concentrations or temperatures below those required for macroscopic changes in lipid organization can facilitate functional membrane processes like fusion (De Kruijff et al., 1985; Batenburg et al., 1987; Ellens et al., 1986; Valtersson et al., 1985; Hope & Cullis, 1981) and lipid transbilayer movement (De Kruijff et al., 1985; Noordam et al., 1981; Van Duijn et al., 1986). These events appear to be mechanistically related and most likely share common intermediates. It will be one of the future challenges to understand and characterize the nature and origin of these intermediates at a molecular level.

ACKNOWLEDGEMENTS

We gratefully acknowledge the contributions of F.J. Aranda, O.R. Bokking, F. Borle, R. Brasseur, K.N.J. Burger, V. Chupin, J. Classen, H. van Dalen, B. Deuticke, C.J.A. van Echteld, C. Fabrie, M. Gasset, W.S.M. Geurts van Kessel, J. de Gier, C.W.M. Haest, D. Hains, P. Hilgers, S. Keur, J. Leunissen-Bijvelt, the late Mrs. J.G. Mandersloot, K.U. Prasad, J.M. Ruysschaert, J. Seelig, J.W. Timmermans, D.W. Urry and A.J. Verkleij to the research described in this article.

REFERENCES

Akashi K, Kurahashi K (1977) Biochem Biophys Res Commun **77**:259-267
Andersen OS (1984) Ann Rev Physiol **46**:531-548
Aranda FJ, Killian JA, De Kruijff B (1987) Biochim Biophys Acta **901**:217-228
Batenburg AM, De Kruijff B (1988) Biosci Rep **8**:299-307
Batenburg AM, Hibbeln JCL, Verkleij AJ, De Kruijff B (1987) Biochim Biophys Acta **903**:142-154
Boni CT, Connolly AJ, Kleinfeld AM (1986) Biophys J **49**:122-123
Brasseur R, Cabiaux V, Killian JA, De Kruijff B, Ruysschaert JM (1986) Biochim Biophys Acta **855**:317-324
Brasseur R, Killian JA, De Kruijff B, Ruysschaert JM (1987) Biochim Biophys Acta **903**:11-17.
Chapman D, Cornell BA, Eliasz AW, Perry A (1977) J Mol Biol **113**:517-538
Chupin V, Killian JA, De Kruijff B (1987) Biophys J **51**:395-405
Classen J, Haest CWM, Tournois H, Deuticke B (1987) Biochemistry **26**:6604-6612
Cullis PR, De Kruijff B (1979) Biochim Biophys Acta **559**:399-420

Cullis PR, Hope MJ, De Kruijff B, Verkleij AJ, Tilcock CPS (1985) In: Phospholipids and Cellular Regulation, Vol I. Kuo KF (ed) CRC Press Inc, Boca Raton, pp 1-60
De Kruijff B (1987) Nature **329**:587-588
De Kruijff B, Cullis PR, Verkleij AJ, Hope MJ, Van Echteld CJA, Taraschi TF, Van Hoogevest P, Killian JA, Rietveld A, Van der Steen ATM (1985) In: Progress in Protein-Lipid Interactions, Watts EA, De Pont A (eds), Elsevier Sci Publ BV, Amsterdam, pp 89-142
Doody MC, Pawnall HJ, Kao YJ, Smith LC (1980) Biochemistry **19**:108-116
Ellens H, Bentz J, Szoka F (1986) Biochemistry **25**:4141-4147
Elliott JR, Needham D, Dilger JP, Haydon DA (1983) Biochim Biophys Acta **735**:95-103
Gasset M, Killian JA, Tournois H, De Kruijff B (1988) Biochim Biophys Acta **939**:79-88
Glickson JD, Mayers DF, Settine JM, Urry DW (1972) Biochemistry **11**:477-486
Goldfine H, Johnston NC, Mattai J, Shipley GG (1987) Biochemistry **26**:2814-2822
Israelachvili JN, Mitchell DJ, Ninham BW (1976) J Chem Soc Faraday Trans II **72**:1525-1568
Hawkes GE, Lian L, Randall EW, Sales KD, Curzon EH (1987) Eur J Biochem **166**:437-445
Hope MJ, Cullis PR (1981) Biochim Biophys Acta **640**:82-90
Jung G, Brückner H, Schmidt H (1981) In: Structure and Activity of Natural Peptides, Voelter W, Weitzer G (eds), pp 75-114
Jung G, Katz E, Schmidt H, Voges KP, Winter W (1983) Biopolymers **22**:241-246
Kachar B, Reese TS (1982) Nature, London **296**:464-466
Katz E, Demain AL (1977) Bact Rev **41**:449-474
Killian JA, De Kruijff B (1985a) Biochemistry **24**:7890-7898
Killian JA, De Kruijff B (1985b) Biochemistry **24**:7881-7890
Killian JA, De Kruijff B (1986) Chem Phys Lipids **40**:259-284
Killian JA, De Kruijff B (1988) Biophys J **53**:111-117
Killian JA, Urry DW (1988) Biochemistry **27**:7295-7301
Killian JA, De Kruijff B, Van Echteld CJA, Verkleij AJ, Leunissen-Bijvelt J, De Gier J (1983) Biochim Biophys Acta **728**:141-144
Killian JA, Timmermans JW, Keur S, De Kruijff B (1985) Biochim Biophys Acta **820**:154-156
Killian JA, Borle F, De Kruijff B, Seelig J (1986a) Biochim Biophys Acta **854**:133-142
Killian JA, Van den Berg CW, Tournois H, Keur S, Slotboom A, Van Scharrenburg GJM, De Kruijff B (1986b) Biochim Biophys Acta **857**:13-27
Killian JA, Burger KNJ, De Kruijff B (1987) Biochim Biophys Acta **897**:269-284
Killian JA, Prasad KU, Hains D, Urry DW (1988) Biochemistry **27**:4848-4855
Killian JA, Prasad KU, Urry DW, De Kruijff B (1989) Biochim Biophys Acta **978**:341-345
Kolb HA, Bamberg E (1977) Biochim Biophys Acta **464**:127-141
Langs DA (1988) Science **241**:188-191
Lewis BA, Engelman DM (1983) J Mol Biol **166**:211-217
Lipmann F (1980) Adv Microbiol Phys **21**:227-266
Macdonald PM, Seelig J (1988) Biochemistry **27**:2357-2364
Madden TD, Cullis PR (1982) Biochim Biophys Acta **684**:149-153
Masotti L, Spinsi A, Urry DW (1980) Cell Biophys **2**:241-251
Mayer LO, Hope MJ, Cullis PR (1986) Biochim Biophys Acta **858**:161-168
Meulendijks GHWM, Sonderkamp T, Buck HJ (1988) Biochim Biophys Acta **979**:321-330
Noordam PC, Van Echteld CJA, De Kruijff B, De Gier J (1981) Biochim Biophys Acta **646**:483-487.
Pasquali-Ronchetti I, Spisni A, Casali E, Masotti L, Urry DW (1983) Biosci Rep **3**:127-133
Prasad KU, Trapane TL, Busath D, Szabo G, Urry DW (1983) Int J Pept Prot Res **22**:341-347
Rice D, Oldfield E (1979) Biochemistry **18**:3272-3279
Seelig J (1977) Quat Rev Biophys **10**:353-418
Shungu DC, Hinton JF, Koeppe II RE, Millett FS (1986) Biochemistry **25**:6103-6108
Siegel DP (1987) In: Cell Fusion (Sowers AE, ed), Plenum Press, New York, pp 181-207
Spisni A, Pasquali-Ronchetti I, Casali E, Lindner L, Cavatorta P, Masotti L, Urry DW (1983) Biochim Biophys Acta **732**:58-68
Stark G, Strassle M, Tachar Z (1986) J Membr Biol **89**:23-27
Stutzin A (1986) FEBS Lett **197**:274-280
Sychev SV, Nevskaya NA, Jordanov S, Shepel EN, Miroshnikow AI, Ivanov VT (1980) Bioorg Chem **9**:121-151
Tank DW, Wu ES, Meers PR, Webb WW (1982) Biophys J **40**:129-135
Tournois H, Fabrie ChHJP, Burger KNJ, Mandersloot J, Hilgers P, Van Dalen H, De Gier J and De Kruijff B (1989) Manuscript in preparation
Tournois H, Henseleit U, De Gier J, De Kruijff B, Haest CWM (1988) Biochim Biophys Acta **946**:173-177

Tournois H, Killian JA, Urry DW, Bokking OR, De Gier J, De Kruijff B (1987a) Biochim Biophys Acta **905**:222-226
Tournois H, Leunissen-Bijvelt J, Haest CWM, De Gier J, De Kruijff B (1987b) Biochemistry **26**:6613-6621
Urry DW (1985) In: The Enzymes of Biological Membranes, Vol I. Martonosi AN (ed) Plenum Press, New York, pp. 229-257
Urry DW, Glickson JD, Mayers DF, Haider J (1972) Biochemistry **11**:487-493
Urry DW, Spisni A, Khaled MA, Long MM (1979) Int J Quant Chem: Quant Biol Symp **6**:289-303
Urry DW, Trapane TL, Prasad KU (1983) Science **221**:1064-1067
Valtersson C, Van Duijn G, Verkleij AJ, Chojnacki T, De Kruijff B, Dallner G (1985) J Biol Chem **260**:2742-2751
Van Duijn G, Valtersson C, Chojnacki T, Verkleij AJ, Dallner G, De Kruijff B (1986) Biochim Biophys Acta **861**:211-223
Van Echteld CJA, De Kruijff B, Verkleij AJ, Leunissen-Bijvelt J, De Gier J (1982) Biochim Biophys Acta **692**:126-138
Veatch WR, Blout ER (1974) Biochemistry **13**:5257-5264
Veatch WR, Fossel ET, Blout ER (1974) Biochemistry **13**:5249-5256
Wallace BA (1986) Biophys J **49**:295-306
Wallace BA, Ravikumar K (1988) Science **241**:182-187
Wieslander Å, Rilfors L, Johansson LB-A, Lindblom G (1981) Biochemistry **20**:730-735

INFLUENZA VIRUS MEDIATED MEMBRANE FUSION: THE IDENTIFICATION OF FUSION INTERMEDIATES USING MODERN CRYOTECHNIQUES

Koert N.J. Burger (1), Gerd Knoll (2), Peter M. Frederik (3), and Arie J. Verkleij (1).

(1) Institute of Molecular Biology & Medical Biotechnology
University of Utrecht
Padualaan 8, 3584 CH Utrecht
The Netherlands

INTRODUCTION

Higher life-forms have evolved in the course of evolution by virtue of (cellular) compartmentalization. Each cellular compartment (nucleus, endoplasmic reticulum, Golgi, etc..) is surrounded by one or two membranes and has a unique structure and composition along with a unique cellular function. Since most of the constituents of the different compartments are synthesized centrally in the cell, a highly efficient sorting and targeting machinery must exist in order to create and maintain the differences in composition between the individual cellular compartments. An important part of this sorting and targeting machinery is vesicular transport, in which membrane fusion is the key event.

Membrane fusion has to occur in a very specific and highly controlled fashion, and therefore specific modulators have to be present determining exactly when and where membrane fusion occurs. The role of certain lipids in determining the overall fusogenicity of a membrane is well established, i.e. membrane lipids with low head group hydration and non-bilayer preferring lipids promote membrane fusion in model systems (De Kruijff et al., 1985a; Rand and Parsegian, 1986). However, based on the great variety and high specificity of biological fusion events, proteins are the most likely candidates for determining the high specificity of a fusion reaction. Proteins probably play an important role not only in the recognition event preceding membrane fusion but also in the regulation of the fusion event itself. Proteins could either play a passive role by sterically hindering the close apposition of membranes and preventing membrane fusion, or an active role by inducing membrane fusion (Verkleij, 1984; De Kruijff et al., 1985b; Blumenthal, 1987).

Taking this as a starting point, two important questions must be answered. The first question concerns the identification of the putative proteins involved in the regulation of intracellular membrane fusion. This kind of research critically depends on the development of in vitro recon-

(2) Fakultät für Biologie, Universität Konstanz, Postfach 5560, D-7750 Konstanz, FRG; (3) EM-unit, Department of Pathology, Univ. of Limburg, P.O. Box 616, 6200 MD Maastricht, NL.

NATO ASI Series, Vol. H 40
Dynamics and Biogenesis of Membranes
Edited by J.A.F. Op den Kamp

stitution systems and sensitive fusion assays. Using this approach Rothman and co-workers have recently succeeded in identifying and purifying a protein involved in intra-Golgi vesicular transport (Block et al., 1988; Weidman et al., 1989; Wilson et al., 1989). The second question concerns the working mechanism of a fusion protein.

In order to get insight into the working mechanism of a biologically relevant fusion protein we turned to one of the best characterized protein-modulated membrane fusion reactions, the fusion of enveloped animal viruses with their target membranes.

THE MODEL SYSTEM: INFLUENZA VIRUS MEDIATED MEMBRANE FUSION

During cellular infection enveloped animal viruses fuse with either the plasma membrane or the endosomal membrane. Membrane fusion is accomplished through the specific action of viral glycoproteins, or spike proteins (White et al., 1983).

The influenza virus is one of the best studied enveloped animal viruses and enters the cell by receptor-mediated endocytosis (Matlin et al., 1981). The low pH of the endosomal compartment results in activation of the influenza virus hemagglutinin glycoprotein (HA) leading to fusion of viral and endosomal membrane and release of the viral nucleocapsid into the cytoplasm (Maeda et al., 1981; White et al., 1982a). The influenza hemagglutinin glycoprotein is involved in the initial attachment of the virus to sialic acid-containing receptors on the target membrane and in the fusion process itself. The molecular structure of the hemagglutinin glycoprotein has been extensively studied both at neutral and low pH (Wilson et al., 1981; Wharton, 1987; Gething et al., 1988).

At low pH the interactions between the HA monomers in the trimeric spike protein complex are partially lost. This irreversible conformational change leads to the exposure of the extremely hydrophobic and highly conserved N-terminal part of subunit-2 of HA (Wiley and Skehel, 1987). Almost any change in the amino acid sequence of this part of the HA molecule (by site-directed mutagenesis; Gething et al., 1986) drastically influences the fusion activity. Therefore this N-terminal fusion sequence is thought to be directly involved in the fusion reaction. Following common sense the N-terminal fusion sequence could interact with the target membrane forcing viral and endosomal membranes together and inducing them to fuse. Although detailed information on the low pH-induced conformational change of the HA glycoprotein is accumulating, the mechanism by which HA induces membrane fusion after its conformational change remains unsolved.

An ultrastructural approach could possibly shed some light on the fusion mechanism in general, and on possible intermediates of the actual joining of the two lipid bilayers in particular. Therefore, we characterized influenza HA-induced membrane fusion morphologically using modern cryotechniques. Three model systems were used: the fusion of large unilamellar vesicles (eggPC/eggPE/Cholesterol/Gd1a, 1:1:2:0.2 molar ratio) with the apical plasma membrane of influenza virus-infected epithelial cells (Knoll et al., 1988), the fusion of influenza virus with resealed human erythrocyte ghost (Burger et al., unpublished results), and the fusion of influenza virus with large unilamellar vesicles (composition as mentioned above; Burger et al., 1988). Different fusion assays were used to allow biochemical characterization in parallel to the morphological characterization of the fusion process (Van Meer and Simons, 1983; Knoll et al., 1988; Burger et al., 1988).

THE TECHNIQUES: FREEZE FRACTURE, FREEZE SUBSTITUTION, AND CRYO ELECTRON MICROSCOPY

Before electron microscopical analysis is feasible the intrinsically instable structure of a sample has to be stabilized using chemical fixation, physical (or cryo-) fixation, or a combination of both. Chemical fixation critically depends on the penetration of chemical cross-linkers into the sample and consequently offers only low time-resolution (minute-range). In addition, membrane fusion events seem to be extremely susceptible to chemical pretreatments. Intramembrane particle clearance and the formation of naked membrane diaphragms during membrane fusion have been shown to be preparation artifacts (Heuser et al., 1979). Only pure cryofixation avoids these preparation artifacts and might in addition offer sufficient time-resolution to capture membrane fusion intermediates (both freezing-time and life-time of fusion intermediates are thought to be in the msec-range; for discussion see Knoll et al., 1987). Of course pure cryofixation should be combined with a reliable visualization technique.

We used controlled induction of fusion by lowering of the pH followed by synchronized fast-freezing. Subsequent analysis was performed using freeze fracture electron microscopy, freeze substitution low-temperature plastic embedding, and cryo electron microscopy. These techniques are particularly well suited to study membrane dynamics. Freeze fracture offers an extended view of the membrane interior and has been extensively used to study membrane fusion. Using this technique fusion intermediates have been described in both biological and artificial (pure lipid) systems (Schmidt et al., 1983; Verkleij, 1984). Freeze substitution low-temperature plastic embedding avoids the artifacts caused by chemical pretreatments during classical chemical fixation, cryoprotection and dehydration procedures, and offers a reliable cross-sectional view of membrane systems (Steinbrecht and Muller, 1987). Since phospholipids are not extracted during freeze substitution, the redistribution of radioactive phospholipids after membrane fusion can be followed (Knoll et al., 1988), and even pure lipid systems can be visualized using this technique (Verkleij et al., 1985). In cryo electron microscopy the sample is not fractured and replicated or dehydrated and embedded in plastic, but fast-frozen and visualized at low-temperature in its original state. Cryo electron microscopy has recently shown its great potential in the examination of membrane structure in pure lipid systems (Frederik et al., 1989).

RESULTS: MORPHOLOGICAL CHARACTERIZATION AND MOLECULAR INTERPRETATION OF FUSION INTERMEDIATES

Influenza virus hemagglutinin-mediated membrane fusion is strictly pH-dependent. Binding at neutral pH and induction of fusion by a rapid lowering of the pH thus offers both an excellent synchronization and an optimal density of fusion events, and maximizes the chance of catching short-lived fusion intermediates.

Indeed, specific morphological features were observed on the fracture faces of the liposomal membrane and the plasma membrane during fusion of liposomes with influenza virus-infected cells (Fig. 1; Knoll et al., 1988). The observed particles and pits are strictly correlated to the interaction of liposomes with the apical plasma membrane induced by low pH. They were never seen in samples containing liposomes and cells at neutral pH nor in liposomes or cells incubated separately at low or neutral pH. Furthermore, preliminary experiments on the interaction of influenza virus with resealed erythrocyte ghosts have also shown the presence of

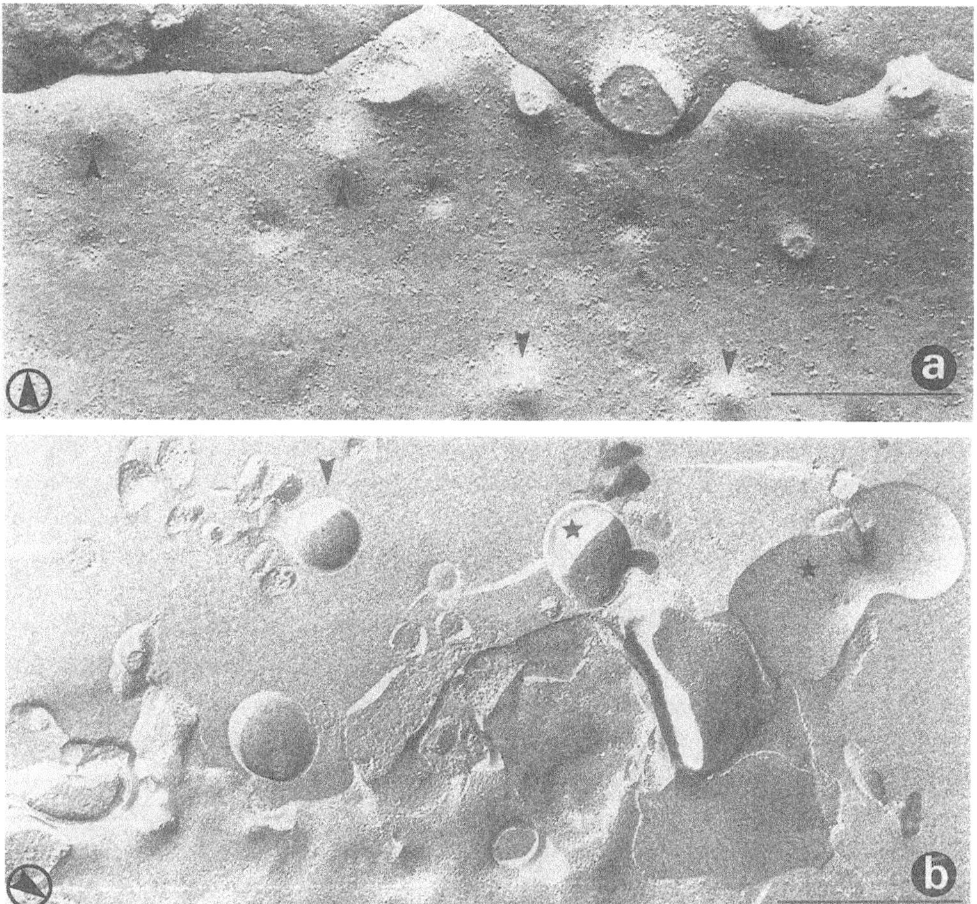

Figure 1. Freeze fracture of the apical plasma membrane during fusion with associated liposomes. Cells were fast-frozen after a 30 s incubation at pH 5, 37°C. Protrusions of the protoplasmic fracture face with a central particle or pit are marked by arrowheads (a). In b the exoplasmic fracture face of the apical membrane is exposed. A liposome probably just fusing with a microvillus is visible (arrowhead). Some liposomes (marked by stars) reveal single particles, presumably in areas in contact with the plasma membrane. Direction of Pt/C shadow indicated by encircled arrowheads. Bars, 0.5 μm (taken from Knoll et al., 1988).

fusion-correlated particles and pits on the fracture faces of the erythrocyte ghost membrane (Fig. 2b; Burger et al., unpublished results). Unfortunately, no technique is currently available to determine whether a fusion-correlated particle or pit represents a protein, a lipid, or a lipid-protein complex. Nevertheless, arguments can be given in favour of a lipidic nature and against a proteinaceous nature. The arguments against a proteinaceous nature of the fusion-correlated particles are based on the size of the particles and the location of the particles on either (or both) fracture face(s).

Let us first consider the partition behavior of integral membrane proteins in general. Intramembrane particles normally arise from the partitioning of a substantial part of an integral membrane protein on one of the fracture faces during freeze fracture (Verkleij and Ververgaert, 1978). Broadly speaking, membrane proteins that partition on the protoplasmic fracture face have either a large intracellular domain or interact with the cytoskeleton; whereas proteins that partition on the exoplasmic fracture face have a large extracellular domain or interact with the extracellular matrix (e.g. Edwards et al., 1979).

During HA-mediated membrane fusion, protein particles could arise from the cytoplasmic part of the HA molecule, or from the N-terminal part of HA subunit-2 after penetration into the target membrane. Because the HA glycoprotein is almost entirely exposed at the extracellular surface, intramembrane particles can only arise from partitioning of the cytoplasmic part of HA on the exoplasmic fracture face of the plasma membrane of virus-infected cells. Since fusion-correlated particles are also observed on the protoplasmic fracture face of the plasma membrane during liposome-cell fusion (Fig. 1a), these particles are not very likely to be of a proteinaceous nature. On the other hand one could argue that the (hypothetical) penetration of the N-terminal fusion sequence of HA subunit-2 into the target membrane at low pH could cause (protein) particles on the fracture faces of the target membrane. For the same reasons as mentioned above the N-terminal fusion sequence would partition on the concave fracture face of the liposomal membrane (liposome-cell system) or on the exoplasmic fracture face of the erythrocyte ghost membrane (virus-ghost system). However, the size of the N-terminal fusion sequence would be far too small to allow visualization by freeze fracture electron microscopy. In agreement with this, preliminary experiments on low pH-induced binding of isolated HA spikes (by bromelain treatment according to Skehel et al., 1982) to pure lipid vesicles did not reveal the presence of intramembrane particles (Burger et al., unpubl. results). Consequently, the fusion-correlated particles on both fracture faces of the liposomal membrane (Fig. 1b) and erythrocyte ghost membrane (e.g. the protoplasmic fracture face, Fig. 2b) are again not very likely to be of a proteinaceous nature.

Moreover the presence of fusion-correlated pits, the often observed complementarity of the fracture faces, and the similarity in size and shape of the fusion-correlated particles to those observed in pure lipid systems, all directly favour a lipidic character of the fusion-correlated particles (and pits). Therefore we tend to conclude that the fusion-correlated particles observed during liposome-cell fusion and virus-ghost fusion are lipidic, and represent a structural intermediate in HA-mediated membrane fusion. If so, the fusion-correlated particles (and pits) either represent the adhesion, joining, or fission stage of membrane fusion.

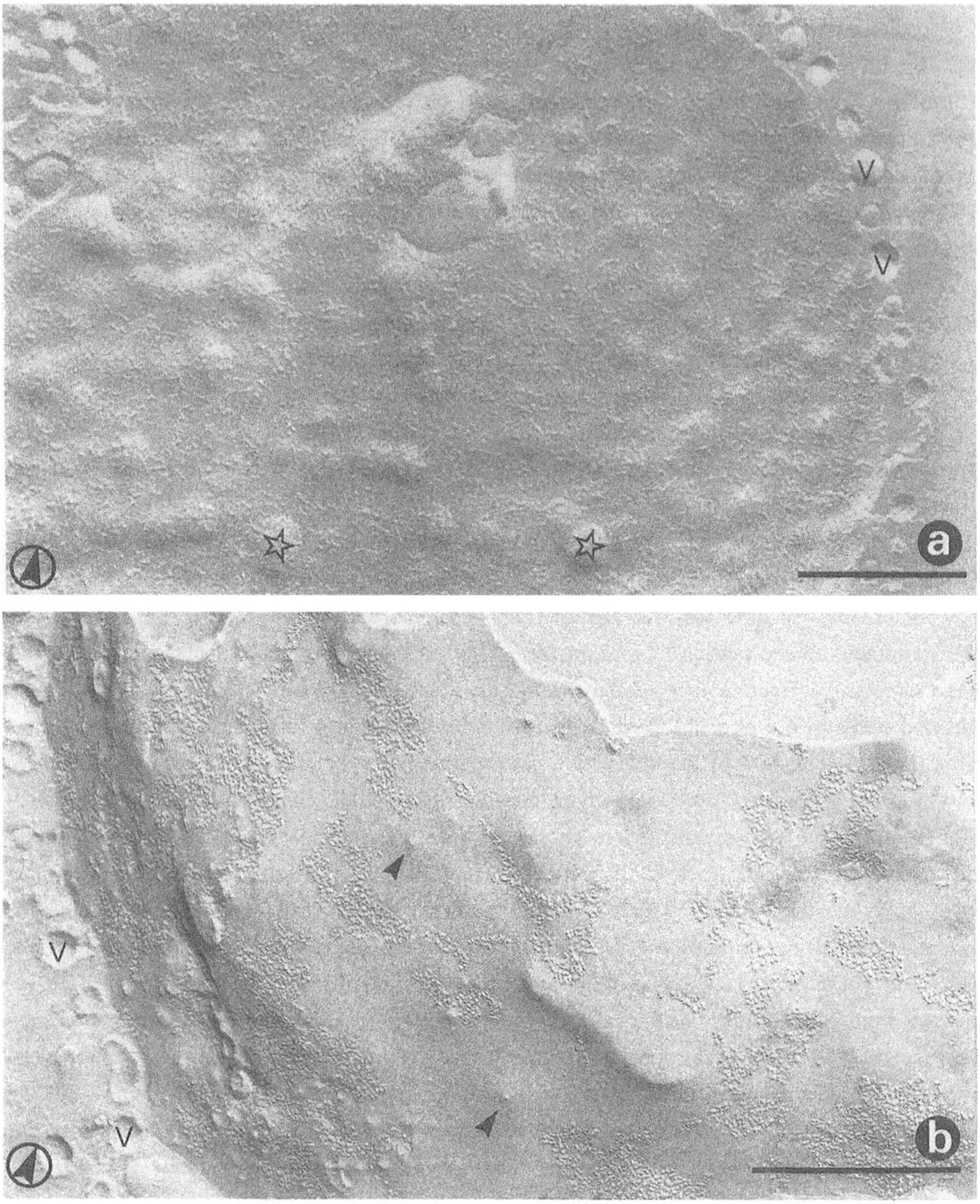

Figure 2. Freeze fracture morphology of influenza virus-erythrocyte ghost interaction. Unsealed erythrocyte ghosts were prepared according to Steck and Kant (1974) and sealed by warming in phosphate-buffered saline. At neutral pH (a) virions (V) are partially engulfed by the ghost membrane and protrusions are visible on the exoplasmic fracture face (marked by stars). Initiation of fusion by lowering of the pH (b) leads to aggregation of integral membrane proteins (as described earlier by Yoshimura et al., 1985), and the appearance of fusion-correlated particles on the protoplasmic fracture face of the ghost membrane (marked by arrowheads; sample incubated for 2 min at pH 5.5 and 37°C). Direction of Pt/C shadow indicated by encircled arrowheads. Bars, 0.5 μm.

A local point contact between two membranes can give rise to particles and pits without the necessity of actual membrane fusion (Miller, 1980). Local point adhesion sites would result in relatively large and ill-defined particles on the protoplasmic or convex fracture face and pits on the exoplasmic or concave fracture face (e.g. during virus-liposome interaction at neutral pH, see later; also see Verkleij, 1984). In the present study the possibility of local point adhesion sites is largely ruled out because both particles and pits were found on the protoplasmic fracture face of the plasma membrane (Fig. 1a) and particles were present on both fracture faces of the liposomal membrane (Fig. 1b) and erythrocyte ghost membrane (not shown).

In fact, the fusion-correlated particles are quite similar in size and shape to the well-defined spherical particles often observed in pure lipid systems. These particles have been interpreted as being inverted micellar intermediates at the joining stage of two fusing membranes (Verkleij, 1984). Recent theoretical work supports the involvement of inverted lipid micelles in membrane fusion but predicts an extremely short lifetime of the inverted micellar intermediate; inverted micellar intermediates are thought to escape detection by freeze fracture electron microscopy due to their low stability and short lifetime (Siegel, 1986). Instead it has been suggested that the well-defined lipidic particles in a pure lipid system (and consequently, the fusion-correlated particles in the current system) might well represent post-fusion structures with a minute (not detectable) aqueous channel already formed ('interlamellar attachment sites'or ILA's, Siegel, 1987). In cubic lipid phases, direct evidence has been presented in favour of this interpretation. Although freeze fracture electron microscopy shows well-defined spherical particles apparently without an aqueous channel (next to vulcano-like protrusions which clearly do possess an aqueous channel; most cubic lipid phases are heterogeneous, see e.g. Verkleij, 1984), NMR measurements do not seem to detect the expected discontinuity in the diffusion of lipids (and water) within the cubic phase (Lindblom and Rilfors, 1989); it is concluded that cubic phases are built up from either open structures (ILA's) or closed type I structures, but not from inverted lipid micelles. The possible absence of inverted lipid micelles in (meta)stable lipid phases, however, does not exclude their presence during lipid phase transitions (lamellar to hexagonal H_{II} phase) or membrane fusion. On the other hand, one has to realize that no direct evidence has been obtained to date, proving the presence (or absence) of inverted lipid micelles during dynamic membrane events. In conclusion, the morphological data on HA-mediated liposome-cell fusion and virus-ghost fusion strongly suggest a local point fusion mechanism with a lipidic particle as a specific structural fusion intermediate. Irrespective of the interpretation of the lipidic particle as an inverted micellar intermediate or an interlamellar attachment site, an inverted lipid micelle is thought to be present at the joining stage of the two fusing membranes.

At first glance, a model system composed of highly purified influenza virus and unilamellar vesicles of a defined lipid composition would seem particularly attractive both for biochemical and morphological characterization of the fusion process. Indeed, fusion is easily demonstrated by using biochemical fusion assays (Stegmann et al., 1985, 1986; Burger et al., 1988) and the fusion process can be followed quite conveniently by different cryotechniques (Fig. 3b-d). Freeze fracture allows an extended view of the membrane interior; virus-liposome fusion is de-

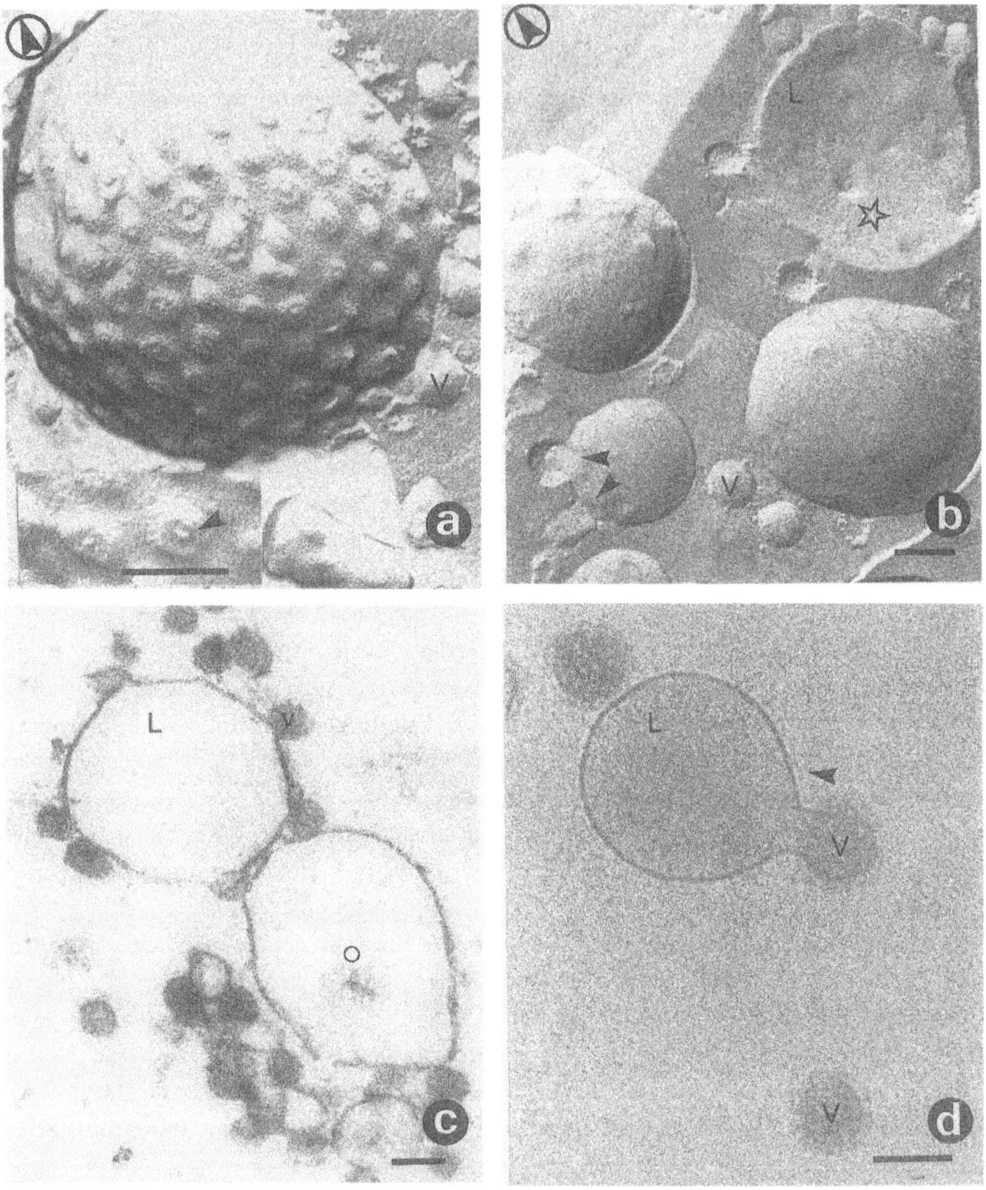

Figure 3. Morphology of influenza virus-liposome interaction as observed by freeze fracture (a,b), freeze substitution (c), and cryo electron microscopy (d). Binding of virus (V) to liposomes (L) at neutral pH leads to protrusions on the convex fracture face of the liposomal membrane (a); occasionally a central particle is observed (a, insert). Lowering the pH (1-2 min, pH 5.5, 37°C [b,c] or 20°C [d]) initiates membrane fusion. Fusion is demonstrated by continuity between viral and liposomal membrane (marked by arrowheads in b), lateral diffusion of viral intramembrane particles (marked by star in b), the presence of viral nucleocapsid in the liposomal interior (marked by circle in c), and the diffusion of viral spike proteins into the liposomal membrane (marked by arrowhead in d). No fusion-correlated particles were seen (see text). Direction of Pt/C shadow indicated by encircled arrowheads. Bars, 0.1 μm (a, b and c taken from Burger et al., 1988).

monstrated by continuity between viral and liposomal membrane, and by the appearance of viral intramembrane (protein) particles in the liposomal membrane (Fig. 3b). Freeze substitution offers the possibility to obtain a reliable cross-sectional view of membrane systems; membrane continuity is again observed, as is the presence of the nucleocapsid in the liposomal interior (Fig. 3c). Finally cryo electron microscopy visualizes a specimen in its original composition; membrane fusion is accompanied by membrane continuity and by the diffusion of viral spike glycoproteins into the liposomal membrane (Fig. 3d).

However, the morphological characterization of virus-liposome interaction revealed marked differences as compared to HA-mediated liposome-cell and virus-ghost interaction.

First of all, binding of influenza virus to liposomes at neutral pH resulted in local protrusions of the convex liposomal fracture face, often bearing an irregularly shaped central particle (Fig. 3a). Since fusion at neutral pH can be virtually excluded, this central particle is interpreted as a prefusion structure. A local point contact between viral and liposomal membrane apparently exists in spite of the high density of spike glycoproteins in the intact virion.

Secondly and again in sharp contrast to HA-mediated liposome-cell and virus-ghost fusion, samples frozen under early fusion conditions did not reveal the presence of fusion-correlated particles or pits. This is possibly caused by a shorter lifetime of fusion intermediates during virus-liposome fusion. One of the major differences between the liposome-cell and virus-ghost system, and the virus-liposome system, is the absence of the cytoskeleton in the latter system; and one might speculate that membrane-cytoskeleton interactions could impose certain steric constraints upon liposome-cell and virus-ghost fusion increasing the lifetime of fusion intermediates.

CONCLUSIONS AND PROSPECTS: THE ROLE OF NON-BILAYER STRUCTURES AND NON-BILAYER PREFERRING LIPIDS IN BIOLOGICAL MEMBRANE FUSION

The results obtained in all three model systems support the idea that influenza virus HA-mediated membrane fusion involves local point adhesion followed by local point fusion with a lipidic particle as a specific structural fusion intermediate. An inverted lipid micelle is thought to be present at the joining stage of the two fusing membranes (irrespective of the interpretation of the lipidic particle as an inverted micellar intermediate or an interlamellar attachment site).

The likely involvement of an inverted micellar intermediate in HA-mediated membrane fusion suggests an important role for hexagonal H_{II} phase preferring lipids. Indeed, influenza HA-mediated membrane fusion is promoted by the presence of hexagonal H_{II} phase preferring lipids (White et al., 1982b; Van Meer et al., 1985). Furthermore, the fact that virus-membrane fusion also occurs in complete absence of H_{II} preferring lipids in the target membrane does not exclude the role of these lipids in HA-mediated membrane fusion, nor does it exclude the possibility of an inverted micellar intermediate. Enveloped viruses have a lipid composition very similar to that of the host cell membrane and contain a significant amount of H_{II} phase preferring lipids (Van Meer and Simons, 1982).

Non-bilayer lipid structures have probably also been observed during other biological membrane fusion events such as influenza virus-budding (Knoll et al., 1988) and exocytosis in chromaffin cells (Schmidt et al., 1983) and blood platelets (Hols et al., 1985). In the case of HA-mediated membrane fusion, the appearance of non-bilayer lipid structures (fusion-correlated particles) is strictly correlated with the low pH-induced conformational change of HA, and the exposure of the hydrophobic and highly conserved N-terminal fusion sequence of subunit-2 of the HA molecule. In analogy to a number of peptides and proteins that are capable of triggering a bilayer-to-non-bilayer phase transition and induce membrane fusion (melittin, cardiotoxin, myelin basic protein and gramicidin, De Kruijff et al., 1985b; also see contribution by Killian in this volume), the N-terminal fusion sequence of HA might actively and locally induce a bilayer-to-non-bilayer phase transition leading to membrane fusion.

To test this hypothesis, the behavior of synthetic peptides with an amino acid sequence identical to the N-terminus of HA subunit-2 should be studied in detail; both the fusion activity and a possible influence on lipid phase behavior should be determined. Future research should also focus on the low pH-induced changes in the intact virion to determine the number of HA spike proteins directly involved in the induction of membrane fusion, and to determine whether their N-termini penetrate the viral or the target membrane.

It is often suggested that viral fusion proteins have evolved from a preexisting intracellular fusion protein (e.g. Lucy, 1984). Unravelling the mechanism of HA-mediated membrane fusion might give important clues on the mechanism by which proteins regulate intracellular membrane traffic. Recently a protein involved in intra-Golgi vesicular transport has been purified (Block et al., 1988). If this protein turns out to be a real fusogen, it will be particularly interesting to study its fusion mechanism and find out whether the fusion mechanism resembles that of viral fusion proteins.

REFERENCES

Block MR, Glick BS, Wilcox CA, Wieland FT, Rothman (1988) Purification of an N-ethylmaleimide-sensitive protein catalyzing vesicular transport. Proc Natl Acad Sci USA 85: 7852-7856

Blumenthal R (1987) Membrane Fusion. In Current Topics in Membranes and Transport (Klausner RD, Van Renswoude J, Kempf C, eds) Vol 29, pp 203-254. Academic Press Orlando

Burger KNJ, Knoll G, Verkleij AJ (1988) Influenza virus-model membrane interaction. A morphological approach using modern cryotechniques. Biochim Biophys Acta 939: 89-101

De Kruijff B, Cullis PR, Verkleij AJ, Hope MJ, Van Echteld CJA, Taraschi TF (1985a) Lipid polymorphism and membrane function. In The Enzymes of Biological Membranes (Martonosi AN, ed) pp 131-204. Plenum Press New York

De Kruijff B, Cullis PR, Verkleij AJ, Hope MJ, Van Echteld CJA, Taraschi TF, Van Hoogevest P, Killian JA, Rietveld A, Van Der Steen ATM (1985b). In Progress in Protein-Lipid Interactions (Watts A, de Pont JJHHM, eds) pp 89-142. Elsevier Amsterdam

Edwards HH, Mueller TJ, Morrison M (1979) Distribution of transmembrane polypeptides in freeze fracture. Science 203: 1343-1346

Frederik PM, Stuart MCA, Verkleij AJ (1989) Intermediary structures during membrane fusion as observed by cryo-electron microscopy. Biochim Biophys Acta 979: 275-278

Gething MJ, Doms RW, York D, White J (1986) Studies on the mechanism of membrane fusion: site-specific mutagenesis of the hemagglutinin of influenza virus. J Cell Biol 102:11-23

Gething MJ, Henneberry J, Sambrook J (1988) Fusion activity of the hemagglutinin of influenza virus. In Current Topics in Membranes and Transport (Düzgünes N, Bronner F, eds) Vol 32, pp 337-364. Academic Press San Diego

Heuser JE, Reese TS, Dennis MJ, Jan Y, Jan L, Evans L (1979) Synaptic vesicle exocytosis captured by quick freezing and correlated with quantal transmitter release. J Cell Biol 81: 275-300

Hols H, Sixma JJ, Leunissen-Bijvelt J, Verkleij AJ (1985) Freeze-fracture studies on human blood platelets activated by thrombin using rapid freezing. Trombosis and Haemostasis 54: 574

Knoll G, Burger KNJ, Bron R, van Meer G, Verkleij AJ (1988) Fusion of liposomes with the plasma membrane of epithelial cells: fate of incorporated lipids as followed by freeze fracture and autoradiography of plastic sections. J. Cell Biol 107: 2511-2521

Knoll G, Verkleij AJ, Plattner H (1987) Cryofixation of dynamic processes in cells and organelles. In Cryotechniques in Biological Electron Microscopy (Steinbrecht RA, Zierold K, eds) pp 258-271. Springer Verlag Berlin

Lindblom G, Rilfors L (1989) Cubic phases and isotropic structures formed by membrane lipids-possible biological relevance. Biochim Biophys Acta 988: 221-256

Lucy JA (1984) Do hydrophobic sequences cleaved from cellular polypeptides induce membrane fusion reaction in vivo? FEBS 166: 223-231

Maeda T, Kawasaki K, Ohnishi SI (1981) Interaction of influenza virus hemagglutinin with target membrane lipids is a key step in virus-induced hemolysis and fusion at pH 5.2. Proc Natl Acad Sci USA 78: 4133-4137

Matlin KS, Reggio H, Helenius A, Simons K (1981) Infectious entry pathway of influenza virus in a canine kidney cell line. J Cell Biol 91: 601-613

Miller RG (1980) Do 'lipidic particles' represent intermembrane attachment sites? Nature 287: 166-168

Rand RP, Parsegian VA (1986) Mimicry and mechanism in phospholipid models of membrane fusion. Ann Rev Physiol 48: 201-212

Schmidt A, Patzak A, Lingg G, Winkler H, Plattner H (1983) Membrane events in adrenal chromaffin cells during exocytosis: a freeze-etching analysis after rapid cryofixation. Eur J Cell Biol 32: 31

Siegel DP (1986) Inverted micellar intermediates and the transition between lamellar, cubic, and inverted hexagonal lipid phases. II. Implications for membrane-membrane interactions and membrane fusion. Biophys J 49: 1171-1183

Siegel DP (1987) Membrane-membrane interactions via intermediates in lamellar-to-inverted hexagonal phase transitions. In Cell Fusion (Sowers AE, ed) pp 181-208. Plenum Press New York

Skehel JJ, Bayley PM, Brown EB, Martin SR, Waterfield MD, White JM, Wilson IA, Wiley DC (1982) Changes in the conformation of influenza virus hemaglutinin at the pH optimum of virus-mediated membrane fusion. Proc Natl Acad Sci USA 85: 7852-7856

Steck TL, Kant JA (1974) Preparation of impermeable ghosts and inside-out vesicles from human erythrocyte membranes. Meth Enzymol 31:172-180

Stegmann T, Hoekstra D, Scherphof G, Wilschut J (1985) Kinetics of pH-dependent fusion between influenza virus and liposomes. Biochem 24: 3107-3113

Stegmann T, Hoekstra D, Scherphof G, Wilschut J (1986) Fusion activity of influenza virus. A comparison between biological and artificial target membrane vesicles. J Biol Chem 261: 10966-10969

Steinbrecht RA, Müller M (1987) Freeze-substitution and freeze-drying. In Cryotechniques in Biological Electron Microscopy (Steinbrecht RA, Zierold K, eds) pp 149-172. Springer Berlin

Van Meer G, Simons K (1982) Viruses budding from either the apical or the basolateral plasma membrane domain of MDCK cells have unique phospholipid compositions. EMBO J 1: 847-852

Van Meer, Simons K (1983) An efficient method for introducing defined lipids into the plasma membrane of mammalian cells. J Cell Biol 97: 1365-1374

Van Meer G, Davoust J, Simons K (1985) Parameters affecting low pH-mediated fusion of liposomes with the plasma membrane of cells infected with influenza virus. Biochem 24: 3593-3602

Verkleij AJ (1984) Lipidic intramembranous particles. Biochim Biophys Acta 779: 43-63

Verkleij AJ, Humbel B, Studer D, Müller M (1985) 'Lipidic particle' systems as visualized by thin-section electron microscopy. Biochim Biophys Acta 812: 591-594

Verkleij AJ, Ververgaert PHJTh (1978) Freeze-fracture morphology of biological membranes. Biochim Biophys Acta 515: 303-327

Weidman PJ, Melancon P, Block MR, Rothman JE (1989) Binding of an N-ethylmaleimide-sensitive fusion protein to golgi membranes requires both a soluble protein(s) and an

integral membrane receptor. J Cell Biol 108: 1589-1596
Wharton SA (1987) The role of influenza virus haemagglutinin in membrane fusion. Microbiol Sc 4: 119-124
White J, Helenius A, Gething MJ (1982a) Haemagglutinin of influenza virus expressed from a closed gene promotes membrane fusion. Nature 300: 658-659
White J, Kartenbeck J, Helenius A (1982b) Membrane fusion activity of influenza virus. EMBO J. 1: 217-222
White J, Kielian M, Helenius A (1983) Membrane fusion proteins of enveloped animal viruses. Quart Rev Biophys 16: 151-195
Wiley DC, Skehel JJ (1987) The structure and function of the hemaglutinin membrane glycoprotein of influenza virus. Ann Rev Biochem 56: 365-394
Wilson DW, Wilcox CA, Flynn GC, Chen E, Kuang WJ, Henzel WJ, Block MR, Ullrich A, Rothman JE (1989) A fusion protein required for vesicle-mediated transport in both mammalian cells and yeast. Nature 339: 355-359
Wilson IA, Skehel JJ, Wiley DC (1981) Structure of the haemagglutinin membrane glycoprotein of influenza virus at 3 Å resolution. Nature 289: 366-373
Yoshimura A, Yamashina S, Ohnishi SI (1985) Mobilization and aggregation of integral membrane proteins in erythrocytes induced by interaction with influenza virus at acidic pH. Exp Cell Res 160: 126-137

CELL-FREE SYSTEMS FOR STUDYING THE PATHWAY OF RECEPTOR-MEDIATED ENDOCYTOSIS

Graham Warren, Marc Pypaert, Elizabeth Smythe and Philip Woodman
Imperial Cancer Research Fund
PO Box 123
Lincoln's Inn Fields
London WC2A 3PX
England

Introduction

Receptor-mediated endocytosis is a process used by cells to take up nutritionally important molecules such as transferrin (Octave et al., 1983). Transferrin receptors provide the means of taking transferrin into cells because they both bind transferrin tightly and are themselves located in coated pits, presumably because of certain features in the cytoplasmic tail (see Glickman et al., 1989). Coated pits invaginate, presumably by re-arrangement of the clathrin coat (Pearse and Crowther, 1987), and eventually undergo membrane scission to form coated vesicles (Hopkins and Trowbridge, 1983). After uncoating (Rothman and Schmid, 1986) they deliver their contents to endosomes (Helenius et al., 1983) and the low pH within these organelles causes the iron to dissociate from the transferrin (Aisen and Listowsky, 1980). Apo-transferrin, still bound to the receptors, then recycles back to the cell surface (Klausner et al., 1983; Dautry-Varsat et al., 1983) where it is released allowing further molecules of transferrin to bind and continue the cycle bringing iron into the cell.

The components which underlie the endocytic mechanism are located, for the most part, on the cytoplasmic side of the membrane. They are not, therefore, accessible to direct biochemical manipulation. Cells must be broken to gain access to these components and cell-free assays are needed to identify the different proteins involved in the different steps. Here we describe two assays which allow us to look for these endocytic components.

Budding Assay

The assay is outlined schematically in figure 1. A431 cells are incubated at 4°C with ^{125}I-

NATO ASI Series, Vol. H 40
Dynamics and Biogenesis of Membranes
Edited by J.A.F. Op den Kamp

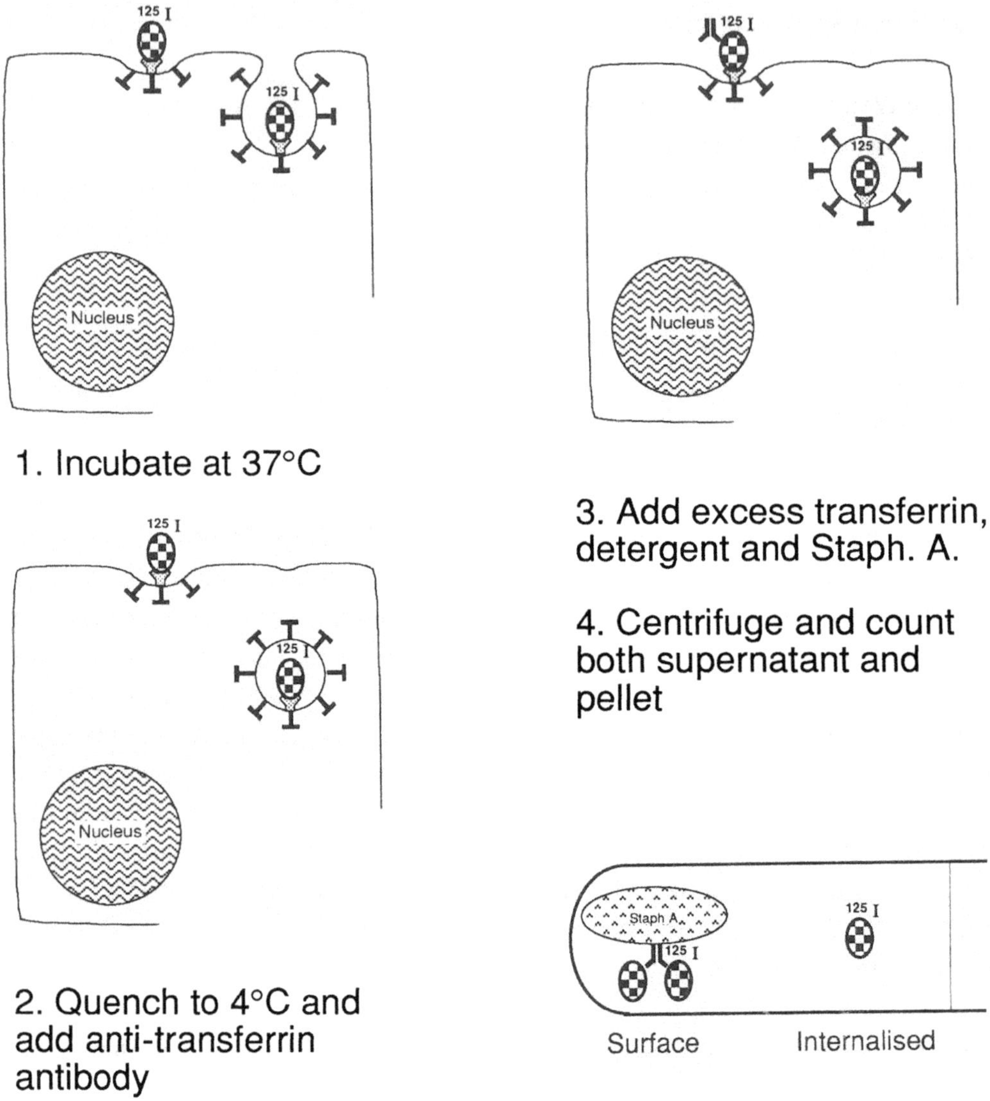

Figure 1. Outline of the Budding Assay described in more detail in the text. Adapted with permission from Warren et al. (1988)

transferrin and washed to remove the excess. The cells are then broken by scraping with a rubber policeman. Breakage is measured by the release of a cytoplasmic marker, lactate dehydrogenase. Typically, more than 90% of the cells are broken by this procedure yet the overall morphology of the cells remains remarkably intact (Figure 2). Broken cells are then incubated at 37°C to allow internalization to occur. The cells are

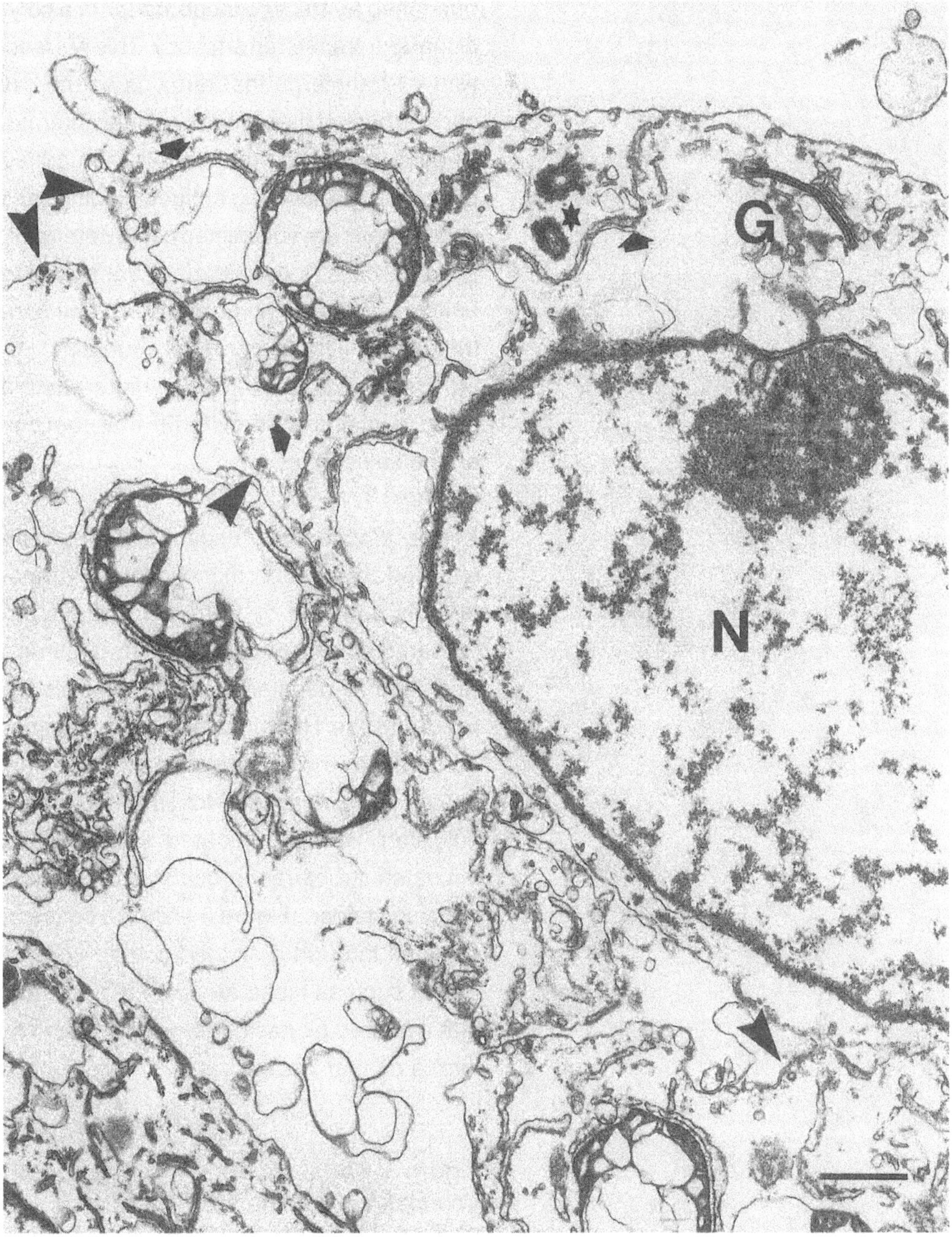

Figure 2. **A431 Cells Broken by Scraping.** Note that the morphology is remarkably intact despite breakage of >90% of the cells. Arrowheads point to deeply-invaginated coated pits and vesicles. Also visible are endoplasmic reticulum (arrows), centrioles (star), Golgi (G), and nucleus (N). Bar = 1μm.

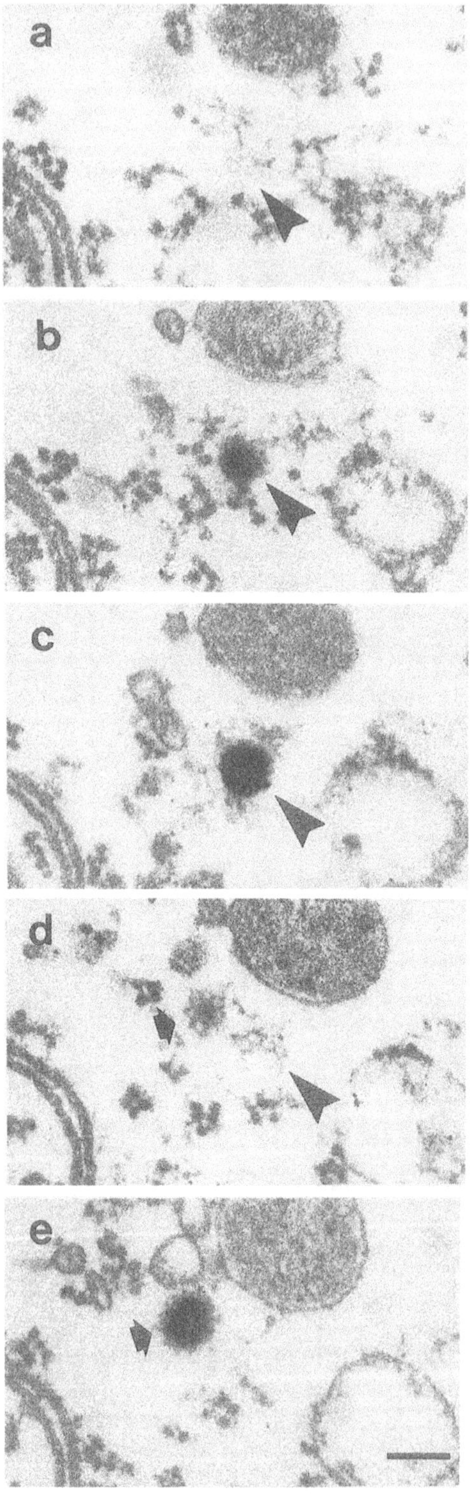

quenched at 4°C and incubated with a polyclonal anti-transferrin antibody. This will react with ^{125}I-transferrin that remains on the cell surface but not that which has been internalized. Excess cold transferrin is then added to block any remaining antigen binding sites and the cells are solubilised using detergent. *Staphylococcus Aureus* cells are added to bind the antibody and are pelleted by centrifugation. The supernatant contains ^{125}I-transferrin that is internalized whereas the pellet contains ^{125}I-transferrin that remains on the cell surface.

These biochemical experiments can be checked by electron microscopy. ^{125}I-transferrin is replaced by transferrin coupled to horseradish peroxidase (HRP-transferrin). At different time after warming, the cells are fixed and the HRP located using diaminobenzidine as the substrate. At zero time only coated pits are labelled for HRP-transferrin but after warming, coated vesicles and uncoated structures (probably endosomes) are also stained. Figure 3 shows a series of sections through a stained coated vesicle. Series such as these are essential to show that a coated pit has undergone scission to form a coated vesicle.

Figure 3. Serial sections through coated vesicles containing HRP-transferrin. (a-e) Five consecutive sections (30nm thick) through an entire coated vesicle (arrowheads) formed during incubation of broken A431 cells. The arrows point to a stained coated vesicle present in the last two sections (d,e). Bar = 0.1µm.

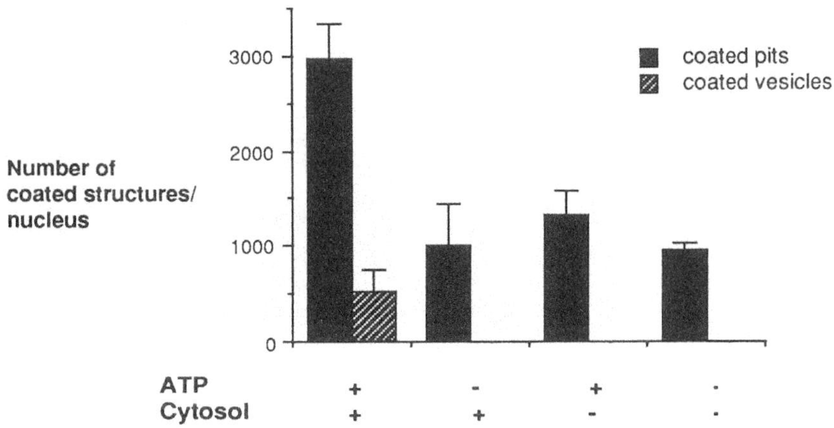

Figure 4. **Effect of ATP and Cytosol on the Number of Coated Pits and Vesicles.** Broken A431 cells were incubated in the presence or absence of ATP or cytosol for 15 min. at 31 °C. The number of coated pits and vesicles stained for HRP-transferrin was determined using the disector (Sterio, 1984).

The combination of these two methods was used (Smythe et al., 1989) to dissect the earliest part of the pathway into three distinct steps:

(a) **Nucleation**. Broken A431 cells typically had 1000-1500 coated pits/cell but when incubated in the presence of ATP and cytosol, this figure increased 2-3 fold to 3000 coated pits/cell (Figure 4). Since the size of the coated pits in profile did not change significantly this meant that new coated pits had been formed. The reason probably lies in the large number of receptors found on the surface of A431 cells. These receptors would be found exclusively in coated pits in other cells but the number and variety of receptors on A431 cells means that they have to compete for the available coated pits, and many end up outside them. The added cytosol presumably contains all the components needed to bind these excluded receptors and generate new coated pits.

(b) **Invagination**. More than 75% of the shallow coated pits became deeply-invaginated within 5 min. of incubation, a rate similar to that in intact cells (Smythe et al., 1989). Invagination occurred at 31 °C but not 0°C and neither ATP nor cytosol were required (Figure 5). We cannot, however, exclude the possibility that some cytosolic components and essential nucleotides remain tightly bound to the pits during the

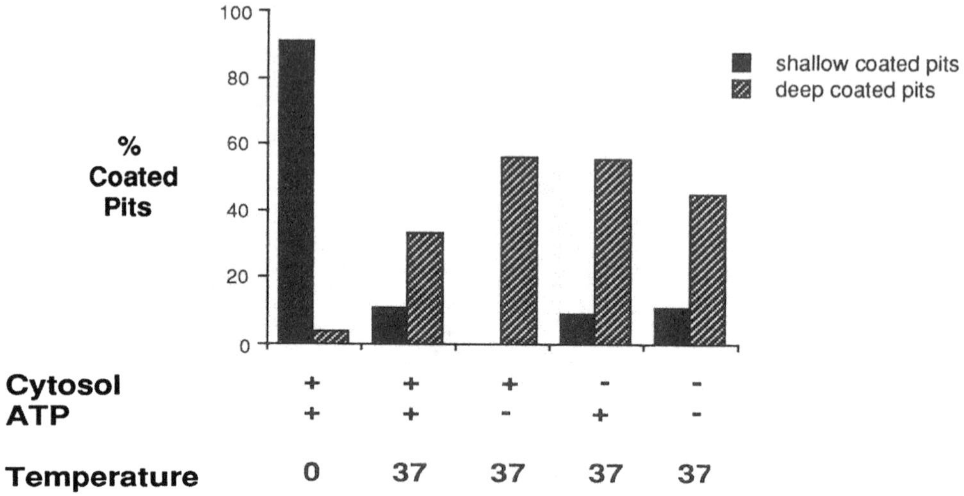

Figure 5. **Effect of ATP and Cytosol on the Invagination of Coated Pits. Broken** A431 cells were incubated in the presence or absence of ATP and/or cytosol for 15 min at 31°C or at 0°C in the presence of both. The percentage of shallow and deep coated pits was determined from profile measurements (Pypaert et al., 1987).

washing procedures.

(c) **Scission**. In the absence of either ATP or cytosol or both, deeply-invaginated coated pits did not undergo scission to form coated vesicles (Figure 4). In the presence of both ATP and cytosol only about 25% of the coated pits underwent scission suggesting that some components of the process were either missing or inactivated.

The requirements for each of these steps are summarized schematically in figure 6.

Fusion Assay

This is shown schematically in figure 7. Two sets of membrane fractions are prepared, which we term donor and acceptor membranes. Donor membranes are prepared from A431 cells to which ^{125}I-labelled transferrin has been bound and internalized. After internalization, the cells are quenched at 0°C to stop further internalization, homogenized, centrifuged and the post-nuclear supernatants are frozen and stored in liquid

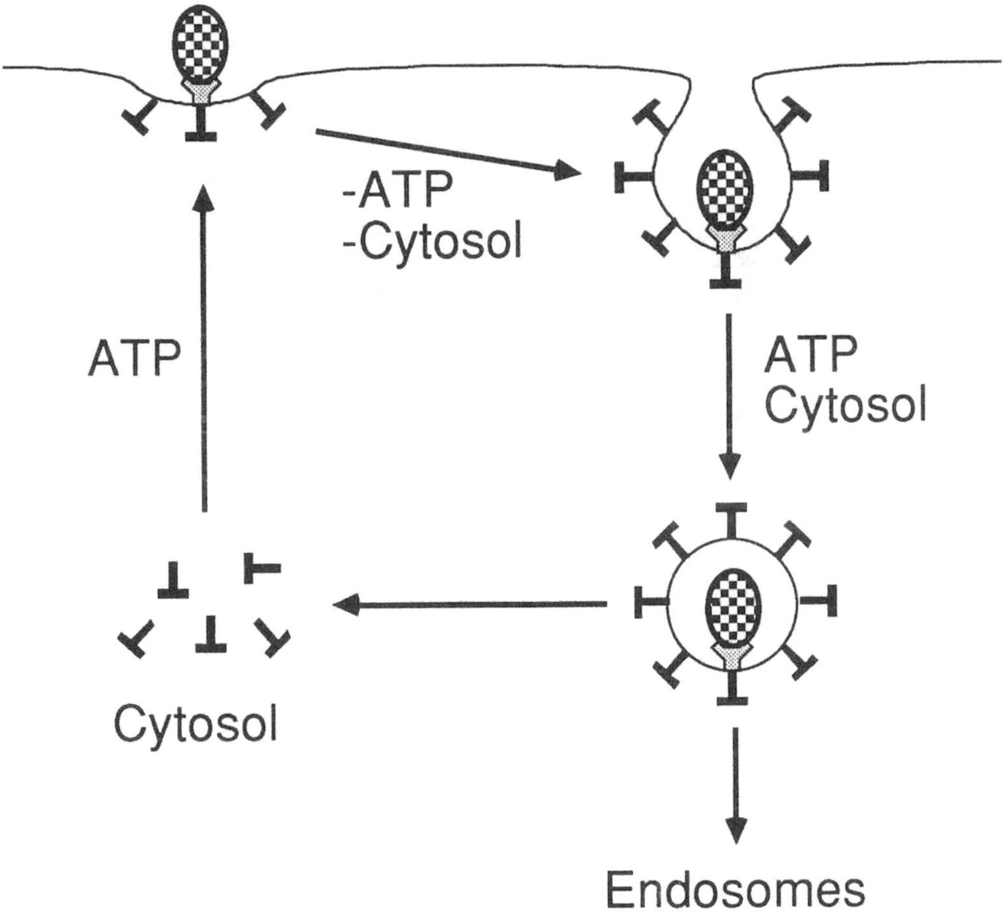

Figure 6. Schematic showing the requirements of the early steps on the endocytic pathway. See text for discussion.

nitrogen.

Acceptor membranes are prepared from A431 cells to which transferrin has been bound, followed by excess anti-transferrin antibody. Excess antibody ensures monovalent binding of antibody to transferrin leaving a free binding site for another transferrin molecule. After internalization at 37°C, the cells are quenched at 0°C and post-nuclear supernatants are prepared as described for donor membranes.

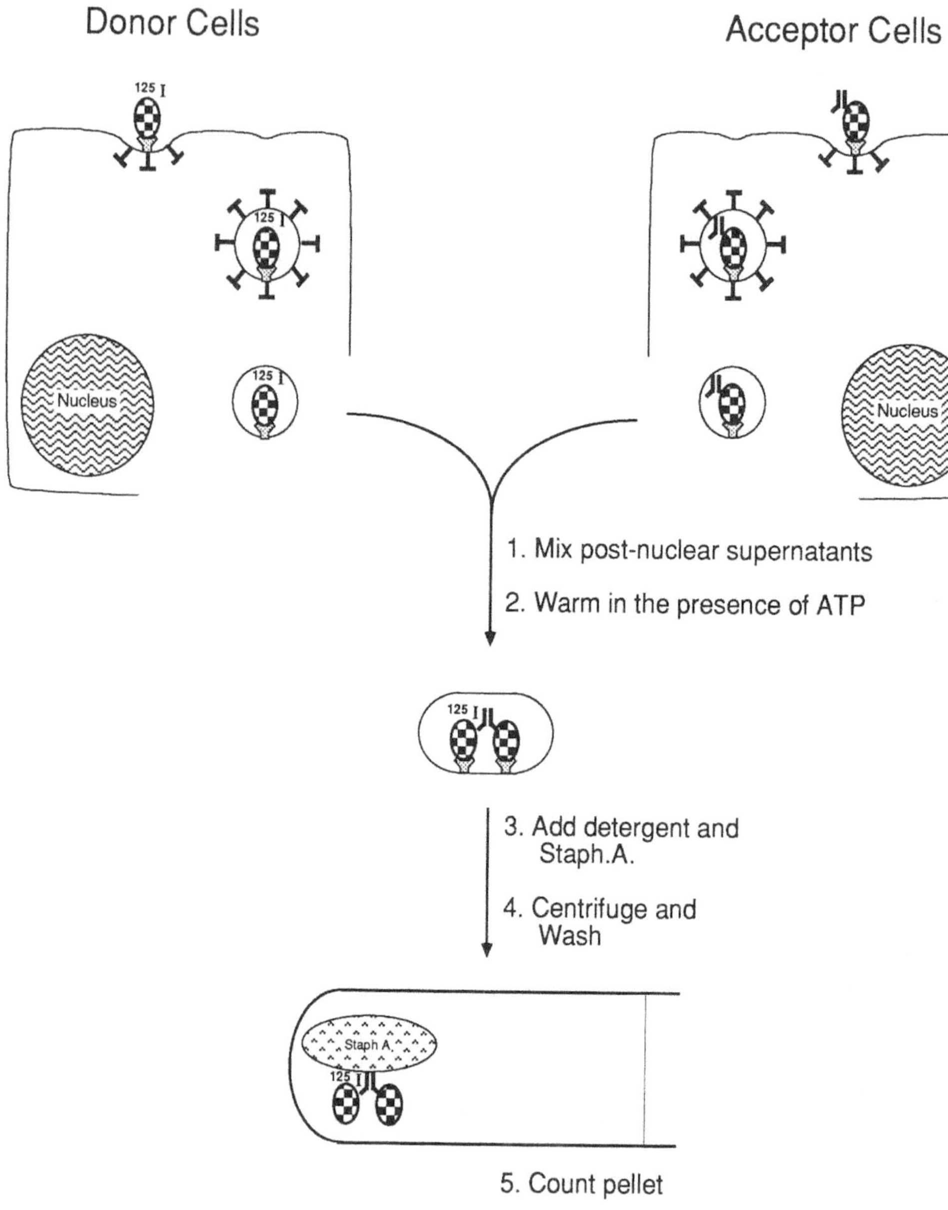

Figure 7. Outline of the Fusion Assay described in more detail in the text. Adapted with permission from Warren et al. (1988).

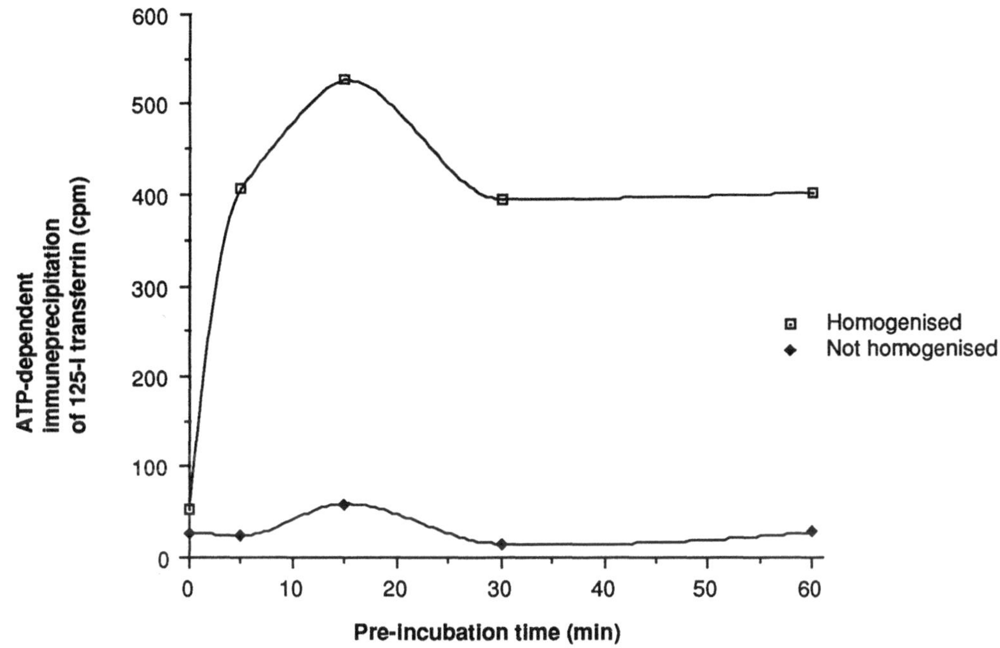

Figure 8. Homogenization is required for Donor Activity. Donor A431 cells were broken and pre-incubated at 37°C in the presence of ATP and cytosol for varying times. The cells were washed and then either added directly to a standard acceptor preparation or first homogenized. Further incubation was for 60 min. at 37°C in the presence of both ATP and cytosol.

Donor and acceptor membranes are then mixed in the presence of ATP and, because the cells are broken open, it is now possible for endocytic vesicles containing ^{125}I-transferrin to fuse with endosomes containing antibody bound to transferrin (Figure 7). In consequence, ^{125}I-transferrin can bind to the free antibody binding sites and the complex can be isolated, after detergent solubilisation, using *Staphylococcus Aureus.* The binding of ^{125}I-transferrin to antibody therefore provides an indirect measure of the fusion event.

Using this assay we have shown (Woodman & Warren, 1988) that the fusion event has the following characteristics:

(a) **Energy.** Fusion requires an energy source which we add in the form of ATP.

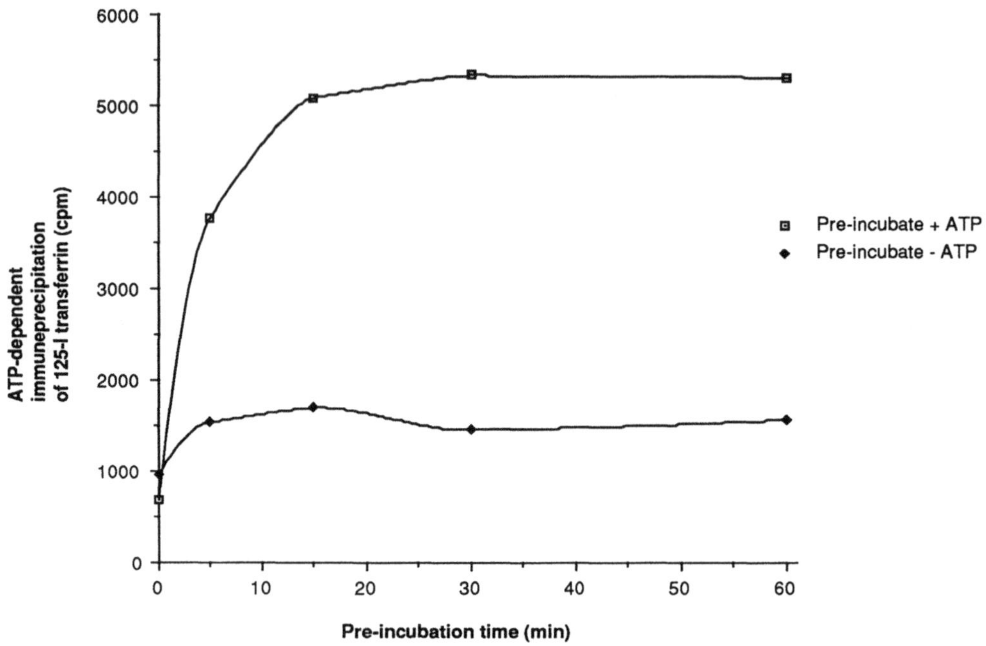

Figure 9. **Donor Activity in Broken Cells requires ATP.** Donor A431 cells were broken and pre-incubated in cytosol at 37°C for varying times in the presence or absence of ATP. The cells were washed, homogenized and then added directly to a standard acceptor preparation for a further 60 min. in the presence of ATP and cytosol

(b) **Specificity**. Only donor and acceptor vesicles from the early part of the endocytic pathway fuse. If internalization is prevented before the post-nuclear supernatants are prepared, fusion does not occur.

(c) **Protein requirements**. Pre-treatment of donor or acceptor membranes with proteases inhibits fusion suggesting that proteins present on the cytoplasmic side of the membrane are required. Cytoplasmic proteins are also required since isolated donor and acceptor membranes are inactive on their own. At least one of the required proteins is sensitive to N-ethyl-maleimide (NEM). This protein is present both in the cytosol and on the membranes and may be related to the NEM-sensitive protein involved in intra-Golgi traffic (Wilson et al., 1989).

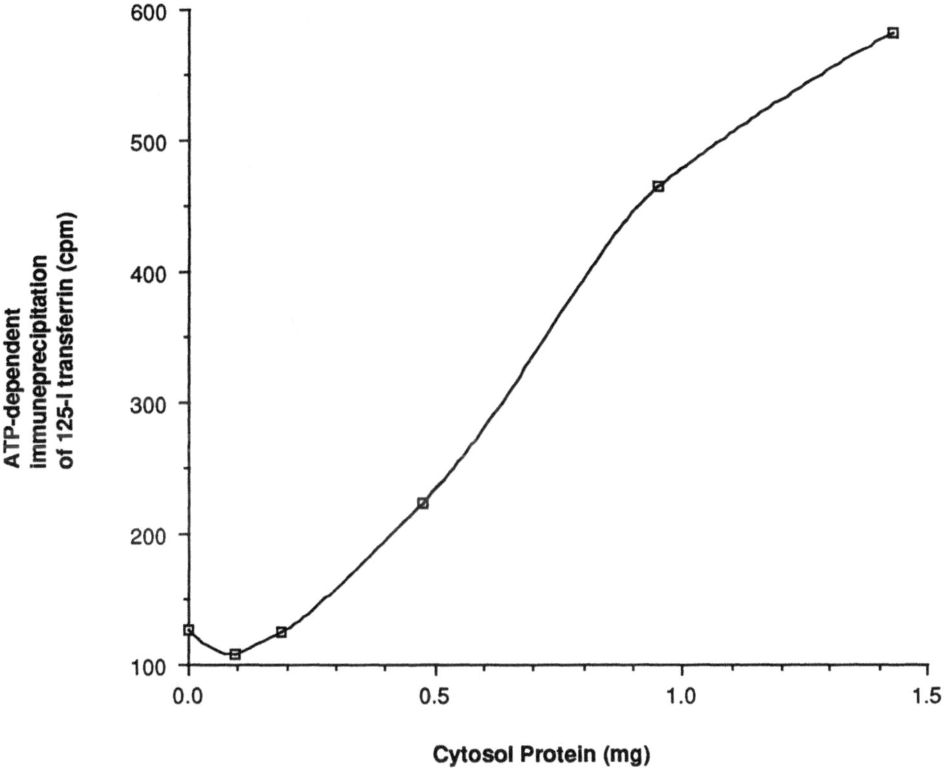

Figure 10. **Donor Activity in Broken Cells requires Cytosol.** Donor A431 cells were broken, washed to remove cytosol and pre-incubated for 60 min. at 37°C in the presence of ATP and increasing amounts of cytosol. The cells were then washed, homogenized and added directly to a standard acceptor preparation for a further 60 min. in the presence of ATP and cytosol.

Linking the Two Assays

The coated vesicles produced in the budding assay (Figure 1) should act as donors in the fusion assay (Figure 7). To test this, broken A431 cells with pre-bound ^{125}I-transferrin were incubated at 37°C in the presence of both ATP and cytosol (to ensure scission of coated pits) and were then added to a standard preparation of acceptor membranes. To our surprise (Figure 8) there was little if any fusion activity. If, however, the broken cells were further homogenized before addition to the acceptor membranes, fusion activity was readily detected (Figure 8). This suggests either that the coated vesicles formed within the broken cells cannot escape or that acceptor membranes cannot enter. This re-inforces our view that scraping is an effective means of breaking cells without extensive damage to the morphology and explains why we were able to account for all forms of membrane containing HRP-transferrin in our budding experiments (Smythe et al., 1989). Presumably, the cytoskeleton is still intact in these broken cells and can act as a filter, keeping structures as large as vesicles within the broken cell.

Figure 8 also shows that maximal donor activity is present within 5 min. of incubation suggesting that coated vesicles (or their uncoated products) are active in the fusion assay. More surprising perhaps is the fact that prior incubation up to 60 min. did not diminish the donor activity. This either means that the coated vesicles in the budding assay do not find endogenous endosomes with which to fuse or that, after fusion, the endosomes remain competent to fuse (with vesicles or endosomes in the acceptor preparations) for at least an hour.

The results from our budding assay (Figure 4) show clearly that the scission of coated pits to give coated vesicles requires both ATP and cytosol. We would not, therefore, expect to be able to prepare donor vesicles from broken A431 cells in the absence of either ATP or cytosol. The experiments presented in figures 9 and 10 bear this out. Figure 9 shows that broken cells, pre-incubated in the absence of ATP, do not generate donor activity. The same is true if the cells are incubated in the absence of cytosol (Figure 10).

Future Prospects

These assays make it possible to look at all of the early steps on the endocytic pathway under conditions where complete control of the environment is possible. Using such assays it should be possible to isolate and characterize the components which constitute the endocytic mechanism.

References

Aisen, P and Listowsky, I (1980) Iron transport and storage proteins. Ann Rev Biochem 49: 357-393

Dautry-Varsat, A, Ciechanover, A and Lodish, H (1983) pH and the recycling of transferrin receptors during receptor-mediated endocytosis. Proc Natl Acad Sci USA 80: 2258-2262

Glickman, JN, Conibear, E and Pearse, BMF (1989) Specificity of binding of clathrin adaptors to signals on the mannose-6-phosphate/insulin-like growth factor II receptor. EMBO J 8: 1041-1047

Helenius, A, Mellman, I, Wall, D and Hubbard, A (1983) Endosomes. Tr Biochem Sci 8: 245-250

Hopkins, CR and Trowbridge, IS (1983) Internalization and Processing of Transferrin and the Transferrin Receptor in Human Carcinoma A431 Cells. J Cell Biol 97: 508-521

Klausner, R, Ashwell, G, Van Renswoude, J, Harford, JB and Bridges, KR (1983) Binding of apo-transferrin to K562 cells: Explanation of the transferrin cycle. Proc Natl Acad Sci USA 80: 2263-2266

Octave, JN, Schneider, YJ, Trouet, A and Crichton, RR (1983) Iron uptake and utilization by mammalian cells. I: Cellular uptake of transferrin and iron. Tr Biochem Sci 8: 217-220

Pearse, BMF and Crowther, RA (1987) Structure and assembly of coated vesicles. Annu Rev Biophys Biophys Chem 16: 49-68

Pypaert, M, Lucocq, JM and Warren, G (1987) Coated pits in interphase and mitotic cells. Eur J Cell Biol 45: 23-29

Rothman, JE and Schmid, SL (1986) Enzymatic recycling of clathrin from coated vesicles. Cell 46: 5-9

Smythe, E, Pypaert, M, Lucocq, JM and Warren, G (1989) Formation of coated vesicles from coated pits in broken A431 cells. J Cell Biol 108: 843-853

Sterio, DC (1984) The unbiased estimation of number and sizes of arbitrary particles using the disector. J Microscop 134: 127-136

Warren, G, Woodman, P, Pypaert, M and Smythe, E (1988) Cell-free assays and the mechanism of receptor-mediated endocytosis. Tr Biochem Sci 13: 462-465

Wilson, DW, Wilcox CA, Flynn GC, Chen E, Kuang W–J, Henzel W J, Block MR, Ullrich A, and Rothman JE (1989) A Fusion Protein Needed for Transport From the Endoplasmic Reticulum and Within the Golgi Stack in Both Animal Cells and Yeast. Nature in press.

Woodman, PG and Warren, G (1988) Fusion between vesicles from the pathway of receptor-mediated endocytosis in a cell-free system. Eur J Biochem 173: 101-108

Plasma Membrane Biogenesis in Polarized Rat Hepatocytes

A.L. Hubbard
Department of Cell Biology and Anatomy
Johns Hopkins University School of Medicine
725 North Wolfe Street
Baltimore, Maryland 21205

NATO ASI Series, Vol. H 40
Dynamics and Biogenesis of Membranes
Edited by J.A.F. Op den Kamp

INTRODUCTION

One of the major goals in cell biology today is to understand the pathways and mechanisms by which newly-synthesized molecules reach their sites of function and some recycle periodically through the cell. How is a protein or lipid correctly targeted? What are the signals, the sorting machineries, and the compartments of the cell involved in sorting? What are the consequences of missorting? These questions are part of the sorting issue and are being actively studied in many labs.

We are studying the biogenesis of the plasma membrane (PM) in mammalian cells. The targeting of transmembrane PM proteins in non-polarized cells, such as fibroblasts, is currently thought to be by "default". That is, these molecules contain no positive sorting or retention signals other than the signal sequence(s) that initially directs their correct insertion into the membrane of the rough endoplasmic reticulum (RER) during translation. Thereafter, they travel through the well-known "secretory" pathway and end up in the PM, because they were neither retained in nor sorted to another compartment.

PM biogenesis in polarized epithelial cells is more complicated than in non-polarized cells, because these cells have specialized PM domains, each facing a different environment and containing a distinct set of proteins. Therefore, proteins destined for these domains must contain positive sorting (or retention) signals to target them to their sites of function. Where and how this domain-specific sorting occurs are central questions to those studying PM biogenesis in epithelial cells.

PM BIOGENESIS IN SIMPLE EPITHELIA

The organization of a simple epithelial cell's surface is illustrated in Figure 1. The basal and lateral domains face the internal milieu and are functionally analogous to the PM of nonepithelial cells. The basal domain is also specialized for adhesion to the basal lamina and the lateral domain for adhesion to other cells. These two domains are often collectively called the basolateral domain. The apical domain faces the external medium and is considered the differentiated surface, since it displays the specialized structures and carries out the specialized functions that largely define each epithelium.

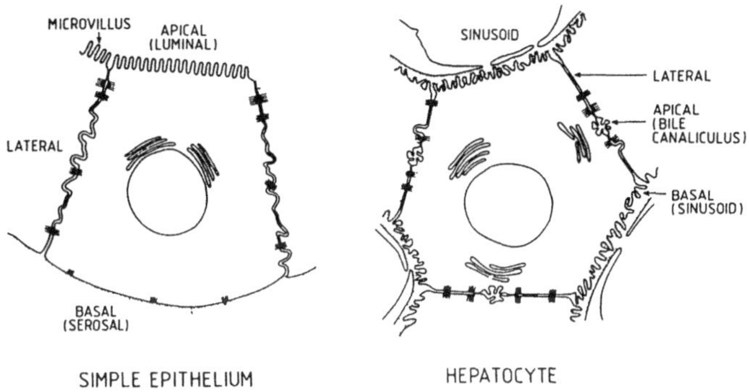

FIGURE 1

The MDCK (Madin-Darby canine kidney) cell line has been the predominant in vitro model for studies of domain-specific sorting for two reasons. First, it forms a simple polarized monolayer in culture, and second, enveloped viruses have been shown to infect these monolayers and mature in a polarized fashion, thus providing investigators with well-characterized membrane glycoproteins whose biogenesis can be studied with little interference by endogenous proteins. These studies have convincingly established that after synthesis, transport and glycosylation within the RER and Golgi, influenza HA (a model apical plasma membrane protein) and VSV G protein (a model basolateral PM protein) reach the trans-portion of the Golgi together, are sorted intracellularly, and are each shipped directly to the correct PM domain. The evidence has been reviewed by others (Matlin, 1986; Salas et al., 1985; Simons & Fuller, 1985).

PM BIOGENESIS IN LIVER HEPATOCYTES IN VIVO

We chose to study PM biogenesis in the major cell of the liver, the hepatocyte. It is also an epithelial cell, but its architecture and surface organization are different from that of a simple bipolar epithelial cell (Figure 1). The hepatocyte is polygonal and has multiple basal (sinusoidal) domains and multiple apical (bile canalicular) domains.

In our initial studies we asked: "Where do newly-synthesized PM proteins first appear at the cell surface?" The possibilities were: 1) direct insertion into the correct domain (the result in MDCK cells); 2) random insertion followed by retrieval and correct delivery; and 3) insertion of all molecules into one domain with those destined for the other domain subsequently segregated and delivered there. We followed five endogenous PM proteins--two that are restricted to the basolateral surface at steady state and three that are restricted to the apical surface (Table I). All are integral membrane sialo-glycoproteins and are synthesized in the RER.

TABLE I
THE BIOGENESIS OF FIVE HEPATOCYTE PLASMA MEMBRANE PROTEINS

DOMAIN LOCALIZATION	MOLECULE (Mab)	CHARACTERISTICS				
		Function	Glycosylated Mr	Degly'd Mr	Number N-linked Chains Complex	High-Mannose
BASOLATERAL (Sinusoidal-Lateral)	CE9	?	48 kD	32 kD	+(4-8)	-
	HA116	Asialo-glycoprotein receptor	42 kD	33 kD	+(1-2)	+(1-2)
APICAL (Bile canalicular)	HA4	Ecto-ATPase	115,105 kD	58,48 kD	+(16)	+(1-2)
	HA301	Dipeptidyl dipeptidase IV	110 kD	80 kD	+(8-15)	+/- (<1)
	HA201	Amino-peptidase N	150 kD	130 kD	+(5-10)	-

TABLE I
See Bartles et al. (1985) for methods and a more detailed description of these molecules. HA4 has recently been identified as an ecto-ATPase (Lin & Guidott, 1989; Margolis et al., in preparation).

Because adult hepatocytes neither proliferate nor form simple polarized monolayers in vitro, our initial studies were conducted in vivo and consisted of metabolic labeling (pulse-chase) experiments in conjunction with subcellular fractionation (Bartles et al., 1987). Of course, subcellular fractionation never yields a pure preparation of any organelle, thus it is important to consider likely contaminants that could influence the results. Furthermore, yields of a particular organelle in the fractions that are finally analyzed are always less than 100% and often less than 10%, making extrapolation to the intact cell problematic.

The above caveats notwithstanding, at selected times after administration of ^{35}S-methionine, we immunoprecipitated each of the five proteins from detergent extracts of a highly-enriched preparation of PM sheets containing known amounts of the two membrane domains and extracts of fractions enriched in apical or basolateral PM vesicles. The newly synthesized, mature (terminally glycosylated), apical and basolateral PM proteins were found to reach the hepatocyte PM at approximately the same rate, with maximal specific radioactivity reached after 45 minutes of chase. In light of the results in virally infected MDCK cells, we were surprised to find that <u>all</u> of the newly synthesized proteins--apical and basolateral--were present at 45 minutes of chase in vesicles that appeared to be bona fide basolateral PMs by several independent criteria. (a) They fractionated as basolateral PM in sucrose density gradients and in free-flow electrophoresis. (b) They could be separated from the bulk of the likely organellar contaminants, including vesicles derived from the trans-Golgi cisternae and endosomes. (c) They contained proven basolateral constituents as judged from vesicle immunoadsorption experiments. (d) Finally, they were oriented in an ectoplasmic-side-out fashion, as expected for vesicles derived from the PM but not for those derived from intracellular organelles. At chase times greater than 45 minutes, the newly synthesized apical proteins began to appear in the apical plasma membrane vesicles, but they did so at different rates. Thus, in hepatocytes, sorting occurs at or after the basolateral surface (route c above) rather than at an intracellular location as in MDCK cells.

A UNIFYING HYPOTHESIS

How can the apparently different domain-specific PM sorting routes in two different epithelial cells be reconciled? Examination of the pathways taken by constitutively secreted proteins in MDCK cells and hepatocytes may provide some clues (Figure 2). MDCK cells constitutively secrete distinct sets of endogenous proteins from both their apical and basolateral surfaces. In addition, a variety of exogenous secretory proteins introduced into MDCK cells by transfection appear to be secreted constitutively from both apical and basolateral surfaces, as though they were not recognized by the MDCK cells' sorting machinery (Gottlieb et al., 1986; Kondor-Koch et al., 1985).

Hepatocytes constitutively secrete a large number of plasma proteins that are synthesized in the RER, modified and packaged

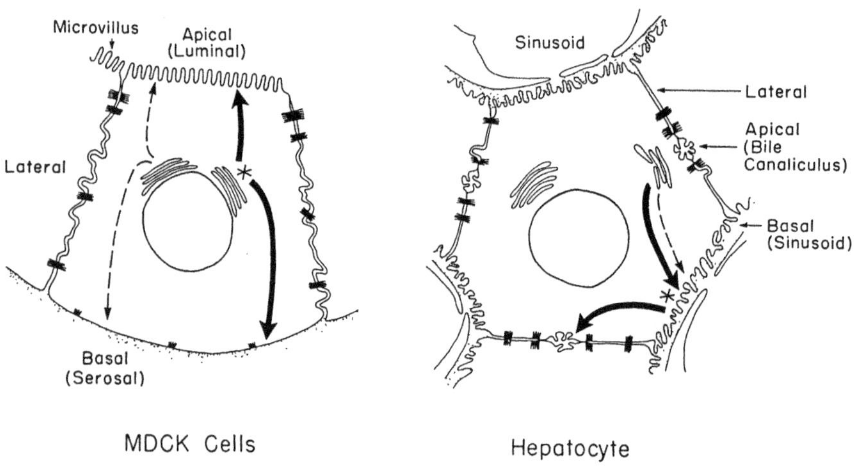

FIGURE 2

Schematic diagrams of MDCK cell and hepatocyte illustrating the differences in the pathways taken by constitutive secretory proteins (dashed arrows) integral plasma membrane proteins (solid arrows) in these two epithelial cell types. The postulated sites of sorting of apical and basolateral plasma membrane proteins are designated by asterisks. In the hepatocyte, the sorting occurs after the proteins arrive at the basolateral plasma membrane; whereas, in the MDCK cell, the proteins appear to be sorted intracellularly, probably in the trans-Golgi network, before they ever arrive at the plasma membrane. This difference may relate to the observation that the hepatocyte, unlike the MDCK cell, does not have an apically directed pathway for constitutive protein secretion. (From Trends Biochem. Sci. 13: 181-184 (1988)) in the Golgi and released at the basal surface into the circulation. However, current evidence indicates that hepatocytes do not have a direct vesicle route through these same intracellular compartments to the apical surface. The proteins present in intrahepatic bile are not secreted directly into bile, but are derived from the plasma either by paracellular leakage through the intercellular space and the hepatocytes' tight junctions or by fluid-phase or receptor-mediated transcytosis (Coleman, 1987; Kloppel et al., 1986).

There appears to be an interesting correlation between the sorting pathways for PM proteins and constitutive secretory proteins within different polarized epithelial cell types (Bartles & Hubbard, 1988). The MDCK cell, which can constitutively secrete in either the apical or basolateral direction, ships its apical PM proteins directly from the Golgi complex to the apical PM. In contrast, the hepatocyte, which has no apically directed pathway for constitutive secretion, employs a basolateral-to-apical transcytotic route for the delivery of its apical PM proteins. Recent in vitro studies on the human colonic carcinoma cell line, Caco2, suggest that it has a predominant basolateral secretory pathway with little or no apical secretion (Traber et al., 1987; Eilers et al., 1989; Rindler & Traber, 1988). However, the sorting pathway for PM proteins is not yet clear.

Theoretically, the relationship between constitutive secretion and domain-specific PM expression could be expressed at any of a number of levels (Figure 3). At one extreme, both classes of proteins, PM and constitutive secretory, may be sorted by the same mechanism and delivered in the same post-Golgi transport vesicles (Gumbiner & Kelly, 1982). Even though constitutive secretory proteins (albumin and transferrin) and an exogenous PM protein (VSV G protein) have been colocalized in putative secretory vesicles in the Hep G2 human hepatoma cell line (Strous et al., 1983), more recent morphological evidence suggests that albumin may be sorted from an endogenous PM protein (the asialoglycoprotein receptor) in the trans-portion of the Golgi in rat hepatocytes in vivo (Geuze et al., 1987). The problem with both of these morphological studies is that they fail to distinguish newly synthesized secretory and membrane proteins from the preexisting ones. This distinction is crucial because there is a large intracellular pool of asialoglycoprotein receptors (Wall & Hubbard, 1985) whose localization might obscure those of the newly synthesized molecules.

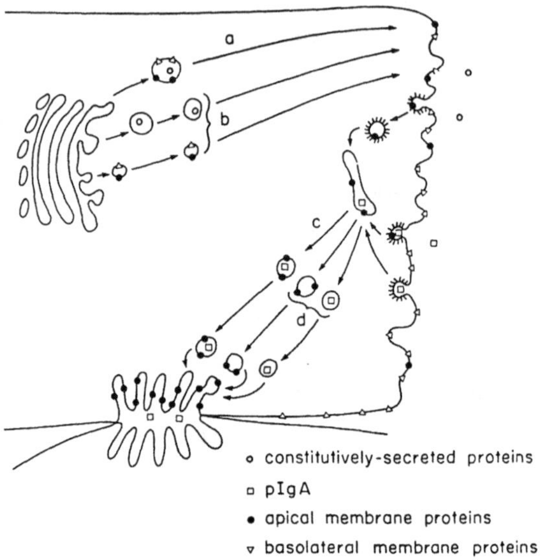

FIGURE 3

Schematic diagram of possibilities for vesicle transport of soluble and PM proteins in hepatocytes. In the route from the Golgi to sinusoidal front, both PM and secretory proteins could travel together (a) or separately (b). Along the transcytotic route, polymeric IgA (pIgA) and apical PM proteins could travel together (c) or separately (d).

Alternatively, this apparent relationship between the pathways of constitutive exocytotic vesicular traffic operating within a single epithelial cell type could be more indirect. There could be separate sorting mechanisms and transport vesicles for PM and constitutive secretory proteins in polarized epithelial cells, but both pathways might be subject to certain generalized directional constraints, such as those that could result from restricting the spatial distributions of microtubules or perhaps other cytoskeletal elements. Microtubules have been postulated to serve as tracks that guide vesicular traffic within cells (Burgess & Kelly, 1987; Vale, 1987). It is interesting to note that microtubule-disrupting drugs, such as colchicine, appear to impair sorting of apical PM proteins in MDCK cells (Rindler et al., 1987; but see Salas et al., 1986), rat hepatocytes (Durand-Schneider et al., 1987; B. Stieger and A.L. Hubbard, unpublished), and rat small intestine (Achler et al., 1989; Eilers et al., 1989; Ellinger et al., 1983; Pavelka et al., 1983; Quaroni et al., 1979).

REMAINING QUESTIONS

We are still far from a complete molecular understanding of PM protein sorting in epithelial cells. Consequently, numerous questions remain, of which only a few are mentioned here.

1. Where are the sorting sites? Since the routes taken by apical proteins in the rat hepatocyte in vivo and MDCK cells in vitro are different, the sorting sites must also be different. Where are they? In MDCK cells, morphological colocalization of the model PM glycoproteins, VSV G and influenza HA, throughout the Golgi (Rindler et al., 1984) and their biochemical co-accumulation in the trans-portion of the organelle by incubation at 20°C (Fuller et al., 1985) have focused attention on this subcompartment as the site where segregation of apical from basolateral PM proteins most likely occurs. However, no morphological separation has yet been observed.

Our results indicate that the hepatocyte sorts its apical and basolateral PM proteins after their initial insertion into the basolateral domain of the plasma membrane (Bartles et al., 1987). Therefore, we predict that one type of vesicular carrier would first deliver all PM molecules to the basolateral membrane (Figure 3). Then, either (a) the apical proteins could be selectively incorporated into endocytic vesicles formed at the basolateral surface and transported to the apical surface; or (b) both basolateral and apical glycoproteins could be internalized and sorted intracellularly, with the basolateral proteins being recycled back to the basolateral surface and the apical proteins being delivered to the apical surface. In the first case, the sorting would occur at the PM. In the second case, the sorting would presumably occur in an endocytic compartment, not unlike the one that has been postulated to function in the sorting of recycling basolateral receptors (e.g. the asialoglycoprotein receptor) and transcytotic receptors (e.g. the polymeric IgA receptor) following endocytosis in rat hepatocytes in vivo (Geuze et al., 1984).

2. Which vesicles transport PM proteins? Very little is known about the vesicles involved in the transport of constitutively expressed PM proteins from the Golgi complex to the surface of either polarized or nonpolarized cells. These vesicles would appear to be very short-lived, since VSV G protein trapped in the trans-Golgi network by incubation at 20°C begins to appear at the PM within just a few minutes following warming to 32°C (Griffiths et al., 1985). Those vesicles delivering exogenous PM proteins, such as VSV G protein in Hep G2 cells (Strous et al., 1983) or in CHO cells, (Bergmann & Singer, 1983) or influenza HA in AtT-20 cells or in pancreatic B cells (Orci et al., 1987), are relatively small electronlucent structures with diameters in the range of 100-300 nm, which are presumably derived by budding off of the trans-Golgi cisterna (Griffiths et al., 1985; Orci et al., 1987). The release of putative exocytic vesicles from virally-infected MDCK cells has been reconstituted using permeabilized cells (Bennett et al., 1988). This approach can now be used as a starting point for the isolation of apical and basolateral transport vesicles.

In the case of the hepatocyte, it would appear that the most likely vesicles to carry the newly synthesized apical and basolateral PM proteins from the trans-Golgi to the basolateral PM are engaged in transport of the hepatocyte's constitutive secretory proteins, such as albumin and transferrin, to its basolateral surface. These vesicles also typically fall within the range of 100-300 nm in diameter and display an electronlucent core, which often contains what appear to be very low density lipoprotein particles (Geuze et al., 1987; Yokota & Fahimi, 1981). Likewise, the most likely candidates for mediating the transcytotic movement of the apical proteins would be the 100-160 nm diameter endosomal vesicles that have been implicated in the transcytosis of the polymeric IgA and its receptor in hepatocytes in vivo (Geuze et al., 1984; Hoppe et al., 1985).

3. What are the sorting mechanisms? As discussed earlier, domain-specific signals must exist on at least one set of epithelial cell PM proteins that serve to segregate them from those destined for the other domain. Despite prodigious efforts, the definitive identification of a sorting signal on any PM protein has proven to be difficult (Breitfeld et al., 1989). Perhaps this is due not only to signals that are conformational rather than linear, but also to the existence of a hierarchy of signals (or receptors) that operates differently in different cellular compartments. That is, one signal (or recognition molecule) may be dominant in a particular site (owing to environment conditions) but another signal becomes dominant in a second site. In this regard, the sorting of the liver pIgA receptor in both hepatocytes and MDCK cells serves as a useful model (Mostov et al., 1987; Mostov et al., 1986).

REFERENCES:

Achler C, Filmer D, Merte C and Drenckhahn D (1989. Role of microtubules in polarized delivery of apical membrane proteins to the brush border of the intestinal epithelium. J. Cell Biol. 108: 179-190.

Bartles JR, Braiterman LT and Hubbard AL (1985) Biochemical characterization of domain-specific glycoproteins of the rat hepatocyte plasma membrane. J. Biol. Chem. 260: 12792-12802.

Bartles JR, Feracci HM, Stieger B and Hubbard AL (1987) Biogenesis of the rat hepatocyte plasma membrane in vivo: comparison of the pathways taken by apical and basolateral proteins using subcellular fractionation. J. Cell Biol. 105: 1241-1251.

Bartles JR and Hubbard AL (1988) Plasma membrane protein sorting in epithelial cells: do secretory pathways hold the key? Trends Biochem. Sci. 13: 181-184.

Bennett MK, Wandinger-Ness A and Simons K (1988) Release of putative exocytic transport vesicles from perforated MDCK cells. EMBO J. 7: 4075-4085.

Bergmann JE and Singer SJ (1983) Immunoelectron microscopic studies of the intracellular transport of the membrane glycoprotein (G) of vesicular stomatitis virus in infected Chinese Hamster ovary cells. J. Cell Biol. 97: 1777-1787.

Breitfeld PP, Casanova JE, Simister NE, Ross SA, McKinnon WC and Mostov KE (1989) Sorting signals. Current Opin. Cell Biol. 1: 617-623.

Burgess TL and Kelly RB (1987) Constitutive and regulated secretion of proteins. Annu. Rev. Cell Biol. 3: 1235-1247.

Coleman R (1987) Biochemistry of bile secretion. Biochem. J. 244: 249-261.
Durand-Schneider A-M, Maurice M, Dumont M and Feldmann G (1987) Effect of colchicine and phalloidin on the distribution of three plasma membrane antigens in rat hepatocytes: comparison with bile duct ligation. Hepatology 7: 1239-1248.
Eilers U, Klumperman J and Hauri H-P (1989) Nocodazole, a microtubule-active drug, interferes with apical protein delivery in cultured intestinal epithelial cells (Caco-2). J. Cell Biol. 108: 13-22.
Ellinger A, Pavelka M and Gangl A (1983) Effect of colchicine on rat small intestinal absorptive cells. II. Distribution of label after incorporation of [^{3}H]fucose into plasma membrane glycoproteins. J. Ultrastruct. Res. 85: 260-271.
Fuller SD, Bravo R and Simons K (1985) An enzymatic assay reveals that proteins destined for the apical or basolateral domains of an epithelial cell line share the same late Golgi compartments. EMBO J. 4: 297-307.
Geuze HJ, Slot JW and Schwartz AL (1987) Membranes of sorting organelles display lateral heterogeneity in receptor distribution. J. Cell Biol. 104: 1715-1723.
Geuze HJ, Slot JW, Strous GJAM, Peppard J, Von Figura K et al. (1984) Intracellular receptor sorting during endocytosis: comparative immunoelectron microscopy of multiple receptors in rat liver. Cell 37: 195-204.
Gottlieb TA, Beaudry G, Rizzolo L, Colman A, Rindler M et al. (1986) Secretion of endogenous and exogenous proteins from polarized MDCK cell monolayers. Proc. Natl. Acad. Sci. USA 83: 2100.
Griffiths G, Pfeiffer S, Simons K and Matlin K (1985) Exit of newly synthesized membrane proteins from the trans cisterna of the Golgi complex to the plasma membrane. J. Cell Biol. 101: 949-964.
Gumbiner B and Kelly RB (1982) Two distinct intracellular pathways transport secretory and membrane glycoproteins to the surface of pituitary tumor cells. Cell 28: 51-59.
Hoppe CA, Connolly TP and Hubbard AL (1985) Transcellular transport of polymeric IgA in the rat hepatocyte: biochemical and morphological characterization of the transport pathway. J. Cell Biol. 101: 2113-2123.
Kloppel TM, Brown WR and Reichen J (1986) Mechanisms of secretion of proteins into bile: studies in the perfused rat liver. Hepatology 6: 587-594.
Kondor-Koch C, Bravo R, Fuller SD, Cutler D and Garoff H (1985) Exocytotic pathways exist to both the apical and the basolateral cell surface of the polarized epithelial cell MDCK. Cell 43: 297-306.
Lin S-H and Guidotti G (1989) Cloning and expression of a cDNA coding for a rat liver plasma membrane ecto-ATPase. J. Biol. Chem. 264: 14408-14414.
Matlin KS (1986) The sorting of proteins to the plasma membrane in epithelial cells. J. Cell Biol. 103: 2565-2568.

Mostov KE, Breitfeld P and Harris JM (1987) An anchor-minus form of the polymeric immunoglobin receptor is secreted predominantly apically in Madin-Darby Canine Kidney cells. J. Cell Biol. 105: 2031-2036.
Mostov KE, de Bruyn Kops A and Deitcher DL (1986) Deletion of the cytoplasmic domain of the polymeric immunoglobulin receptor prevents basolateral localization and endocytosis. Cell 47: 359-364.
Orci L, Ravazzola M, Amherdt M, Perrelet A, Powell SK et al. (1987) The trans-most cisternae of the Golgi copmlex: a compartment for sorting of secretory and plasma membrane proteins. Cell 51: 1039-1051.
Pavelka M, Ellinger A and Gangl A (1983) Effect of colchicine on rat small intestinal absorptive cells. I. Formation of basolateral microvillus borders. J. Ultrastruct. Res. 85: 249-259.
Quaroni A, Kirsch K and Weiser MM (1979) Synthesis of membrane glycoproteins in rat small-intestinal villus cells. Effect of colchicine on the redistribution of L-[1,5,6-^{3}H]fucose-labelled membrane glycoproteins among Golgi, lateral basal and microvillus membranes. Biochem. J. 182: 213-221.
Rindler MJ and Traber MG (1988) A specific sorting signal is not required for the polarized secretion of newly synthesized proteins from cultured intestinal epithelial cells. J. Cell Biol. 107: 471-480.
Rindler MJ, Ivanov, IE, Plesken H, Rodriguez-Boulan E and Sabatini DD (1984) Viral glycoproteins destined for apical or basolateral plasma membrane domains traverse the same Golgi apparatus during their intracellular transport in doubly infected Madin-Darby Canine Kidney cells. J. Cell Biol. 98: 1304-1319.
Rindler MJ, Ivanov IE and Sabatini DD (1987) Microtubule-acting drugs lead to the nonpolarized delivery of the influenza hemagglutinin to the cell surface of polarized Madin-Darby Canine Kidney cells. J. Cell Biol. 104: 231-241.
Salas PJI, Misek DE, Vega-Salas DE, Gundersen D, Cereijido M et al. (1986) Microtubules and actin filaments are not critically involved in the biogenesis of epithelial cell surface polarity. J. Cell Biol. 102: 1853-1867.
Salas PJI, Vega-Salas D, Misek D and Rodriguez-Boulan E (1985) Intracellular routes of apical and basolateral plasma membrane proteins to the surface of epithelial cells. Pflugers Arch. 405: S152-S157.
Simons K and Fuller SD (1985) Cell surface polarity in epithelia. Annu. Rev. Cell Biol. 1: 243-288.
Strous GJAM, Willemsen R, van Kerkhof P, Slot JW, Geuze HJ et al. (1983) Vesicular stomatitis virus glycoprotein, albumin, and transferrin are transported to the cell surface via the same Golgi vesicles. J. Cell Biol. 97: 1815-1822.
Traber MG, Kayden HJ and Rindler MJ (1987) Polarized secretion of newly synthesized lipoproteins by the Caco-2 human intestinal cell line. J. Lipid Res. 28: 1350-1363.
Vale RD (1987) Intracellular transport using microtubule-based motors. Annu. Rev. Cell Biol. 3: 347-378.

Wall DA and Hubbard AL (1985) Receptor-mediated endocytosis of asialoglycoproteins by rat liver hepatocytes: biochemical characterization of the endosomal compartments. J. Cell Biol. 101: 2104-2112.

Yokota S and Fahimi HD (1981) Immunocytochemical localization of albumin in the secretory apparatus of rat liver parenchymal cells. Proc. Natl. Acad. Sci. USA 78: 4970-4974.

Genetic approaches to the study of protein targeting.

Anthony P. Pugsley
Unité de Génétique Moléculaire
Institut Pasteur
25, rue du Dr Roux
Paris 75724 Cedex 15
France
Tel: (1) 45 68 84 94; FAX: (1) 45 68 85 21; EARN/BITNET:MAXàPASTEUR

Genetics and protein targeting. No single methodological approach can lead to a complete understanding of any complex biological system. The emphasis placed upon different approaches depends to a great extent on the ease with which they can be adapted to a particular problem. One area of research which has benefitted enormously from the use of classical and in particular molecular genetics is that of protein targeting. This is not to say, however, that a genetic approach can be used in isolation; instead it provides basis for further studies using biochemical and other approaches. There is, for example, no doubt that the great advances in our understanding of the mechanisms of protein targeting in bacteria which arose from the pioneering work on the genetics of the system by Jon Beckwith, Tom Silhavy, Maurice Hofnung and Maxime Schwartz in the late 1970s only came to fruition when Bill Wickner and others developed the first *in vitro* assays with which to study the process biochemically.

Geneticists either study genome structure and the ways in which genes are regulated and expressed, or use the tools of genetics to study other processes. These tools are numerous and varied; the purpose of this article is to briefly describe some of them and to discuss the results obtained through their use to study protein targeting.

Gene expression systems. The simplest genetic tools used in studies on protein targeting are gene expression systems. Virus-infected cells were often used in early studies on protein targeting in mammalian cells. Major coat proteins of many viruses such as Vesicular Stomatitis virus, Influenza virus or Semliki Forest virus are assembled into viruses at the plasma membrane which they reach via the same secretory pathway (endoplasmic reticulum-transition element-Golgi apparatus-trans Golgi network-secretory vesicles-plasma membrane) as endogenous secretory proteins. Viral genome-encoded proteins are the major products of virus-infected cells, and are therefore readily detectable in simple pulse-chase experiments. Even the pioneering work on protein export in the bacterium *Escherichia coli* by Bill Wickner and his colleagues was based on studies of procoat (precoat), the major component of the

NATO ASI Series, Vol. H 40
Dynamics and Biogenesis of Membranes
Edited by J.A.F. Op den Kamp

bacteriophage M13 which assembles in the cytoplasmic membrane of infected bacteria.

Although viruses continue to be extensively used in studies on protein targeting, they are gradually being replaced by custom-made expression systems in which genes for endogenous proteins are placed under the control of a strong foreign promoter. The more sophisticated, "second generation" systems have regulated promoters which allow the amount of protein being produced to be carefully controlled. The entire package of promoter and cloned gene is usually inserted into high copy number plasmids or defective viruses which can be introduced into novel host cells. Plasmids usually carry selectable markers (eg, antibiotic resistance genes) which allow them to be maintained in the new host. In other cases, the plasmid or defective virus is allowed to integrate into the genome of the host cells, which increases the stability of the recombinant cell lines but reduces the copy number of the cloned gene (and hence the amount of encoded protein which can be produced). These expression systems have opened the way to studying targeted proteins in specially adapted cell lines such as certain epithelial cells or other cells which can be readily grown in culture.

Gene fusions and hybrid proteins. The premise that targeted proteins contain one or more signals which direct them into particular targeting pathways and ensure that they are correctly sorted is the basis of much of the present work on protein targeting. In the most complex situation, that of plant cells, a minimum of five distinct routing signals are required to distinguish cytoplasmic proteins from proteins which are routed into the secretory pathway (cell surface, secretory organelles and vacuole), mitochondria, chloroplasts or other plasmids, the nucleus or the peroxisome. In fact, different proteins routed to the same target may have different routing signals and use different pathways, so that the actual number of routing signals which have to be "decoded" by the cell is probably somewhat larger than five. Additional sorting signals are required for the correct localization of proteins once they have entered a particular route. Most if not all routing signals are linear sequences of amino acids. While some sorting signals are also linear amino acid sequences, many may be composed of amino acids which are brought together only in the folded polypeptide (patch signals) and others are known to include prosthetic groups (eg the mannose-6-phosphate residues which are the sorting signals on lysosomal hydrolases).

Gene fusions have proven particularly powerful tools for detecting and identifying routing signals and sorting signals composed of short polypeptide segments. The basic principle of the gene fusion approach is that a passenger or reporter protein devoid of any routing and/or sorting signals can be targeted to any location in the cell by fusing it to routing and/or sorting signals derived from another protein. Such is the power of *in vitro* recombinant DNA technology that such techniques have become routine practice in many laboratories. The principle is best illustrated by a simple, hypothetical example in which DNA coding for a normally cytoplasmic protein is fused in-frame to DNA coding for different segments of a secretory protein which is assumed to have a secretory routing signal (signal peptide or signal sequence). The recombinant DNA molecules are cloned into a suitable expression vector and introduced into an appropriate host which is then examined for the presence of the normally cytoplasmic protein in the secretory pathway. Once "positive" clones have been identified, the DNA fragment which is assumed to encode the routing signal can be reduced in size to determine the minimum coding sequence required. Until recently, the *in vitro* manipulation of DNA based on the so-called cut and patch approach was limited by the availability of suitable restriction endonuclease cleavage sites in the DNA being tested. This limitation has now been overcome with technical improvements in the synthesis of and more widespread use of synthetic oligonucleotides.

While gene fusions have been particularly valuable in defining routing and sorting signals, they are not without their limitations. One of the major disadvantages of the gene fusion approach is that reporter proteins are rarely "inert". For example, many cytoplasmic proteins are intrinsically unable to cross certain membranes (eg, the bacterial cytoplasmc membrane) when fused to routing signals. Another disadvantage is that the context in which routing or sorting signals are placed may affect their activity. Many such signals are located at the extreme amino- or carboxy-termini, and do not function correctly when placed elsewhere in reporter proteins. Conversely, karyophilic signals which route proteins into the nucleus are often located in the centre of the protein, and yet may not function correctly if they become buried within the reporter protein once it is folded.

A second major disdadvantage of the gene fusion approach is that many routing signals are "degenerate", ie many different sequences of amino acids can be substituted for an authentic routing signal provided certain sequence and structural constraints are respected. Numerous studies have shown that secretory and mitochondrial routing signals seen to be particularly degenerate. It is therefore not unreasonable to expect that routing signal-like sequences might exist within the mature part of many secretory or mitochondrial proteins. These "pseudorouting signals" would not normally influence the localization of the protein concerned, but might become "active" when fused to the amino-terminus of a reporter protein. Thus, results obtained in gene fusion studies should be treated with caution, particularly if the kinetics and the specificity of the targeting of the hybrid protein are not fully characterized. Finally, it should be be noted that patch signals are not amenable to analysis by gene fusion techniques.

Hunting for mutants. The major contribution made by geneticists in many fields of biological research has been in the selection and analysis of mutants. Mutations affecting protein targeting have been selected in a variety of different cell lines, ranging from the bacterium *E.coli* through the yeast *Saccharomyces cerevisae* to plant and mammalian cell lines. These mutations can be divided into two groups depending on the protein or process affected:

i. Mutations affecting targeted proteins. Once routing or sorting signals have been identified by techniques such as gene fusions (see above), their sequences can be altered by mutagenesis in order to determine the extent to which they can be changed without affecting their activity. Some of the most elegant early experiments were carried out by Jon Beckwith, Phil Bassford, Tom Silhavy and their colleagues who based their method for selecting mutations affecting signal peptide function on the observation that certain hybrids containing a signal peptide fused to the normally cytoplasmic protein β-galactosidase were inactive and/or toxic to producing bacteria. Toxicity was found to be relieved or β-galactosidase activity restored by mutations which prevented the entry of the hybrid protein into the secretory pathway. The vast majority of these mutations caused alterations in the signal peptide part of the hybrid such that its sequence now differed significantly from the consensus signal peptide structure (ie, the hydrophobic domain was considerably shortened or contained charged or helix-breaking residues). The malfunction of these altered signal peptides was confirmed by reintroducing the corresponding DNA sequence into the gene from which it was originally derived: the resulting mutated genes encoded proteins which were only inefficiently exported out of the cytoplasm.

Changes which bring about more subtle alterations in signal peptide function cannot be selected by these *in vivo* procedures. Such changes, which might affect the signal peptidase processing site or the generally positively charged amino terminus of signal peptides, can be obtained by site-directed mutagenesis. The experiments are usually designed so that specific amino acid substitutions can introduced, and their individual influence on signal function assessed. Such experiments have demonstrated for example, that certain residues in the karyophilic signal of SV40 virus large T antigen are critical for routing into the nucleus, and that the periodically arranged Arg residues in the routing signal of mitochondrial precursor proteins (the prepeptide or presequence) are critical for efficient routing to the mitochondrion.

It is difficult to envisage ways in which such precise information on the sequence constraints on routing or sorting signals could have been obtained by any other technique. The major limitation of the mutagenesis technique is that it is difficult to predict the effects of sequence changes which occur outside regions of the polypeptide which are directly involved in routing or sorting. Some sequence changes may induce such drastic conformational changes that essential routing or sorting signals are no longer correctly exposed on the surface of the polypeptide, and thus will not function correctly.

ii. Mutations affecting components of the targeting pathway. Routing and sorting signals must somehow be recognized or decoded. Mutations which affect the synthesis or function of proteins involved in signal recognition will thus prevent protein targeting. Likewise, mutations which prevent the correct functioning of proteins which form part of the machinery responsible for protein translocation across membranes or the transport of proteins within or between organelles will also block protein targeting. Since many different and essential proteins usually use the same pathway to reach the same target, mutations blocking such a pathway will eventually be lethal to the cell. For this reason, it has become expedient to select temperature-conditional lethal mutations, ie, mutations which have little or no effect at the "permissive" temperature, but which block protein targeting at the "nonpermissive" temperature. Temperature- conditional mutants may be either temperature-sensitive (growing only at 20-30°) or cold-sensitive (growing only at 37-42°). Temperature- conditional mutations may belong to either of two distinct groups distinguished by whether the expression of the affected gene is blocked at the nonpermissive temperature or whether the gene product itself becomes inactive (due to conformational changes) at the nonpermissive temperature. The vast majority of mutations affecting protein targeting belong to the latter group.

Mutations affecting any one of a large variety of different processes may produce a temperature conditional phenotype. Therefore, an additional, simple screening procedure has to be included to enrich for mutations affecting protein targeting. The many procedures which have been designed to select such mutations are too numerous to describe here, and only some general examples can be given. One of the most productive selection procedures was that devised by Randy Schekman and his colleagues, who observed that *S.cerevisiae* mutants in which the secretory pathway was blocked were more dense than wild-type cells. This simple screening procedure led to the isolation of several classes mutants, each blocked at different stages in the secretory pathway. Genes which encode the wild type homologue of the protein whose activity is defective in these mutants are now being cloned by complementation selection procedures (ie, the cloned wild-type gene suppresses the effects of the mutated gene, and the cells can grow at the normally nonpermissive temperature). This allows the sequence of the protein to be determined and the protein itself to be identified, overprodued (in recombinant organisms) and purified for use in *in vitro* reconstitution assays.

Although this selection and screening procedure proved to be particularly productive, it led to the isolation of very few mutations affecting the earliest stage in the secretory pathway, ie, the translocation of secretory proteins across the membrane of the endoplasmic reticulum. A selection procedure specifically designed to overcome this problem is based on a method developed earlier by Jon Beckwith and his colleagues for the selection of mutations affecting the corresponding stage in the secretory pathway in *E.coli.* In brief, the basic idea is to force the cells to compartmentalize a normally cytoplasmic protein which is essential for growth under certain conditions. This was done by fusing DNA coding for a signal peptide from a secretory protein to DNA coding for the cytoplasmic protein. The cytoplasmic protein thus becomes trapped in the endoplasmic reticulum (in yeasts) or in the cytoplasmic membrane (in bacteria) and cannot function correctly. Its function is restored, however, in mutants affected in the routing of secretory proteins into the secretory pathway. In future, the analysis of these and other, similar mutants is likely to provide invaluable information on the identity and function of proteins involved in routing secretory proteins to and across membranes. Similar selection procedures could be devised for selecting mutants affected in the routing of proteins to other organelles.

These and numerous other examples illustrate the power of genetics as a tool for dissecting complex biological processes. There can be no doubt that our understanding of the mechanics of protein targeting, limited though it is, would not have advanced anywhere near as quickly or as far if it were not for the contribution made by geneticists.

Further reading

Beckwith J, Ferro-Novick S (1986) Genetic studies of protein export in bacteria. Curr Topics Microbiol Immunol 125: 5-28.

Benson SA, Hall MN and Silhavy TJ (1985) Genetic analysis of protein export in *Escherichia coli* K12. Ann Rev Biochem 54: 101-134.

Fleischer S, Fleischer B (eds) (1983 and 1986). Methods in Enzymology Vols 96-98 and 125. Academic Press, Orlando, Florida, USA.

Garoff H. (1985) Using recombinant DNA techniques to study protein targeting in the eukaryotic cell. Ann Rev Cell Biol 1: 404-445.

Nicholson DW, Neupert W (1988) Synthesis and assembly of mitochondrial proteins. In Protein Transfer and Organelle Biogenesis (Das RA, Robbins PW, eds) pp 677-746. Academic Press, San Diego, USA.

Pugsley AP (1988) The use of hybrid proteins in the study of protein targeting signals. In Membrane Biogenesis (Op den Kamp JAP, ed) pp 399-418. Springer-Verlag, Berlin, West Germany.

Schekman R (1985) Protein localization and membrane traffic in yeast. Ann Rev Cell Biol 1:115-143.

Silver PA, Hall MN (1988) Transport of proteins into the nucleus. In Protein Traffic and Organelle Biogenesis (Das RA, Robbin PW, eds) pp 747-770. Academic Press, San Diego, USA.

Walter P, Lingappa VR (1986) Mechanism of protein translocation across the endoplasmic reticulum membrane. Ann Rev Biol 2:499-516.

Signals and components involved in the translocation of proteins across the endoplasmic reticulum membrane

Tom A. Rapoport

Central Institute of Molecular Biology
Academy of Sciences of the GDR
1115 Berlin-Buch

The endoplasmic reticulum (ER) is a major site of protein translocation in eukaryotic cells. Secretory proteins, lysosomal enzymes, lumenal proteins of the ER, plasma membrane proteins, integral ER membrane proteins, to name a few classes which have been studied extensively, are all translocated across or inserted into the ER membrane. Signals contained in the protein sequence target the protein to the ER and determine whether or not it is integrated in the membrane. The signals have to be recognized in the cell and the protein must be transferred at least in part across the membrane. The present review summarizes our current knowledge on some aspects of these problems, with special emphasis on the work carried out in our laboratory.

Signal, stop-transfer and insertion sequences

Proteins targeted to the ER membrane have signal sequences which are typically 15-25 residues long (von Heijne, 1985). They do not share any sequence homology but contain a core of at least 6 consecutive hydrophobic amino acids. The existence

NATO ASI Series, Vol. H 40
Dynamics and Biogenesis of Membranes
Edited by J.A.F. Op den Kamp

of an accessible hydrophobic segment seems to be a characteristic feature of signal sequences (Finkelstein et al. 1983). In many cases, signal sequences are located at the N-terminus of precursor molecules and are cleaved off at the lumenal side of the ER membrane. However, among membrane proteins it is quite common that the signal sequence is not cleaved off and functions as a membrane-spanning segment.

It has been proposed that stop-transfer sequences in a membrane protein terminate the translocation process previously initiated by a signal sequence (Blobel, 1980). According to this model, a multi-spanning membrane protein is inserted into the membrane in a sequential manner by the alternating function of signal and stop-transfer sequences. Recently, evidence has been provided by Wessels and Spiess (1988) that multi-spanning membrane proteins require only a single signal sequence. As predicted in the amphipathic tunnel hypothesis (Rapoport, 1986), the most N-terminal hydrophobic sequence functions as a signal sequence, and all succeeding hydrophobic segments are merely membrane-spanning segments. Obviously, the membrane orientation of the most N-terminal hydrophobic sequence will determine that of all other membrane-spanning segments (see next section).

Whereas signal sequences are probably directed towards a translocation apparatus, insertion sequences trigger the spontaneous incorporation of proteins into membranes. For example, cytochrome b5 is inserted into any membrane with a C-terminal domain (insertion sequence) (Bendzko et al. 1985).

Predicting the orientation of membrane proteins

Membrane proteins synthesized on the ER membrane can have different orientations. If they have a cleavable signal sequence, their N-terminus will be generally located on the lumenal side. However, if they have an uncleaved signal sequence, it may have both orientations (NexoCcyt and Ncytexo). The mechanism which determines the orientation is unknown as yet. However, the following procedure can be used to predict the orientation (Hartmann et al. 1989a).

The algorithm for predicting the orientation of the most N-terminal membrane onchor sequence is outlined in Fig.1. The hydrophobic sequence is searched for by standard computer programs beginning with the N-terminus of the protein. The net charge is calculated for segments of 15 residues each, starting with the charged residue closest to the hydrophobic sequence on either side. The more positive side will be located in the cytosol (Hartmann et al. 1989a).

```
- -   + -   +  -  -           +      -+   +   +  -+
EVETPIRNEWGCRCNDSSDPLVVAASIIGLHLILILDRLFFKCIYRFFEHG..
   |<sum: -1.0  >|        |  |<sum: +1.5  >|
                          |
                          |
         middle of the hydrophobic segment
```

Figure 1: Explanation of the procedure for prediction of the orientation of a membrane protein (see text). The hydrophobic sequence is underlined. The right hand side is predicted to be located in the cytosol.

The "charge difference" rule has been tested on a large set of membrane proteins with both orientations. Among 51 proteins with a NexoCcyt-orientation of the most N-terminal membrane anchor, all corresponded to the rule. For the opposite orientation we found 3 exceptions among 40 different cases. It therefore appears that factors other than the charge difference also contribute to the NcytCexo-orientation. Nevertheless, the charge difference is a major determinant of the membrane orientation and has been successfully used for predictions (Hartmann et al. 1989a).

Recognition of signal sequences in the cytosol

Two mechanisms of signal recognition have been proposed. Based on the fact that signal sequences are hydrophobic and do not share any homology, it has been suggested that they simply dissolve into the lipid bilayer of the target membrane (von Heijne and Blomberg (1979); Engelman and Steitz (1981)). According to an alternative model, signal sequences are recognized by common specific protein receptors (Blobel and Sabatini, 1971).

For the majority of proteins translocated across the ER membrane, it is clear now that the first step of signal recognition occurs already in the cytosol by the signal recognition particle (SRP) (Walter and Blobel, 1981). As soon as the signal sequence of a nascent polypeptide chain has completely emerged from the ribosome, it is bound by SRP through its 54 kDa-polypeptide component (Kurzchalia et al. 1986; Krieg et al. 1986). Identification of this interaction has become possible by a crosslinking approach. A photoactivatable chemical group was incorporated into the signal sequence of nascent preprolactin and was crosslinked to the 54 kDa polypeptide of SRP after irradiation. Similar results were obtained with other secretory proteins (Rapoport et al. 1989) indicating that the 54 kDa polypeptide is a general signal sequence interacting component.

The SRP

SRP is a cytoplasmic ribonucleoprotein particle consisting of six polypeptide chains (72, 68, 54, 19, 14, 9 kDa) and a 7S RNA (Walter and Blobel, 1982). The 9 and 14 kDa polypeptides form a functional dimer in the particle that appears to be required only for a slow-down in translation (elongation arrest) on interaction of the signal sequence with the SRP (Siegel and Walter, 1985). The 68/72 kDa dimer is involved in the interaction of the SRP with its cognate receptor in the ER membrane. The 54 kD-polypeptide which interacts with the signal sequence is bound through the 19 kDa polypeptide and not directly to the RNA (Siegel and Walter, 1988).

The SRP-receptor (docking protein)

SRP interacts with the ER membrane through the SRP-receptor (docking protein) (Meyer et al. 1982; Gilmore et al. 1982). The latter consists of two different polypeptide chains (subunits alpha (69 kDa) and beta (30 kDa) (Tajima et al. 1986). The alpha-subunit has a cytosolic domain that is responsible for the interaction with the SRP. The receptor has recently been shown to be a GTP-binding protein (Conolly and Gilmore, 1989).

According to this new information, the SRP-cycle can be depicted as shown in Fig.2. SRP interacting with the ribosome and the nascent chain binds to the receptor carrying GDP. GDP is exchanged for GTP and SRP is released from both the ribosome (Gilmore et al. 1983) and from the signal sequence (Wiedmann et al. 1987a) The translocation process continues without participation of SRP and its receptor, which are therefore only needed for its initiation. For SRP to be released from the SRP-receptor, GTP-hydrolysis must occur.

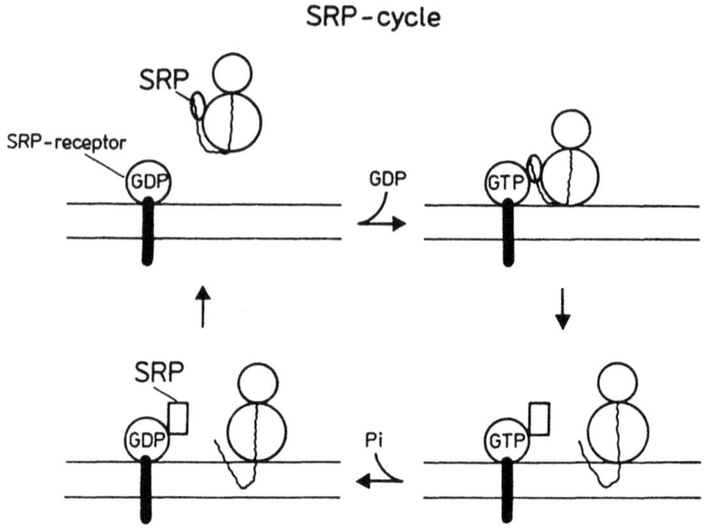

Figure 2: Scheme of the SRP cycle and the involvement of GTP (for details see text).

SRP-independent translocation

Not all translocated proteins strictly depend on SRP. Some small polypeptides can be translocated in the absence of SRP and SRP-receptor across the mammalian ER membrane (Schlenstedt and Zimmermann, 1987). The secretory protein alpha-factor (Rothblatt and Meyer, 1986; Hansen et al. 1986; Waters and Blobel, 1986) and the membrane protein cytochrome P-450 (Wiedmann et al. 1987b) can be translocated posttranslationally, presumably without the need of SRP, across yeast membranes.

Cytoplasmic factors have been found to stimulate the posttranslational translocation. These include 70 kDa heat shock proteins and a N-ethylmaleimide-sensitive factor (Deshaies et al. 1988; Chirico et al. 1988). The heat shock proteins appear to retard folding of the polypeptide chain so that it remains in a translocation competent state. They are unspecific with respect to the destination of the protein.

The signal sequence receptor (SSR)

The signal sequence is not only recognized by its cytosolic receptor, the SRP, but most likely also by membrane proteins of the ER (Prehn et al. 1981; Austen and Ridd, 1983). By use of the crosslinking approach mentioned before, Wiedmann et al. (1987c) provided evidence that the signal sequence of nascent preprolactin, after being released from the SRP, is in close proximity to an integral membrane glycoprotein of ca. 35 kDa, named signal sequence receptor (SSR). It was speculated that SSR represents a basic targetting function, needed by all proteins, including those not requiring SRP, for translocation. The role of SRP was seen in preventing the folding of the polypeptide chain, keeping the signal sequence accessible for the interaction with the SSR (see Wiedmann et al. 1987 a). It is also possible that the function of two signal receptors in sequence may enhance the specificity of recognition.

Transfer of the polypeptide across the membrane

It has been suggested that the nascent polypeptide chain crosses the membrane in an aqueous environment formed by the transient association of membrane proteins (Blobel and Dobberstein, 1975). However, the existence of a protein tunnel has remained elusive as yet. We have recently applied the crosslinking approach to probe the environment of nascent preprolactin at different stages of the translocation process (Wiedmann et al. 1989). A ca. 35 kDa integral membrane glycoprotein, probably identical with the SSR, was found in close proximity to the signal sequence at an early stage, and to the mature part of preprolactin at a later stage. Moreover, the same protein was found in the vicinity of nascent preprolactin after solubilization with detergent of the membranes. It therefore appears that the translocation site consists of a rather stable protein complex (translocon (Walter and Lingappa, 1986), a constituent of which is the ca. 35 kDa protein.

Isolation and characterization of a 34 kDa ER protein that may be essential for protein translocation

We have isolated a major 34 kDa glycoprotein from canine microsomes (Hartmann et al. 1989b) that is probably identical with the one found in the crosslinking experiments (Wiedmann et al. 1987c; 1989): both have about the same size, both are glycosylated, integral membrane proteins, from both proteases split off about 5-7 kDa, both are present in amounts at least equivalent to membrane-bound ribosomes. Direct proof for the identity, however, is lacking as yet.

The 34 kDa protein appears to be essential for protein translocation since antibodies directed against it and Fab-fragments prepared from the antibodies inhibit the in vitro translocation (Hartmann et al. 1989b). In agreement with the assumed general function, the 34 kDa protein was found in various types of cells and organisms.

The amino acid sequence of the 34 kDa protein, determined recently on the protein level and deduced from the nucleotide sequence of cloned cDNA (Prehn et al. unpublished results), indicates the existence of one classical membrane-spanning segment. Most strikingly, a segment with highly negative charge was found close to the N-terminus.

The 34 kDa protein is phosphorylated at its cytoplasmic

tail by a cytosolic protein kinase both in vitro and in intact cells (Kurzchalia et al., unpublished results). It is, however, uncertain as yet if this phosphorylation is connected with protein translocation.

We have recently found that the 34 kDa protein is in tight association with another glycosylated polypeptide of 23 kDa (D. Goerlich et al, unpublished results). From the size of the complex, it appears that it is a hexamer, consisting of three copies each of the 34 kDa and 23 kDa polypepdides.

On the basis of the evidence summarized above, we suggest that the glycoprotein complex is part of the translocation complex of the ER membrane.

Other integral ER membrane proteins involved in translocation

The Table gives a list of membrane proteins that have been implicated in protein translocation.

The signal sequence receptor and the glycoprotein complex have been discussed.

The signal peptidase has been purified from canine microsomes as a complex containing polypeptides of 25, 23, 22, 18 and 12 kDa (Evans et al. 1986). The 22 and 23 kDa subunits are glycosylated and appear to have the same amino acid sequence (Baker and Lively, 1987). Most likely, only one or a few of the polypeptides make up the enzymatic activity. The others might be involved in protein translocation. The signal peptidase complex has also been purified from yeast and is similarly complex (YaDeau and Blobel, 1989).

The ribophorins (Kreibich et al. 1978) may be essential for protein translocation. Antibodies to ribophorin I inhibit translocation (Yu et al. 1988). However, it is unlikely that the ribophorins are involved in ribosome binding (Hortsch et al. 1986).

Table: Integral ER-proteins implicated in translocation

Protein	Mol.mass (kDa)	Source	Amounts	Function
SRP-receptor				
alpha-subunit	69	dog pancreas	low	targetting, GTP-binding
beta -subunit	30	dog pancreas	low	? (anchoring of alpha-subunit)
Signal peptidase complex	25, 23, 22 21, 18, 12	dog pancreas	high	signal peptide cleavage
Glycoprotein complex	34	dog pancreas	high	? (signal sequence receptor, channel proteins)
	23	dog pancreas	high	
Ribophorins I	65	dog pancreas	high	?
II	63	dog pancreas	high	?
Mp 30	30	dog pancreas	high	?
Sec 61	53	saccharomyces	?	? (signal sequence receptor)
Sec 62	32	saccharomyces	?	?
Sec 63	?	saccharomyces	?	?

For references and discussion see text.

An abundant membrane protein of 30 kDa unrelated to the 34 kDa glycoprotein has been found that has an affinity to SRP (Tajima et al. 1986).

In yeast, several proteins involved in translocation were identified by mutations (sec 61-63) (Deshaies and Schekman, 1987). The Sec 61-protein is a multi-spanning membrane protein that has homology to the Sec Y-protein which has been discovered previously in E.coli and which is believed to be involved in signal sequence binding (Emr et al. 1981).

It appears from the Table that a number of different membrane proteins play a role in translocation. Even if one takes into account that except for the SRP-receptor and for the enzymatic activity of the signal peptidase no function has been assigned to any of the polypeptides, at least some of them are likely to be constituents of the translocation complex.

Conclusions

Whereas the initial steps in protein translocation across the ER membrane are reasonably well understood, current efforts concentrate on the process of transfer of the nascent polypeptide across the membrane. Several components of the ER membrane have been implicated in translocation and are currently being studied.

Acknowledgements

I thank D. Bauer, V. Heinze and E. Hartmann for help in preparing the manuscript.

References

Austen, B.M. and Ridd, D.H. (1983) Studies on the binding of a signal peptide to pancreatic rough microsomal membranes. Biochem. Soc. Trans. 11 : 160

Baker, R.K. and Lively, M.O. (1987) Purification and characterization of hen oviduct microsomal signal peptidase. Biochemistry 26 : 8561-8567

Bendzko, P., Prehn, S., Pfeil, W. and Rapoport, T.A. (1982) Different modes of membrane interactions of the signal sequence of carp preproinsulin and of the insertion sequence of rabbit cytochrome b5. Eur. J. Biochem. 123 : 121

Blobel, G. and Sabatini, D. (1971) Ribosome-membrane interaction in eukaryotic cells. In: Manson, L.A. (Ced) Biomembranes. Plenum Press, New York, p. 193

Blobel, G. and Dobberstein, B. (1975) Transfer of proteins across membranes. II. Reconstitution of functional rough microsomes from heterologous components. J. Cell Biol. 67 : 852-862

Blobel, G. (1980) Intracellular protein topogenesis. Proc. Natl. Acad. Sci. USA 77 : 1496-1500

Chirico, W.J., Waters, M.G. and Blobel, G. (1988) 70K heat shock related proteins stimulate protein translocation into microsomes. Nature 332 : 805-810

Connolly, T. and Gilmore, R. (1989) The signal recognition particle receptor mediates the GTP-dependent displacement of SRP from the signal sequence of the nascent polypeptide. Cell, in press

Deshaies, R.J. and Schekman, R. (1987) A yeast mutant defective at an early stage in import of secretory protein precursors in tothe endoplasmic reticulum. J. Cell Biol. 105 : 633-645

Deshaies, R.J., Koch, B.D., Werner-Washburne, M., Craig, E.A. and Schekman, R. (1988) A subfamily of stress proteins facilitates translocation of secretory and mitochondrial precursor polypeptides. Nature 332 : 800-805

Emr, S.D., Hanley-Way, S. and Silhavy, T.J. (1981) Suppressor mutations that restore export of a protein with a defective signal sequence. Cell 23 : 79-88

Engelmann, D.M. and Steitz T.A. (1981) The spontaneous insertion of proteins into and across membranes : The helical hairpin hypothesis. Cell 23 : 411-422

Evans, E.A., Gilmore, R. and Blobel, G. (1986) Purification of microsomal signal peptidase as a complex. Proc. Natl. Acad. Sci. USA 83 : 581-585

Finkelstein, A.V., Bendzko, P. and Rapoport, T.A. (1983) Recognition of signal sequences. FEBS Lett. 161 : 176

Gilmore, R., Walter, P. and Blobel, G. (1982) Protein translocation across the endoplasmic reticulum. II. Isolation and characterization of the signal recognition particle receptor. J. Cell Biol. 95 : 470-477

Gilmore, R. and Blobel, G. (1983) Transient involvement of signal recognition particle and its receptor in the microsomal membrane prior to protein translocation. Cell 35 : 677-685

Hansen, W., Garcia, P.D. and Walter, P. (1986) In vitro protein translocation across the yeast endoplasmic reticulum : ATP-dependent posttranslational translocation of the prepro-alpha-factor. Cell 45 : 397-406

Hartmann, E., Wiedmann, M. and Rapoport, T.A. (1989b) A membrane component of the endoplasmic reticulum that may be essential for protein translocation. EMBO-J. in press

Hartmann, E., Rapoport, T.A. and Lodish, H.F. (1989a) Predicting the orientation of eukaryotic membrane-spanning proteins. Proc. Natl. Acad. Sci. USA, in press

von Heijne, G. and Blomberg, C. (1979) Trans-membrane translocation of proteins : The direct transfer model. Eur. J. Biochem. 97 : 175-181

von Heijne, G. (1985) Structural and thermodynamic aspects of the transfer of proteins into and across membranes. Curr. Top. in Membr. and Transp., 24 : 151-179

Hortsch, M., Avossa, D. and Meyer, D.I. (1986) Characterization of secretory protein translocation : ribosome-membrane interaction in endoplasmic reticulum. J. Cell Biol. 103 : 241-253

Kreibich, G., Czako-Graham, M., Grebenau, R., Mok, W., Rodriguez-Boulan, E. and Sabatini, D.D. (1978) Characterization of the ribosomal binding site in rat liver rough microsomes : ribophorins I and II, two integral membrane proteins related to ribosome binding. J. Supramolec. Structure 8 : 279-302

Krieg, U.C., Walter, P. and Johnson, A.E. (1986) Photocrosslinking of the signal sequence of nascent preprolactin of the 54-kilodalton polypeptide of the signal recognition particle. Proc. Natl. Acad. Sci. USA 83 : 8604-8608

Kurzchalia, T.V., Wiedmann, M., Girshovich, A.S., Bochkareva, E.S., Bielka, H. and Rapoport, T.A. (1986) The signal sequence of nascent preprolactin interacts with the 54K polypeptide of the signal recognition particle. Nature 320 : 634-636

Meyer, D.I., Krause, E. and Dobberstein, B. (1982) Secretory protein translocation across membranes - the role of the docking protein. Nature 297 : 647-650

Prehn, S., Tsamaloukas, A. and Rapoport, T.A. (1980) Demonstration of specific receptors of the rough endoplasmic membrane for the signal sequence of carp preproinsulin. Eur. J. Biochem. 107 : 185-195

Rapoport, T.A. (1986) Protein translocation across and integration into membranes. CRC Crit. Rev. Biochem. 20 : 73-137

Rapoport, T.A., Wiedmann, M., Kurzchalia, T.V. and Hartmann, E. (1989) Signal recognition in protein translocation across the endoplasmic reticulum membrane. Biochem. Soc. Trans. 17 : 325-328

Rothblatt, J.A. and Meyer, D.I. (1986) Secretion in yeast : reconstitution of the translocation and glycosylation of alpha-factor and invertase in a homologous cell-free system. Cell 44 : 619-628

Schlenstedt, G. and Zimmermann, R. (1987) Import of frog prepropeptide Gla into microsomes requires ATP but does not involve doking protein or ribosomes. EMBO-J. 6 : 699-703

Siegel, V. and Walter, P. (1985) Elongation arrest is not a prerequisite for secretory protein translocation across the microsomal membrane. J. Cell Biol. 100 : 1913-1921

Siegel, V. and Walter, P. (1988) Each of the activities of signal recognition particle (SRP) is contained within a distinct domain : analysis of biochemical mutants of SRP. Cell 52 : 39-49

Tajima, S., Lauffer, L., Rath, V.L. and Walter, P. (1986) The signal recognition particle receptor is a complex that contains two distinct polypeptide chains. J. Cell Biol. 103 : 1167-1178

Walter, P. and Blobel, G. (1981) Translocation of proteins across the endoplasmic reticulum. III. Signal recognition protein (SRP) causes signal sequence - dependent and site-specific arrest of chain elongation that is released by microsomal membranes. J. Cell Biol. 91 : 557-561

Walter, P. and Blobel, G. (1982) Signal recognition particle contains a 7S RNA essential for protein translocation across the endoplasmic reticulum membrane. Nature 299 : 691

Walter, P. and Lingappa, V.R. (1986) Mechanism of protein translocation across the endoplasmic reticulum membrane. Ann. Rev. Cell Biol. 2 : 499-516

Waters, M.G. and Blobel, G. (1986) Secretory protein translocation in a yeast cell free system can occur posttranslationally and requires ATP hydrolysis. J. Cell Biol. 102 : 1543-1550

Wessels, H.P. and Spiess, M. (1988) Insertion of a multi-spanning membrane protein occurs sequentially and requires only one signal sequence. Cell 55 : 61-70

Wiedmann, M., Kurzchalia, T.V., Bielka, H. and Rapoport, T.A. (1987a) Direct probing of the interaction between the signal sequence of nascent preprolactin and the signal recognition particle by specific crosslinking. J. Cell Biol. 104 : 201-208

Wiedmann, M., Wiedmann, B., Voigt, S., Wachter, E., Mueller, H.-G. and Rapoport, T.A. (1988b) Post-translational transport of proteins into microsomal membranes of Candida maltosa. EMBO-J. 7 : 1763-1768

Wiedmann, M., Kurzchalia, T.V., Hartmann, E. and Rapoport, T.A. (1987c) A signal sequence receptor in the endoplasmic reticulum membrane. Nature 328 : 830-833

Wiedmann, M., Goerlich, D., Kurzchalia, T.V. and Rapoport, T.A. (1989) Probing by photocrosslinking the environment of nascent preprolactin during its translocation across the endoplasmic reticulum membrane. Submitted for publication.

YaDeau, J. and Blobel, G. (1989) Solubilization and characterization of yeast signal peptidase. J. Biol. Chem. 264 : 2928-2934

Yu, Y.H., Sabatini, D.D. and Kreibich, G. (1988) Antibodies against ribophorin I inhibit the translocation activity of rough microsomal membranes. J. Cell Biol. 107 : 766a

SIGNAL PEPTIDE-LIPID INTERACTIONS AND THEIR SIGNIFICANCE FOR PROTEIN TRANSLOCATION IN *Escherichia coli*

G.J. de Vrije[1], A.M. Batenburg[2], J.A. Killian[+] and B. de Kruijff[o]
[+]Centre for Biomembranes and Lipid Enzymology, and [o]Institute of Molecular Biology and Medical Biotechnology, University of Utrecht, Padualaan 8, 3584 CH Utrecht, NL

INTRODUCTION

During the last years much attention has been focussed on elucidating the process of protein transport in *E. coli* cells. Proteins synthesized in the cytoplasm but destined for the periplasmic space or outer membrane have to cross the plasma membrane. This process can be divided into several steps; efficient targeting to the membrane, translocation across this membrane and processing into the mature form by cleavage of an N-terminal signal sequence. Genetic approaches, biochemical experiments and more recently the use of reconstituted systems of purified precursor proteins and components involved in protein translocation greatly contributed to an understanding of the early events of bacterial protein export. The results of some studies strongly suggest that cytosolic factors act as chaperone proteins and play an important role in stabilizing a translocation competent conformation of precursors (Crooke et al., 1988; Bochkareva et al., 1989; Collier et al., 1988; Gannon et al., 1988; Kusters et al., 1989; Lecker et al., 1989). The presence of SecA protein, a peripheral protein of the *E. coli* inner membrane (Oliver & Beckwith, 1982; Schmidt et al., 1988) is essential for a functional binding of such a complex (Cunningham et al., 1989). In addition, SecA protein couples ATP hydrolysis with precursor protein translocation (Lill et al., 1989). Translocation was earlier shown to be an energy dependent process, which requires ATP hydrolysis (Chen & Tai, 1985) and the presence of an electrochemical membrane potential (Geller et al., 1986; De Vrije et al., 1987). Essential for protein translocation across the inner membrane is the presence of a temporal N-terminal extension of the polypeptide chain: the signal sequence. Analogous polypeptide structures are necessary for protein translocation across endoplasmic reticulum membranes in the secretion pathway of eukaryotes. Although there is almost a complete lack in homology of the primary structure of signal peptides, they contain three characteristic regions: a positively charged N-terminal part, a hydrophobic core of 10-15 amino acids and a more polar C-region, containing the cleavage site (Von Heijne, 1985). Figure 1 shows the primary structure of such a signal peptide.

To understand the functioning of signal peptides in the translocation process, knowledge of the interactions signal peptides have with components of the secretion machinery is of great

[1]Present address: University of Groningen, Dept. of Microbiology, Kerklaan 30, 9751 NN Haren, NL; [2]Present address: Unilever Research Lab., O. van Noortlaan 120, 3133 AT Rotterdam, NL.

NATO ASI Series, Vol. H 40
Dynamics and Biogenesis of Membranes
Edited by J.A.F. Op den Kamp

importance. A direct interaction of the signal peptide with leader peptidase, enzymes which cleave signal peptides (Zwizinski & Wickner, 1980) is supposed to occur. On the other hand, recent results show that signal peptides are not necessarily involved in binding of chaperone to precursor proteins (Lecker et al., 1989) and consequently in the subsequent targeting of the complex to the membrane. However, after dissociation of the precursor from the chaperone proteins, binding of the signal sequence to either a receptor-like proteinaceous component or directly to membrane lipids might initiate membrane translocation of the precursor.

−20 ↓ ... −1 ↓

[n] fMet-Lys$^+$-Lys$^+$-Ser-Thr-Leu-Ala-Leu- Val-Val-Met- Gly-Ile-Val-Ala-Ser-Ala-Ser-Val-Gln-Ala

Fig. 1. Primary structure of the signal peptide of the *E. coli* outer membrane pore protein PhoE.

In this review we will focus on the latter possibility. Due to their basic and amphipathic character, signal sequences seem to be very well suited for a direct interaction with membrane lipids. The present available data on signal interactions between lipids and chemically synthesized signal peptides will be analyzed and together with other studies will be discussed in their relation to initiation of precursor protein translocation.

STRUCTURE OF *E. COLI* INNER MEMBRANE PHOSPHOLIPIDS

The major phospholipids found in *E. coli* inner membranes are phosphatidylethanolamine (PE), phosphatidylglycerol (PG) and diphosphatidylglycerol or cardiolipin (CL). The chemical structures of these lipids are shown in Fig. 2, together with the final steps of their biosynthesis.

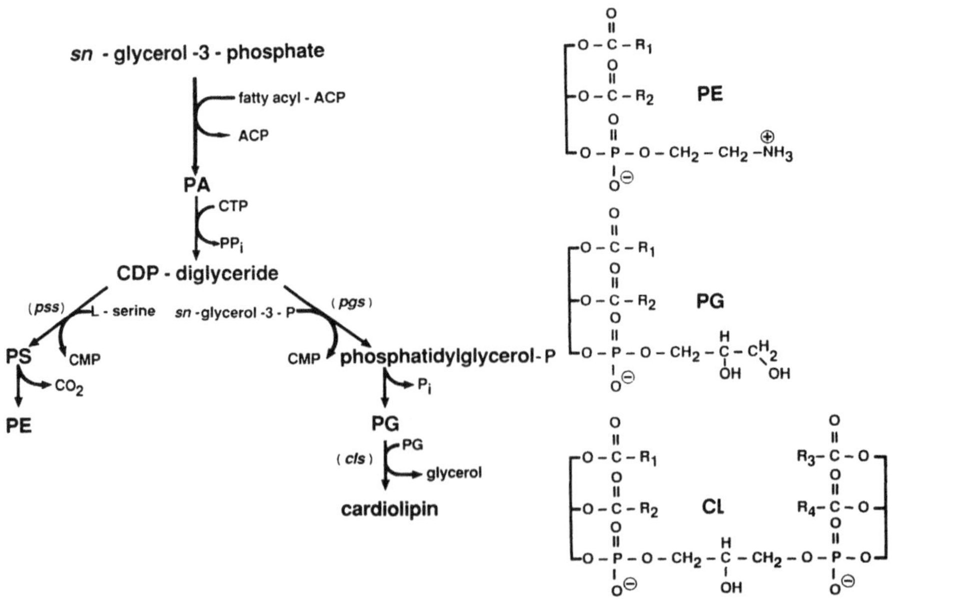

Fig. 2. Biosynthesis and structure of major membrane phospholipids in *E. coli*. The genes coding for some of the enzymes involved in the biosynthesis are indicated within parenthesis.

The phospholipid composition of wild-type *E. coli* strains is 75-80% zwitterionic PE, 20% PG and 1-5% diphosphatidylglycerol or CL, which values depend on the particular conditions (Raetz, 1978). These latter two phospholipids are negatively charged at physiological pH values. The overall organization of the phospholipids in the *E. coli* inner membrane is that of a liquid crystalline bilayer (Burnell et al., 1980). This L_α structure (Fig. 3) is also the dominant organization in the hydrated total phospholipid extract of the inner membrane. However, two alternative lipid organizations, the L_β and H_{II} phase, are relevant for the *E. coli* inner membrane (Fig. 3). Lowering the temperature or incorporation of higher melting fatty acids, using unsaturated fatty acid auxotrophs, results in the formation of the L_β phase in the membrane and derived liposomes (Haest et al., 1972). This disorder -> order phase change in the bilayer has many functional consequences of which inhibition of protein translocation (Kimura & Izui, 1976; Ito et al., 1977; Pagès et al., 1978; DiRienzo & Inouye, 1979) is most relevant in the context of this article.

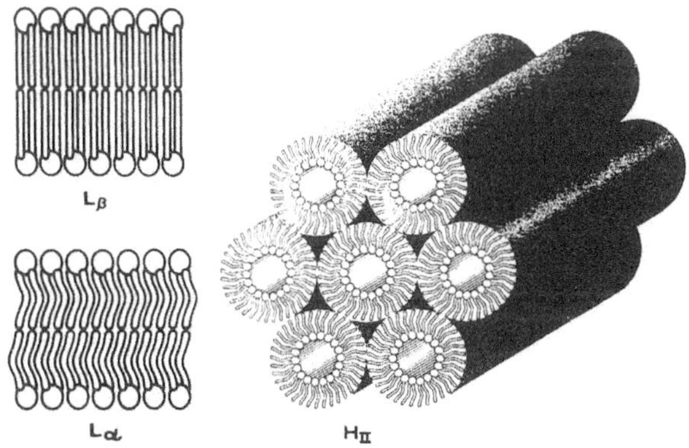

Fig. 3. Major phospholipid organizations adopted by hydrated isolated *E. coli* phospholipids.

Unsaturated PE's in isolation in general favor the hexagonal H_{II} phase (Fig. 3) under physiological conditions. This inverted non-bilayer phase consists of hexagonally organized tubes in which the lipid head groups line the wall of the aqueous channels present within the tubes. The presence of the bilayer preferring lipids such as PG and CL possibly in combination with other membrane components such as proteins or other constraints within the membrane can stabilize the bilayer. Nevertheless, the stability of the bilayer in the inner membrane of *E. coli* is limited as treatment of cells under conditions which more strongly express the hexagonal H_{II} phase-forming capacity of the phospholipids, e.g. divalent cation treatment and elevated temperatures, induces polymorphic changes in membrane organization towards inverted non-bilayer lipid structures (Borovjagin et al., 1987). Local formation of inverted (i.e. type II) non-bilayer lipid structures offers attractive functional abilities to membranes, including protein transport [see De Kruijff (1987) for a short review]. The basis of lipid polymorphism in its most simple form is shown in Fig. 4 as the shape-structure concept. This relates the dynamical shape of the lipid molecule, including hydration and intra- and intermolecular interactions, to its organization in aggregate structures. In case of molecules with a head group area that is smaller than the cross-sectioned area of the hydrocarbon region, such as in unsaturated PE's, the cone shape will favor a concave interface typical of the inverted non-bilayer phases. In the opposite case of lysophospholipids, the convex interface favors formation of type-I aggregate structures such as micelles. In the bilayer organization the two areas are more or less in balance. Local changes in planar --> convex interface curvature (bilayer --> type II non-bilayer) could initiate passage of polar molecules across membranes.

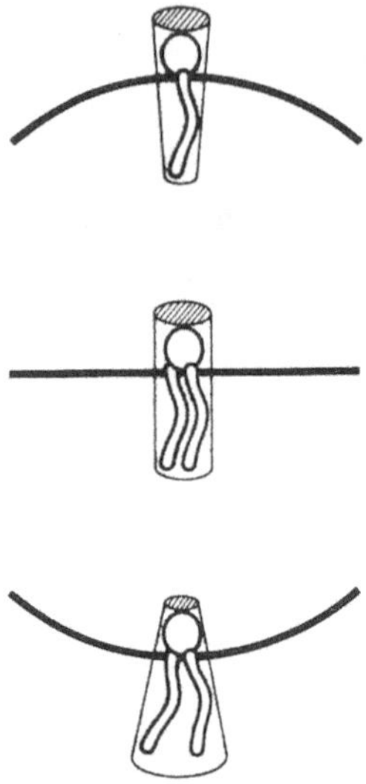

Fig. 4. The shape-structure concept of polymorphism.

With this background information on the structure of the *E. coli* membrane lipids, we now turn to the structure of signal peptides and signal peptide-lipid interactions.

SIGNAL PEPTIDE STRUCTURE AND SIGNAL PEPTIDE-LIPID INTERACTIONS

An excellent review on signal sequences has recently been published (Gierasch, 1989). At this moment nothing is known about the spatial structure of prokaryotic signal sequences in the translocation process, e.g. from its emergence from the ribosome to its cleavage by leader peptidase at the periplasmic side of the inner membrane and subsequent digestion by signal peptide peptidase. The only available conformational information comes from computer predictions and biophysical studies on chemically prepared signal peptides. Structure prediction methods applied to signal peptides in general, reveal a high propensity for either β-structure or α-helix for a major part of the signal peptide (Austen, 1979). In case of the PhoE signal peptide (Fig. 1) the Ala^{-7}-Thr^{-17} sequence has a mean sheet potential of 1.31 and a mean helix potential of 1.10 (Batenburg et al., 1988a). The prediction of β-turns in a number of cases (Austen, 1979) is not unexpected because often in the hydrophobic central region β-turns favoring residues as Pro or Gly (note Gly^{-10} in Fig. 1) are present. Small peptides as the 15-30 long signal peptides can be expected in isolated form to have considerable conformational flexibility. This, together with the strong tendency of these hydrophobic peptides to aggregate in water might contribute to the range of conformations reported for prokaryotic signal peptides in water random coil (Shinnar & Kaiser, 1984; Austen & Ridd, 1981) to β-structure (Batenburg et al., 1988a; Rosenblatt et al., 1980). In non-polar membrane-like environments as trifluoroethanol and SDS micelles the α-helix is the dominant conformation (Batenburg et al., 1988a; Shinnar &

Kaiser, 1984; Austen & Ridd, 1981; Rosenblatt et al., 1980). This ability seems to be a property of functional signal sequences (Briggs & Gierasch, 1984,1986). For the LamB signal peptide an α-helical conformation of the species inserted in a phospholipid monolayer was reported in which the long axis of the helix runs parallel to the acyl chains (Cornell et al., 1989).

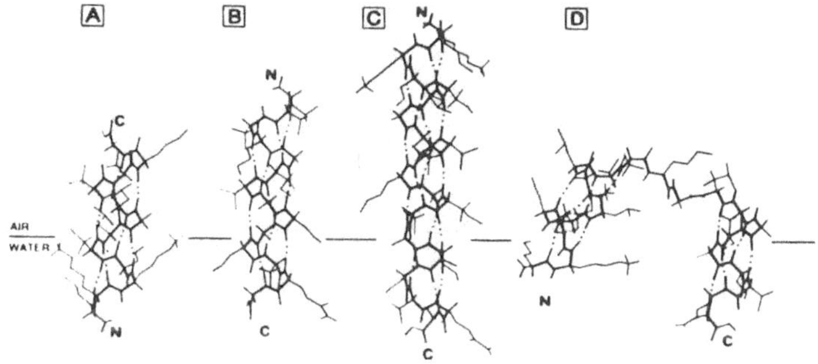

<table>
<tr><td>AREA (Å²)/MOLECULE</td><td></td><td></td><td></td><td></td></tr>
<tr><td>CALCULATED</td><td>158</td><td>125</td><td>172</td><td>370</td></tr>
<tr><td>MEASURED</td><td>175 ± 15</td><td>-</td><td colspan="2">360 ± 20</td></tr>
</table>

Fig. 5. Predicted orientations of the isolated PhoE signal peptide (C, D) and fragments Met^{-21}-Gly^{-10} (A) and Ala^{-15}-Ala^{-1} (B) on an air-water interface and calculated (after monolayer assembly) and measured limiting molecular areas. Reproduced with permission from Batenburg et al. (1988a). For details see this reference.

Figure 5 shows computer predictions of energy minimalized orientations of the PhoE signal peptide and derived fragments in a monolayer at the air-water interface assuming an α-helical conformation. In all cases a preferred orientation of the signal peptide and the C- and N-terminal fragments perpendicular to the interface is predicted to be energetically most favorable (Fig. 5A-C). It is remarkable that the complete sequence, if assumed to consist of 100% helix is predicted to be situated with its N-terminus in the air despite the excess of charged residues at this side. This is due to the decreasing hydrophobicity of the amino acid side chain towards the C-terminus as generally found in signal peptides. The calculated area per molecule occupied in the monolayer after assembly of the peptides is also given in Fig. 5. Assembly of the signal peptide in the conformation of Fig. 5C results in very unfavorable energetics due to electrostatic interactions. The area of 172 A^2/molecule is therefore of the reversed orientation which could be favorable packed (Batenburg et al., 1988a). When the theoretical values are compared to the measured area a striking feature is noted. For the N-terminal fragment (Fig. 5A), the calculated and measured values are in good agreement and fall in the range expected for α-helices. However, the measured area of the complete signal peptide is approximately twice that of the calculated area. Energy minimalization of the conformational space in the sequence Val^{-12}-Ile^{-9}, results mainly due to the presence of Gly^{-10}, in a lowest energy state of the signal peptide after assembly in the monolayer depicted in Fig. 5D. In this looped structure the two α-helices in the N- and C-terminal part are connected by a β-turn-like structure. In this conformation there is a good agreement between the calculated and measured molecular area.

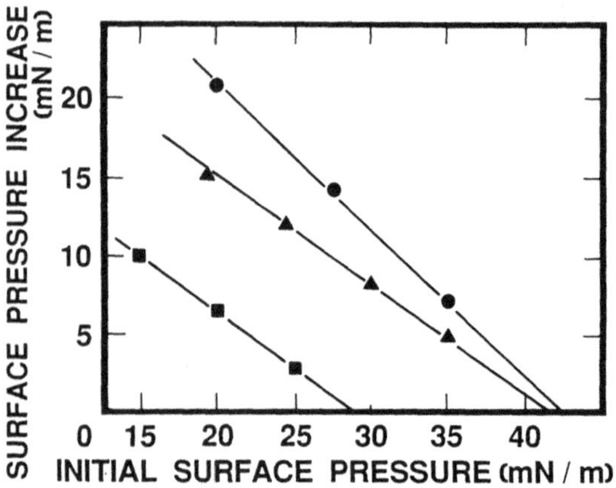

Fig. 6. PhoE signal peptide induced surface pressure increase at 37°C in monolayers of *E. coli* phospholipid extract (▲), isolated PE (■) and PG (●). Reproduced with permission from Batenburg et al. (1988b). For details see this reference.

Signal peptides in general have a very high affinity for membrane lipids (Shinnar & Kaiser, 1989; Fidelio et al., 1987; Briggs et al., 1986; Batenburg et al., 1988b), not unexpected in view of their chemical structure. The nature of the interaction and the specificity for negatively charged phospholipids is illustrated in the monolayer experiment depicted in Fig. 6. Injection of the signal peptide from trifluoroethanol results in case of the total *E. coli* phospholipid extract in a large increase in surface pressure due to penetration of the peptide in between the lipid molecules. At initial surface pressures of around 30 mN/m, which are considered to be relevant for the situation in the biological membrane, the large pressure increase must be mainly due to the PG component as evidenced by the differences in interaction between pure PE and PG monolayers. It can therefore be concluded that the signal peptide-lipid interaction has both electrostatic and hydrophobic components. Under monolayer packing conditions equivalent to the ordered Lβ phase PhoE signal peptide penetration is severely inhibited (Batenburg et al., 1988b).

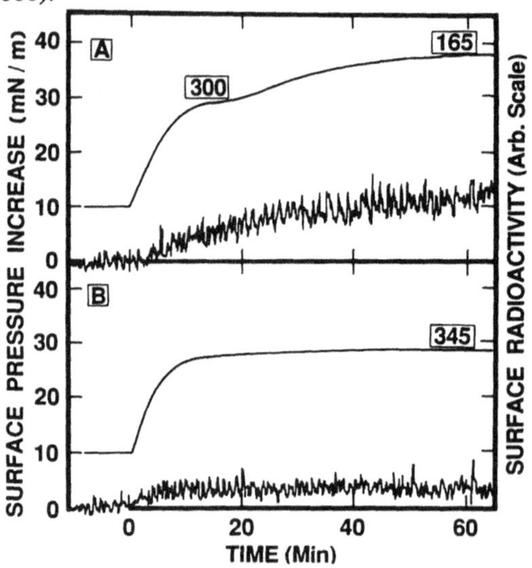

Fig. 7. Kinetics of the increase of surface pressure (upper trace) and surface radioactivity (lower trace) upon injection of [^{14}C]-PhoE signal peptide at time = 0 underneath a CL (A) and phosphatidylcholine (B) monolayer. The numbers indicate the area per molecule (A^2) occupied by the signal peptide in the monolayer. Reproduced with permission from Batenburg et al. (1988b). For details see this reference.

A striking feature of the interaction, reported so far only for the PhoE signal peptide is illustrated in Fig. 7. Using radiolabelled PhoE signal peptide and simultaneous monitoring of surface pressure (penetration) and surface radioactivity (amount of the peptide associated with the monolayer), two phase kinetics is observed for the interaction with monolayers of cardiolipin as a model negatively charged lipid, but only one phase kinetics in case of phosphatidylcholine as a model zwitterionic lipid. The measured areas per molecule occupied by the signal peptide in the lipid monolayer, are indicated in the figure and demonstrate that the initial insertion of the signal peptide in both systems results in a molecular area close to that observed for the pure PhoE signal peptide (compare Fig. 5). However, in case of the negatively charged phospholipid a time-dependent change in the monolayer occurs, finally resulting in a molecular area close to that expected for a signal peptide in an α-helical conformation, oriented perpendicular to the interphase. These observations are an important element in the unlooping model for protein translocation we will postulate this review.

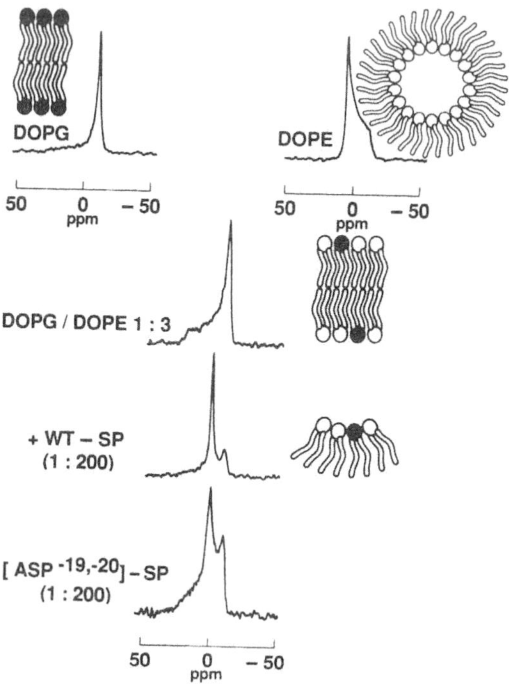

Fig. 8. Influence of PhoE signal peptides (peptide-lipid molar ratio 1:200) on lipid organization of mixed DOPG/DOPE (1:3, molar) bilayers as assessed by ^{31}P NMR.

Another important element of the model comes from the ^{31}P NMR experiments on the PhoE signal peptide-lipid interaction shown in Fig. 8. This technique is a diagnostic method to resolve different macroscopic organizations of phospholipids (Cullis & De Kruijff, 1979). This is because of differences in motional possibilities of phospholipids in various aggregate structures cause differences in averaging of the chemical shift anisotropy, resulting in characteristic ^{31}P NMR line shapes. For instance, dioleoyl-PG (DOPG) forms bilayers in aqueous dispersion as evidenced by the asymmetrical line shape with a high field peak and a low field shoulder separated by approximately 40 ppm. In contrast, the hexagonal H_{II} organization of hydrated DOPE gives rise to a line shape with a reversed symmetry and two-fold reduced width. When the two lipids are mixed in a ratio approximating that of the *E. coli* inner membrane, the bilayer organization prevails due to the bilayer stabilizing activity of the PG component. Addition of small amounts of PhoE signal peptide results in the induction of a new spectral component characteristic of phospholipids undergoing rapid isotropic motion. In conjunction with centrifugation experiments and freeze fracture electron microscopy analysis (Killian et al., 1989), the phospholipid organization in the signal peptide induced new structure can be assigned to a type II inverted organization with a net concave surface curvature. These effects might well be related to the recently (Nagaraj et al., 1987) observed ability of signal peptides to cause aggregation and fusion of lipid vesicles. The loss of bilayer structure induced by the signal peptide is correlated with the functionality level of the signal peptide. Replacement of the two lysines by two aspartic acid residues in the signal peptide of PhoE decreases the rate of translocation across the *E. coli* membrane (Bosch et al., 1989). Similarly, the chemically prepared [Asp^{-19},Asp^{-20}] signal peptide is much less effective in inducing changes in lipid organization upon addition to the mixed DOPE/DOPG dispersion (Fig. 8).

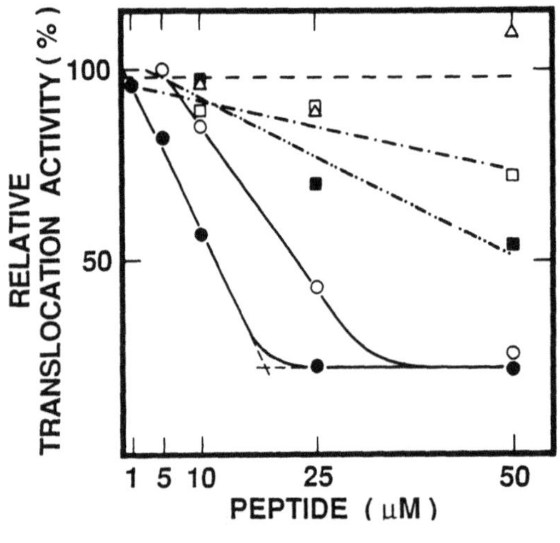

Fig. 9. Inhibition of *in vitro* PhoE translocation across inner membrane vesicles by wild type PhoE signal peptide (●-●) and derived fragments Met^{-21}-Ala^{-15} (Δ), Met^{-21}-Gly^{-10} (■), Ile^{-9}-Ala^{-1} (□) and mutant PhoE signal peptide in which the two lysines are replaced by two aspartic acid residues (o). Reproduced with permission from De Vrije et al. (1989). For details see this reference.

Chemically prepared signal peptides can cause inhibition of *in vitro* protein translocation (Majzoub et al., 1980; Austen et al., 1984; Chen et al., 1987; De Vrije et al., 1989). This is shown in Fig. 9 for PhoE protein translocation in inverted *E. coli* inner membrane vesicles. The inhibition is maximal for the intact signal peptide and is less efficient for the less functional [Asp^{-19}, Asp^{-20}] signal peptide in parallel to the observed differences in signal-peptide lipid interaction. The concentration of signal peptide causing maximal inhibition corresponds to an acidic phospholipid phosphate/peptide molar ratio of 3 (De Vrije et al., 1989). This ratio exactly corresponds to the formation of a charge neutral complex of the peptide (which had a non-formylated N-terminal methionine) with the anionic phospholipids. The high affinity interaction (K_d ~ 0.1 μM) of the PhoE signal peptide with cardiolipin in a monolayer has a similar charge stoichiometry of 2.8 ± 0.4 (lipid phosphate/peptide; Batenburg et al., 1988b).

The data on signal peptide-lipid interactions and the inhibitory effect of signal peptides and other basic polypeptides (De Vrije et al., 1989), which have a high affinity interaction with negatively charged phospholipids on protein translocation are consistent with an important role of signal peptide lipid interactions in protein translocation. The key question is whether such interactions exist in the functional translocation process. The answer to this question is unknown, but many experimental findings suggests that lipids are involved in protein translocation in *E. coli*.

LIPID INVOLVEMENT IN PROTEIN TRANSLOCATION IN *E. COLI*

Here we will summarize all pertinent experimental information on the role of lipids in protein translocation across the inner membrane of the *E. coli* cell.

The extensive studies of the group of Wickner (1979,1988) on M13 procoat translocation has given strong support for the involvement of lipids in protein translocation and form the basis for his membrane trigger hypothesis for protein translocation. This relatively small protein moves across the membrane in a manner independent of the products of the *sec* genes (Wolfe et al., 1985). Moreover, translocation could be reconstituted using vesicles which next to phospholipids only contained the processing enzyme leader peptidase, which is known not to be essential for translocation (Ohno-Iwashita et al., 1984).

Membrane active compounds such as phenethylalcohol and anaesthetics, which preferentially partition in the lipid phase of the membrane and alter lipid packing are long known for their large effects on translocation of a variety of proteins across the inner membrane (see Chen & Tai, 1987, for an example). As indicated above when the amount of ordered (Lβ phase) lipids in the *E. coli* inner membrane increases protein translocation is greatly inhibited (Kimura & Izui, 1976; Ito et al., 1977; Pagès et al., 1978; DiRienzo & Inouye, 1979). Unfortunately a careful parallel analysis of lipid organization and protein translocation in the same system has not yet been carried out.

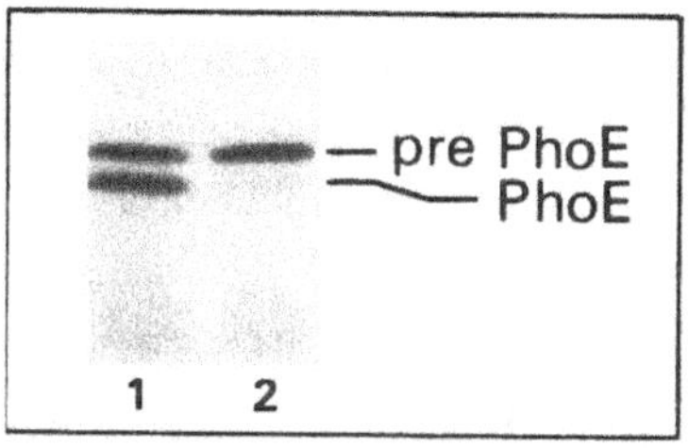

Fig. 10. *In vitro* co-translational translocation of [^{35}S]-labelled prePhoE across *E. coli* wild-type inner membrane vesicles (lane 1) and *pgs*A3 mutant vesicles containing low PG levels (lane 2). Samples were analyzed for translocation by fluorography of SDS-polyacrylamide gels after exposure to protease. For details see De Vrije et al. (1988).

Recently *E. coli* mutants with defects in the biosynthesis of particular phospholipid classes were introduced in the field of protein translocation. In strains carrying a defect in the *pgs* gene (Fig. 2) thereby greatly reducing the amount of PG and CL (Miyazaki et al., 1985; Heacock & Dowhan, 1987) both inhibition of translocation of outer membrane proteins across the inner membrane could be demonstrated as well as a decrease in processing efficiency (De Vrije et al., 1988). Figure 10 illustrates this with an *in vitro* co-translational translocation experiment. In control wild type vesicles with normal PG and CL levels a substantial amount of *in vitro* synthesized precursor (prePhoE) and mature PhoE are protected against externally added proteinase K. In contrast in the *pgs*A3 mutant vesicles which contain only 2.3% PG and 0.8% CL next to 2.8% of negatively charged phosphatidic acid (De Vrije et al., 1988) both total translocation (sum protected prePhoE and PhoE) and processing is inhibited. Using a mutant with a deletion in the N-terminal part of the mature sequence of the PhoE, which results in inhibition of processing (Bosch et al., 1988), it could be proven that translocation itself was inhibited in this *pgs* mutant next to inhibition of processing (De Vrije et al., 1988). The inhibition of processing in the *pgs* mutant suggests that the activity of leader peptidase depends on the level of PG. Fully consistent with this suggestion is the observation that the activity of purified leader peptidase in a detergent system is greatly stimulated by PG and not by the zwitterionic lipids PE and PC (Fig. 11).

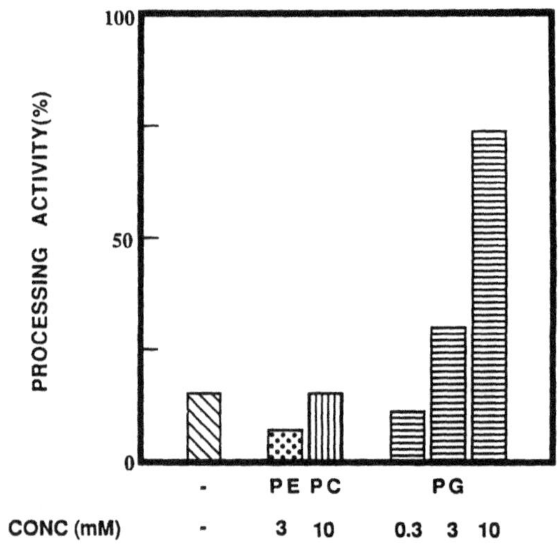

Fig. 11. Stimulation of the processing activity of purified leader peptidase by various concentrations of phospholipids in a mixed micellar β-octylglucoside detergent (30 mM) system. The processing activity is defined as the percentage of total prePhoE synthesized in a cell-free system processed into mature PhoE by 10 μg leader peptidase/ml at 37°C. For details see De Vrije et al. (1988).

SUMMARIZING CONCLUSIONS

Studies on signal peptide-lipid interactions as well as studies on protein translocation processes in *E. coli* and derived *in vitro* and model systems have given important clues as to how lipids could participate in protein export out of the cytoplasm. The main elements of these studies are: 1. insertion of signal peptides in the hydrophobic part of the membrane, which depends on the state of packing of the lipids and which is facilitated by electrostatic interactions between the positively charged N-terminus and the negatively charged membrane phospholipids, 2. conformational and topological changes in the signal peptide upon insertion, and 3. changes in local lipid structure as a result of the signal peptide-lipid interaction.

Although many questions are still unanswered and a direct involvement of lipids in functional protein translocation *in vivo* or *in vitro* has not been demonstrated, we feel that these elements can be rationally incorporated in a working model for PhoE translocation, which is shown and described in Fig. 12. The unlooping model resembles previous models (Inouye & Halegoua, 1980; Engelman & Steitz, 1981), but differs in two important aspects, i.e. a loop in the signal peptide and local changes in lipid bilayer organization (non-bilayer lipids). Energetically an α-helical looped structure spanning half the bilayer seems unrealistic due to constraints in hydrogen bonding in the turn part of the signal peptide.

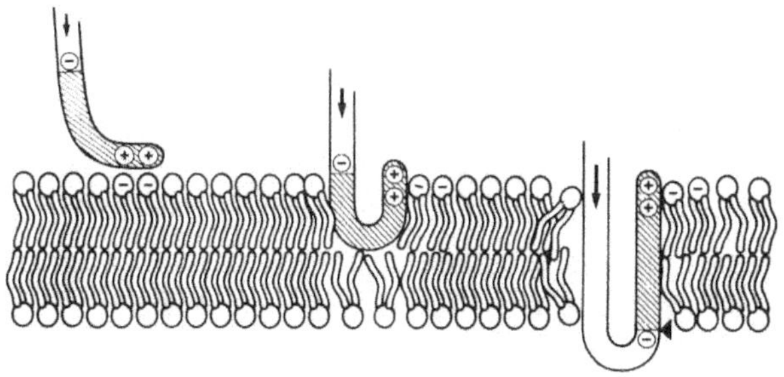

Fig. 12. Unlooping model for PhoE translocation. 1: The positively charged N-terminus of the shaded signal peptide initially interacts electrostatically with the negatively charged phospholipids (PG). 2: Subsequently, the signal peptide inserts as a loop in the hydrophobic core of the bilayer. 3: Unlooping of the signal peptide moves the N-terminus of the mature sequence across the bilayer, which process might be facilitated by electrophoretic movement of the negatively charged amino acid at position +2 of the mature sequence under the influence of membrane potential. During steps 2 and 3 lipid organization is locally perturbed which could facilitate passage of the protein. For simplicity the role of protein components and ATP known to be essential for PhoE translocation are excluded from this model. The bilayer --> type II perturbation of lipid organization shown in the figure is highly schematically.

In this model local changes in lipid organization could overcome this transient state in the translocation process. Future work using purified precursor proteins and defined *in vitro* systems will test the validity of this model.

ACKNOWLEDGEMENTS

We thank other members of the Biomembranes and Model Systems Group of the Centre for Biomembranes and Lipid Enzymology and members of the Section Microbiology of the Department of Molecular Cell Biology of the University of Utrecht, for their important contributions to this work. Dr. W. Wickner (UCLA) is thanked for his kind gift of purified leader peptidase. Drs. W. Dowhan and R. Brasseur are thanked for their collaborations in the studies on the *pgs* mutant vesicles and computer predictions of PhoE signal peptide structure, respectively. This work was supported by the Netherlands Foundation of Chemical Research (SON) with financial aid from the Netherlands Organization for Scientific Research (NWO).

REFERENCES

Austen BA (1979) FEBS Lett 103: 308-313
Austen BA, Ridd DH (1981) Biochem Soc Symp 46, 235-258
Austen BM, Herman-Taylor J, Kaderbhai MA, Ridd DH (1984) Biochem J 224: 317-325
Batenburg AM, Brasseur R, Ruysschaert JM, Van Scharrenburg GJM, Slotboom AJ, Demel RA, De Kruijff B (1988a) J Biol Chem 263, 4202-4207
Batenburg AM, Demel RA, Verkleij AJ, De Kruijff B (1988b) Biochemistry 27: 5678-5685
Bochkareva ES, Lissin NM, Girsohovich AS (1988) Nature 336: 254-257
Borovjagin VL, Sabelnikov AG, Tarahovsky VS, Vasilenko IA (1987) J Membr Biol 100: 229-242
Bosch D, Voorhout W, Tommassen J (1988) J Biol Chem 263, 99152-99157
Bosch D, De Boer P, Bitter W, Tommassen J (1989) Biochim Biophys Acta 979: 69-76

Briggs MS, Gierasch LM (1989) Biochemistry 23: 3111-3114
Briggs MS, Gierasch LM (1986) Adv Prot Chem 38: 109-180
Briggs MS, Cornell DG, Dluhy RA, Gierasch LM (1986) Science 233: 206-208
Burnell E, Van Alphen L, Verkleij AJ, De Kruijff B (1980) Biochim Biophys Acta 597, 492-501
Chen L, Tai PC (1985) Proc Natl Acad Sci USA 82: 4384-4388
Chen L, Tai PC (1987) J Bacteriol 169: 2373-2379
Chen L, Tai PC, Briggs MS, Gierasch LM (1987) J Biol Chem 262: 1427-1429
Collier DN, Bankiatis V, Weiss JB, Bassford PJ Jr (1988) Cell 53: 273-283
Cornell DG, Dluhy RA, Briggs MS, McKnight CJ, Gierasch LM (1989) Biochemistry 28: 2789-2797
Crooke E, Brundage L, Rice M, Wickner W (1988) EMBO J 7: 1831-1835
Cullis PR, De Kruijff B (1979) Biochim Biophys Acta 559: 399-420
Cunningham K, Lill R, Crooke E, Rice M, Moore K, Wickner W, Oliver D (1989) EMBO J 8: 955-959
De Kruijff B (1987) Nature 329: 587-588
De Vrije T, Tommassen J, De Kruijff B (1987) Biochim Biophys Acta 900: 63-72
De Vrije T, De Swart RL, Dowhan W, Tommassen J, De Kruijff B (1988) Nature 334: 173-175
De Vrije T, Batenburg AM, Jordi W, De Kruijff B (1989) Eur J Biochem 180: 385-392
DiRienzo JM, Inouye M (1979) Cell 17: 155-161
Engelman DM, Steitz TA (1981) Cell 23: 411-422
Fidelio GD, Austen BM, Chapman D, Lucy JA (1987) Biochem. J. 244: 295-301
Gannon PM, Li P, Kumamoto CA (1988) J Bact 171: 813-818
Geller BL, Movva NR, Wickner W (1986) Proc Natl Acad Sci USA 83: 4219-4222
Gierasch LM (1989) Biochemistry 28: 923-930
Haest CWM, De Gier J, Van Es GA, Verkleij AJ, Van Deenen LLM (1972) Biochim Biophys Acta 288: 43-53
Heacock PN, Dowhan WJ (1987) J Biol Chem 262: 13044-13049
Inouye H, Halegoua S (1980) CRC Crit Rev Biochem 7: 339-371
Ito K, Sato T, Yura T (1977) Cell 11: 551-559
Killian JA, Bijvelt J, Verkleij AJ, De Kruijff B (1989) Submitted for publication
Kimura K, Izui K (1976) Biochem Biophys Res Commun 70: 900-906
Kusters R, De Vrije T, Breukink E, De Kruijff B (1989) Submitted for publication
Lecker S, Lill R, Georgopoulos C, Bassford PJ Jr, Kumamoto C, Wickner W (1989) EMBO J, in press
Lill R, Cunningham K, Brundage LA, Ito K, Oliver D, Wickner W (1989) EMBO J 8: 961-966
Majzoub JA, Rosenblatt M, Fennick B, Mannus R, Kronenberg HM, Potts JT, Habener JF (1980) J Biol Chem 255: 11478-11483
Miyazaki C, Kuroda M, Ohta A, Shibuya I (1985) Proc Natl Acad Sci USA 82: 7530-7534
Nagaraj R, Joseph M, Reddy GL (1987) Biochim Biophys Acta 903, 465-472
Ohno-Iwashita Y, Wolfe PB, Ito, K, Wickner W (1984) Biochemistry 23, 6178-6184
Oliver DB, Beckwith J (1982) Cell 30: 311-319
Pagès JM, Piovant M, Varenne S, Lazdunski C (1978) Eur J Biochem 86, 589-602
Raetz CRH (1978) Microbiol Rev 42: 614-659
Rosenblatt M, Beaudette NV, Fasman GD (1980) Proc Natl Acad Sci USA 77, 3983-3987
Schmidt MG, Rollo EE, Grodberg J, Oliver DB (1988) J Bact 170: 3400-3417
Shinnar AE, Kaiser ET (1984) J Am Chem Soc 106, 5006-5007
Von Heijne G (1985) J Mol Biol 184: 99-105
Wickner W (1979) Ann Rev Biochem 48: 23-45
Wickner W (1988) Biochemistry 27: 1081-1086
Wolfe PB, Rice M, Wickner W (1985) J Biol Chem 266: 1836-1841
Zwizinski C, Wickner W (1980) J Biol Chem 255: 7973-7977

The Enzymology of Bacterial Protein Export

Bill Wickner

Molecular Biology Institute and
Department of Biological Chemistry
University of California
Los Angeles, California 90024-1570 USA

NATO ASI Series, Vol. H 40
Dynamics and Biogenesis of Membranes
Edited by J.A.F. Op den Kamp

In 1967, as a first-year student in medical school, I attended an elegant series of lectures by Professor Eugene Kennedy on the topic of lipid metabolism. While this topic often fails to quicken the pulse of the audience (and, in fact, is pretty good at flattening brain waves), Dr. Kennedy's lectures were so lucid and witty that I was left wanting to know more of the topic and, frankly, of the man behind the topic. He kindly consented to letting me work in his lab that summer, and, together with Eugene Scarborough, a postdoctoral fellow, we explored the physiology of lactose transport in intact E. coli.

At the end of the summer, I was the proud posessor of a laboratory notebook crammed with data, but had the nagging feeling that it was impossible to understand the fantastically complex machinery of this organism merely by observing sugar uptake rates. With some trepidation, I went to "the boss" and expressed this concern. To my delight, his eyes brightened, and he suggested that I was ready to undertake enzyme purification. Now, before the kind and gentle reader assume that Professor Kennedy was condemning me to the "Coldroom Gulag", I should tell you that there were no pure integral membrane proteins in 1967. In fact, it was still being debated whether the lipid bilayer was really the fundamental substratum of biological membranes, and the elegant experiments of Bretscher which led to the fluid mosaic model were 4 years away. Dr. Kennedy showed me the assay for phosphatidylserine decarboxylase, a membrane-bound enzyme of lipid biosynthesis, and advised that they'd been able to solubilize the enzyme in the nonionic detergent "Cutscum".

I reproduced this result, found that Cutscum had been renamed Triton X100 by Rohm and Haas, and set about trying to purify my first enzyme. I went down the hall to Dr. Charles Richardson, a noted enzymologist, for advice. His first advice was to get rid of the detergent before trying serious enzymology. However, when I did so, the enzyme activity precipitated or was lost. I therefore set about purifying the enzyme in the continued presence of detergent; 4 happy years later, the enzyme was pure. I don't believe that any thrill has ever matched that of seeing a single protein band on that SDS gel.

During the following postdoctoral years, I kept marveling at the wonderful properties of Cutscum, and felt that living cells couldn't conceivably do without detergents! When I set out on my own, I wished to figure out how water-insoluble membrane proteins could be made in the aqueous environment of the cytoplasm. I was sure that the answer was detergents, so sure that I set about trying to isolate a Triton X100 auxotroph of E. coli by classical replica-plating methods. This was, of course, a complete flop, and merely demonstrated my naivete. Still, the field of protein secretion and membrane assembly was young in 1974, and there was time for a beginner to make lots of beginner's mistakes. As Racker would say, there's a lesson here: always pick a sufficiently unpopular or unappreciated problem that you have a chance to try creative experiments without being trampled by the others in the field.

Chastened by these mistakes, I arrived at the decision that I needed the <u>simplest</u> protein with every conceivable technical advantage if I was to make any headway at all. In this reductionist spirit, I settled on studying the membrane assembly of coliphage M13 coat protein, as suggested to me by Doug Brutlag, a friend and colleague at Stanford. This was, for me, a most fortuitous choice. This study (one might even say marriage) occupied me fully for a dozen years, and introduced me to a beatiful protein, leader peptidase, with which I had a serious affair. The study of these proteins taught several lessons which are of general interest for protein insertion into, and across, membranes:

1. Protein transit across a membrane is not coupled to ongoing polypeptide chain growth. This is well-known today, but was, I assure you, heresy when first proposed.

2. Protein conformation has a critical role in its insertion into, or across, a membrane, and this conformation changes as a result of this process.

3. Energy, in the form of an electrochemical potential, is needed for protein translocation. More recently, Tai and colleagues have shown that a second form of energy, ATP, is also essential.

4. Leader sequences can be recognized by enzymes (in our case, leader peptidase).

5. The leader sequence is not enough to support translocation. Additional information is needed, and resides in both polar and apolar domains of the mature protein.

6. Insertion of a protein such as procoat, with a typical leader sequence but an unusually small size, can occur into a liposomal membrane without the assistance of catalytic proteins, such as are encoded by the sec genes.

This work has been reviewed recently (1). During the decade of these studies, sec mutants were being isolated, and we found that the M13 procoat doesn't need them for its membrane assembly (2). To understand what these fascinating sec-encoded proteins do for protein translocation, we had to turn to another protein, the **o**uter **m**embrane **p**rotein A (OmpA), which is made as a precursor, termed proOmpA, with a typical N-terminal leader sequence. My group has adopted a strictly enzymological approach to the study of the transit of this protein across the plasma membrane of E. coli.

We began our studies with the isolation of radiochemically (3), and chemically (4), pure proOmpA. As with many proteins bearing a leader sequence, it is only sparingly water soluble, and thus required urea for its stable solubilization. The purified protein is fully capable of membrane assembly when diluted from urea into aqueous buffer containing inverted plasma membrane vesicles and ATP (5). However, it rapidly denatures unless it can form a complex with an available "chaperone" molecule. Three proteins of E. coli have been identified as chaperones: SecB, trigger factor, and GroEL. Genetic and biochemical studies are underway an many labs to evaluate the _in vivo_ role of each in translocation. Biochemically, each of these chaperones will form a 1:1 stiochiometric complex with proOmpA (6). The proOmpA in this complex is stable for membrane assembly.

As we turned our attention from the cytoplasmic phase to the membrane components needed for translocation, we were faced with a puzzle. Some preparations of membranes were fully competent for protein translocation, while others were less so and needed cytosol to restore their activity. Sedimentation analysis revealed that the active component was approximately 100 kilodaltons, which immediately suggested the SecA protein (7). We entered into a fruitful collaboration with Don Oliver, who had discovered and cloned the _sec_A gene and created a strain which dramatically overproduces this protein. Extracts from Don's strain were highly enriched in this soluble activity, allowing its easy purification.

At this point, we made a most fortunate finding, that membranes which had been inactivated for translocation by exposure to N_3-ATP, as described by Tai, could be fully re-activated for translocation by incubation with fresh SecA protein. Within a few short, exciting weeks, it became clear that SecA protein is an ATPase and that it is the only N_3-ATP sensitive component of the membrane (8). Strikingly, this ATPase activity is enhanced up to 100-fold by the simultaneous presence of proOmpA and inner membrane vesicles. This enhanced activity, which we have called the "translocation ATPase", clearly exhibits all the hallmarks of a central component of what is called the "translocation machinery." While translocation-competent proOmpA (freshly diluted from urea or complexed with chaperone) supports the translocation ATPase, proOmpA which has denatured for translocation will not. Furthermore, neither leader peptide nor OmpA alone will support

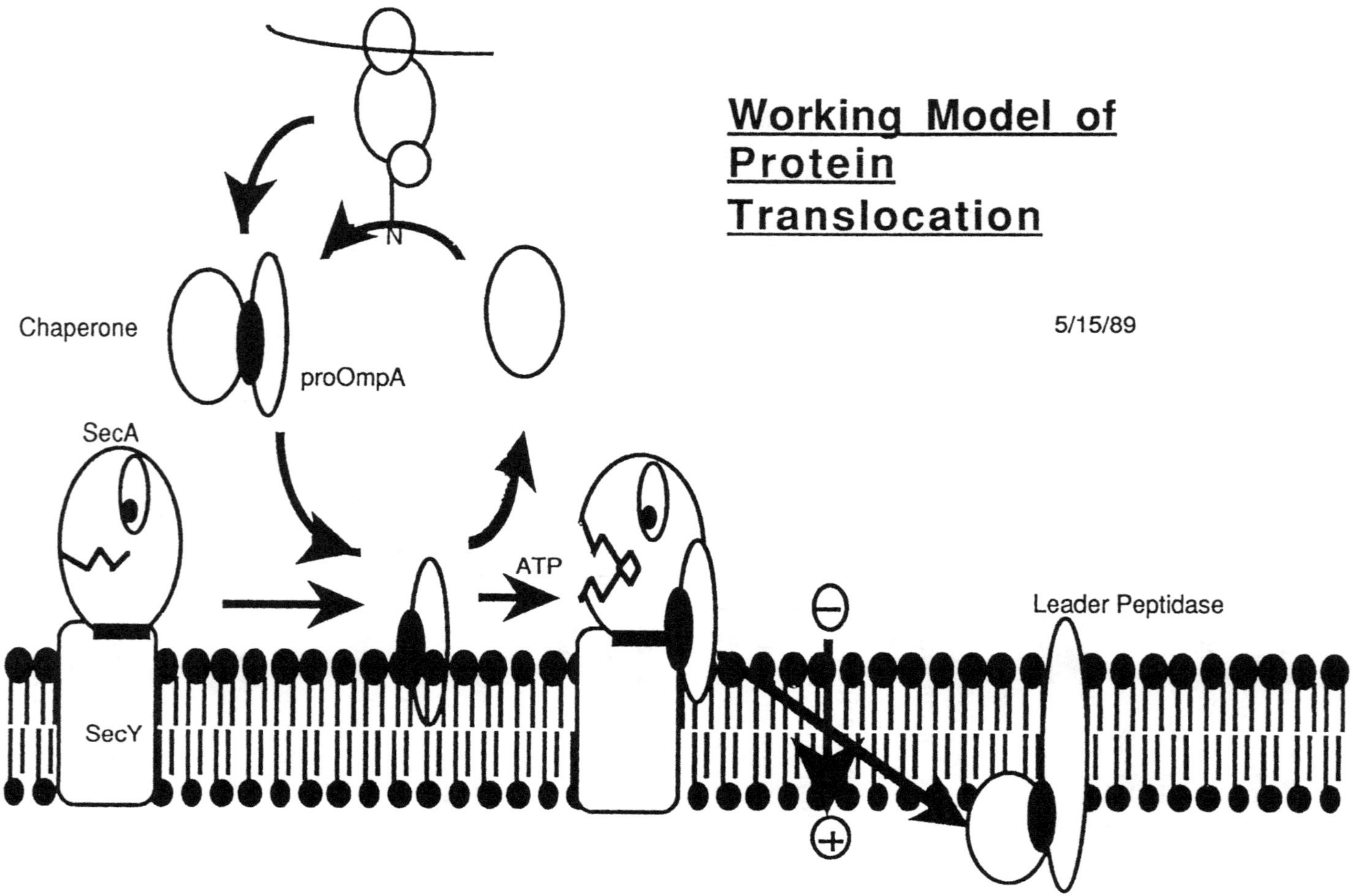
Working Model of
Protein
Translocation
5/15/89
Chaperone
proOmpA
N
SecA
SecY
ATP
Leader Peptidase

this reaction and, in fact, synthetic leader peptide is a classical Michaelis-Menten inhibitor of the translocation ATPase reaction. Thanks to a long-standing collaboration with Koreaki Ito, we were able to show that the membrane component of the reaction must have functional SecY protein. Our current model of protein translocation is shown in Figure 1.

Armed with the translocation ATPase reaction, our work has taken two major directions. One is the solubilization of inner membrane vesicles in detergent and reconstitution of proteoliposomes from the micellar extract. These proteoliposomes are quite efficient at replacing the native inner membrane vesicles in support of the translocation ATPase reaction. This has provided an assay for the isolation of functional SecY protein and, conceivably, could lead to the isolation of other needed membrane components, such as the SecD or SecE proteins, as well. This work is at an early stage.

A second direction is the analysis of the specificity of the SecA protein itself. A key observation was that the high-level ATPase activity characteristic of SecA protein in the presence of proOmpA and membranes becomes far less dependent on the SecY protein of the membranes when the Mg^{2+} concentration is lowered to 0.5mM (9). Under these conditions, SecA protein binds to even liposomes, and this bound protein is virtually inert as an ATPase. However, the addition of proOmpA dramatically stimulates its ATPase activity. This reaction, illustrated in Fig. 2, is referred to as "SecA/lipid ATPase." While either leader peptide alone or OmpA only cause a small stimulation of ATP hydrolysis by SecA bound to liposomes, together they can give up to 70% of the full SecA/lipid ATPase activity seen with intact proOmpA. This study has had several points of interest: 1. The SecA protein itself recognizes precursor molecules, and presumably has a precursor-recognition allosteric site. 2. Though the SecA protein is fully water-soluble, it must have a site which recognizes and binds to phospholipids. This binding has a profound effect on its catalytic site. 3. Recognition of the precursor

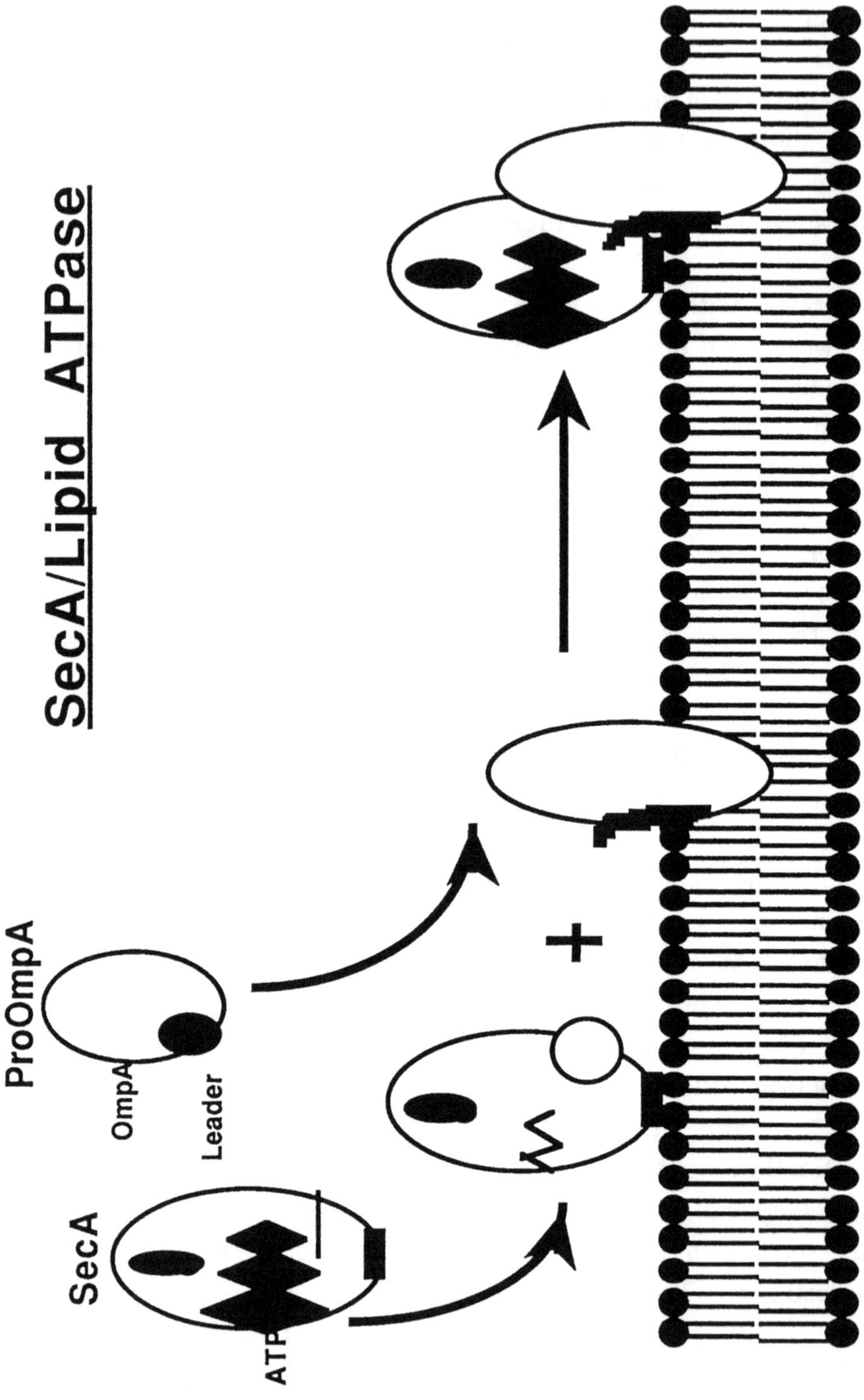
SecA/Lipid ATPase
ProOmpA
SecA
OmpA
Leader
ATP

protein is specific for both the leader *and the mature* region. This is especially suprising, since examination of the 1-dimensional, linear sequences of exported proteins has thus far failed to reveal common themes. 4. We now have a rapid and sensitive assay, using pure and abundant components, for the recognition of the leader and the mature, and may exploit this to attempt to define the important features of preproteins.

It is important to emphasize the problems in our current work as much as the progress. It takes entirely different conditions to reconstitute translocation, as opposed to translocation ATPase, from detergent extracts, and this reconstitution is not that efficient or suitable for fractionation. This is a major challenge! Secondly, our transloction ATPase is clearly "uncoupled", in the sense that far too many ATP molecules are hydrolyzed per proOmpA translocated. Clearly, the major effort on the enzymology of this system has just begun, and will happily fascinate many investigators in many labs for years to come.

References

1. Wickner, W. (1988) Mechanisms of membrane assembly: General lessons from the study of M13 coat protein and Escherichia coli leader peptidase. Biochem. **27**, 1081-1086.

2. Wolfe, P.B., Rice, M., and Wickner, W. (1985). Effects of two sec genes on protein assembly into the plasma membrane of Escherichia coli. J. Biol. Chem. **260**, 1836-1841.

3. Crooke, E. and Wickner, W. (1987). Trigger factor: A soluble protein which folds proOmpA into a membrane assembly competent form. Proc. Natl. Acad. Sci. USA **84**, 5216-5220.

4. Crooke, E., Guthrie, B., Lecker, S., Lill, R., and Wickner, W. (1988) ProOmpA is stabilized for membrane translocation by either purified E. coli trigger factor or canine signal recognition particle. Cell **54**, 1003-1011.

5. Crooke, E., Brundage, L., Rice, M., and Wickner, W. (1988) ProOmpA spontaneously folds in a membrane assembly competent state which trigger factor stabilizes. EMBO J. **7**, 1831-1835.

6. Lecker, S., Lill, R., Ziegelhoffer, T., Bassford Jr., P.J., Kumamoto, C.A., Georgopolis, C., and Wickner, W., manuscript submitted.

7. Cunningham, K., Lill, R., Crooke, E., Rice, M., Moore, K., Wickner, W., and Oliver, D. (1989). SecA protein, a peripheral protein of the Escherichia coli plasma membrane, is essential for the functional binding and translocation of proOmpA. EMBO J., **8**, 955-959.

8. Lill, R., Cunningham, K., Brundage, L.A., Ito, K., Oliver, D., and Wickner, W. (1989). SecA protein hydrolyzes ATP and is an essential component of the protein translocation ATPase of Escherichia coli. EMBO J., **8**, 961-966.

9. Lill, R. and Wickner, W. (1989) Manuscript in preparation.

COMPONENTS AND MECHANISMS INVOLVED IN COLICIN RELEASE AND COLICIN UPTAKE ACROSS THE CELL ENVELOPE IN *E. COLI*.

Claude LAZDUNSKI
Centre de Biochimie et de Biologie Moléculaire du C.N.R.S.
31 Chemin Joseph Aiguier, B.P. 71
13402 Marseille Cedex 9
France

The cell envelope of E. coli

All bacterial organisms have an envelope which has the primary role to constitute a physical barrier between the cytoplasm and the extracellular medium thereby protecting cells from harmful compounds from this medium. However, exchanges with the latter are required since cells must take up nutrients useful for growth. This is the reason why elements responsible for metabolite transport, respiration, etc... are inserted into these membranes which are thus a functional element of the cell.

The cell envelope of *E. coli* consists of two membranes, the cytoplasmic or inner membrane and the outer membrane. The outer layer of the outer membrane is constituted of lipopolysaccharide (LPS) while its inner layer contains phospholipids. This membrane is thus asymetrical. Between the outer membrane and the inner membrane, is located the periplasmic space which contains the bacterial cytoskeleton, the peptidoglycan, conferring the shape and osmotic resistance to the cell envelope. This peptidoglycan is covalently attached to the outer membrane via the so-called major lipoprotein [for a review, see (Lugtenberg and van Alphen, 1983)].

The major proteins in the outer membrane are designated as porins since they form pores through which small hydrophilic solutes with molecular wheights up to 650, in most cases, can pass (Nikaido and Vaara, 1985). The two major porins are named OmpF and OmpC however, another pore protein PhoE is induced when cells are grown under phosphate limitation (Tommassen and Lugtenberg, 1980). In addition to the porins, the outer membrane contains proteins (LamB, BtuB, Tsx) which facilitate the uptake of specific solutes

NATO ASI Series, Vol. H 40
Dynamics and Biogenesis of Membranes
Edited by J.A.F. Op den Kamp

(maltodextrins, vitamin B12, nucleosides).

OmpA is another abundant outer membrane protein, which together with the major lipoprotein appears to be involved in maintaining the structural integrity of the outer membrane and the rod shape of the cell.

Most of the outer membrane porins involved in nutrient uptake are used as colicin receptors and phage receptors (Konisky, 1982; Datta *et al.*, 1977; Braun V. and Hantke, 1977; Kadner *et al.*, 1979). Thus, the polypeptide that serves as the receptor for colicins E1, E2 and E3 functions in uptake of vitamin B12, whereas the colicin K receptor (Tsx) serves as a specific diffusion pathway for nucleosides (Hantke, 1976). Several colicin receptors are involved in iron uptake, serving as siderophore-binding proteins. For example, TonA protein is the receptor for colicin M and ferrichrome (Hantke and Braun, 1975; Wayne and Neilands, 1975) whereas enterochelin and colicins B and D utilize FepA for adsorption (Hantke and Braun, 1975; Hollifield and Neilands, 1978; Pugsley and Reeves, 1977). The colicin Ia, Ib receptor (Cir) is also involved in iron accumulation (Konisky, 1982).

In the following sections, experiments designed to elucidate the mechanism of release and the mechanism of entry of colicins will be described. Emphasis will be laid on colicin A which is the model system in our laboratory.

Colicins release across the cell envelope

Most of the plasmids which encode colicins also encode a small protein, denoted lysis protein (or bacteriocin release protein), which is required for the release of colicins from producing cells to the extracellular medium (for a review, see Lazdunski *et al.*, 1988). The genes for these small, 28-33 amino-acid proteins form an operon with the colicin structural gene and are co-regulated under SOS control (Cole *et al.*, 1985; Hakkart *et al.*, 1981; Lloubès *et al.*, 1988; Toba *et al.*, 1986). However, lysis proteins are produced in lower amounts than the colicins due to the presence of a transcription terminator after the colicin gene (Cole *et al.*, 1985; Ebina and Nakazawa, 1983; Hakkart *et al.*, 1981; Luirink *et al.*, 1986).

The primary structures of many colicin lysis proteins have been determined from the nucleotide sequence of the corresponding colicin operons (de Graaf and Oudega, 1986; Lazdunski *et al.*, 1988). They all contain a signal peptide with a cysteine residue at the cleavage site and in all cases so far examined they have been found to be modified by the addition of lipid, very likely through the same pathway as the major lipoprotein of *E. coli* (Cavard *et al.*, 1987).

The CloDF13 lysis protein has been localized to both inner and outer membranes of *E. coli* (Hakkart *et al.*, 1981; Oudega *et al.*, 1984), while the colicin N lysis protein was found only in the outer membrane (Pugsley, 1988a). In addition, in Citrobacter freundii the mature form of the colicin A lysis protein (Cal) was found to be partly released to the extracellular medium (Cavard *et al.*, 1985).

The induction of lysis proteins has been shown to activate the normally dormant phospholipase A located in the outer membrane (Pugsley and Schwartz, 1984; Cavard *et al.*, 1987; Luirink *et al.*, 1986) thereby increasing the permeability of this membrane. However, it is not yet known whether lysis proteins cause a direct or indirect activation of the phospholipase A or whether this is the primary effect of these proteins. Furthermore as of yet the mechanism by which lysis proteins allow the transfer of colicins across both the inner and outer membranes remains poorly understood. It has however been demonstrated that the release process is non-specific with respect to the colicin itself, which was shown not to contain any topogenic export signal in the case of colicin A (Baty *et al.*, 1987b).

The colicin A lysis protein (Cal) is a 33 amino acid lipoprotein which is responsible for the "quasi-lysis", which involves the release of colicin A and many other proteins from induced colicinogenic cells (Cavard *et al.*, 1985). It is produced as a precursor of 51 amino acids which undergoes a remarkably slow processing, during which the unmodified precursor, modified precursor, mature Cal and the signal sequence are visible for at least 30 min following a pulse-labelling (Cavard *et al.*, 1987).

The aminoacid sequences of almost all lysis proteins are very homologous and even identical for certain of the E colicins (Lazdunski *et al.*, 1988). There is clearly a conserved 18 aminoacids region while the C-terminal region is not conserved at least for colicin A and colicin N lysis proteins (Fig. 1). This pattern of conservation suggested some structural requirement related to the function. In an attempt to address this question, we have used *in vitro* mutagenesis to alter selected aminoacids or regions of the colicin A lysis protein (Cal). We constructed a first series of mutants (Fig. 1) in which the highly conserved amino acids Arg-7, Gly-11 and Val-14 were substituted by using a cassette mutagenesis technique (Howard *et al.*, 1989).

A second series of mutants were designed to test the importance of the C-terminal region of Cal to its function (Fig. 1). Two mutant plasmids coded for truncated Cal proteins which contained only the very homologous first 16 or 18 (S16 and S18) amino acids of the normal 33 aminoacid lysis proteins. Another construct was made to produce a Cal protein in which the C-terminal region (aminoacids 17 to 33) following the conserved amino-terminal portion of Cal was completely altered being different from the wild-type both by in length and in sequence (see FS2 and WT in Fig. 1).

	1	2	3	4	5	6	7	8	9	10	11	12	13	14	15	16	17	18	
WT	Cys	Gln	Val	Asn	Asn	Val	Arg	Asp	Thr	Gly	Gly	Gly	Ser	Val	Ser	Pro	Ser	Ser	
A2							Glu												
A6							Ala												
A11							Gly												
A32							Val												
G4											Glu								
G6											Ala								
G8											Val								
V1														Arg					
V3														Gln					
V8														Leu					
S16	Cys	Gln	Val	Asn	Asn	Val	Arg	Asp	Thr	Gly	Gly	Gly	Ser	Val	Ser	Pro	STOP		
S18	Cys	Gln	Val	Asn	Asn	Val	Arg	Asp	Thr	Gly	Gly	Gly	Ser	Val	Ser	Pro	Ser	Ser	STOP

	16	17	18	19	20	21	22	23	24	25	26	27	28	29	30	31	32	33	
WT	Pro	Ser	Ser	Ile	Val	Thr	Gly	Val	Ser	Met	Gly	Ser	Asp	Gly	Val	Gly	Asn	Pro	STOP
FS2	Pro	Phe	Glu	Tyr	Arg	Tyr	Arg	Ser	Glu	Tyr	Gly	Glu	STOP						

Fig. 1. Primary structure of the Cal point mutants.
The primary structure of the wild-type (WT) Cal protein is indicated, with the consensus amino acids underlined. The numbering of the amino acids is that of the mature form of Cal. For the point mutations only the changed amino acid is shown, while the complete primary sequences of the S16 and S18 mutants are shown. These were constructed by the insertion of a TAA codon after residues 16 and 18 of Cal, respectively. At the bottom of the figure, the sequence of the carboxy-terminal part of the FS2 mutant is presented and compared with that of the wild type (Reprinted from Howard *et al.*, 1989).

The three major phenotypic effects of colicin operon induction which have been shown to be strictly dependent on lysis gene expression are a large but incomplete drop in the optical density of the colicinogenic culture (quasi-lysis), the concurrent release of large quantities of colicin and other soluble cell proteins, and a pronounced degradation of cellular lipids, caused by the activation of phospholipase A of the outer membrane.

Cells containing the mutant plasmids (encoding the mutant Cal proteins) were thus induced with mitomycin C (an inducer of SOS responses) and examined for these effects as a measure of the function of the various mutant proteins.

The results with the point mutants in the conserved region indicated that, despite a very high conservation of the amino-terminal aminoacid sequence among the colicin lysis proteins, the sequence specificity in this region is very relaxed. Not only could each of the conserved amino acids tested be replaced by another amino acid, but they could also be replaced by amino acids of entirely different chemical properties. Thus Arg 7 could be replaced by Val, an amino acid with a large hydrophobic side chain, while Val 14 could be replaced by the positively charged Arg, without detrimental effects on the function of the protein.

Among the substitutions tested, only the replacement of Arg 7 or Gly 11 by Glu had significant effects on the lytic activity of Cal. The pulse-chase analysis of these mutants indicated that the impairement of function was not caused by an inhibition of the normal modification and processing of these mutants (Howard *et al.*, 1989). These results suggest that the charge of the amino terminal region of Cal may play some role in its function. Although there is already an aspartate in this region, the addition of further negative charge appears to have an adverse effect. In contrast, the addition of positive charge in this region in the V1 mutant, and the presence of many positive and negative charges in the carboxy-terminal region of FS2 did not impair the function of Cal. It should also be mentioned that the lysis protein of colicin N contains an additional positive charge in the amino-terminal region relative to the other lysis proteins, and 5 additional positive charges in its carboxy-terminal region (Pugsley, 1988).

A possible role for the leader sequence of lysis proteins in the lytic function has been proposed on the basis of a loose homology with portions of phage lysis proteins (Lau *et al.*, 1987). Indeed the likelihood of this possibility was strengthened on the one hand by the observation that the signal sequences of Cal (Cavard *et al.*, 1985) and the Col E2 lysis protein (Pugsley and Cole, 1987) are apparently very stable, and on the other by the observations that when Cal or the Col E2 lysis protein were modified so that they could not be processed (preventing the release of the stable signal sequence), the lytic events following their induction were sharply attenuated or delayed (Cavard *et al.*, 1987; Pugsley and Cole, 1987). The results presented here however, argue strongly against this hypothesis, since the Cal mutants A2 and G4, while being impaired in the lytic function, were not impaired with respect to their processing, and thus accumulated the signal sequence just as the wild-type did. The question remains as to why the signal sequences of lysis proteins are so stable. It may be that for structural reasons the lysis protein signal sequences are not susceptible to cleavage by signal peptide peptidase, and another possibility is that they are for some reason not able to contact the peptidase in the envelope. A study on the properties of various lipoprotein signal sequences has indicated that the Cal signal sequence is more hydrophobic than those of other lipoproteins (Klein *et al.*, 1988), a factor which could be important in determining its stability.

The fact that the truncated proteins S16 and S18 were neither modified nor processed (Howard *et al.*, 1989) is very interesting. Like the mutant Cal proteins which were altered at the processing and cleavage site that we previously constructed (Cavard *et al.*, 1987), the S16 and S18 proteins were completely inactive. The ability to be modified and processed was entirely restored by the addition of an 11 amino acid carboxy-terminal sequence in the frameshift mutant FS2. This addition also restored the activity to the mutant protein. This result again supports the hypothesis that the modification is the key requirement for the activity of the lysis proteins. Moreover, this suggests that the carboxy-terminal sequence, not conserved among lysis proteins, functions only in the assembly of the nascent lysis protein polypeptide chains.

We do not yet know why the truncated Cal proteins are unable to be post-translationally processed while the full length protein can be. It is possible that the explanation lies in a coupling between synthesis and export of the lysis protein. About 35 amino acids are masked within the large subunit of the ribosome during translation (Bernabeu and Lake, 1982). When the polypeptide chain is truncated, as in S16 and S18, with the signal sequence of 18 amino acids, the nascent chains (totalling 34 or 36 amino acids) would not emerge from the ribosome and thus could not be accessible to a cellular machinery involved in export and modification. As a consequence, these nascent chains may be released into the cytoplasm. In contrast, the addition of a "spacer sequence" of 12 amino acids in FS2 may allow the nascent chain to emerge from the ribosome, and be recognized and inserted into the pathway of export and modification.

All of the results so far obtained indicate that aside from the structural features required for correct assembly, the modification of Cal by the addition of lipid to the amino terminal cysteine seems to be the most important if not the only requirement for quasi-lysis and the release of colicin A. In this context, it is interesting that a family of lipopeptides extracted from the culture media of various microorganisms have been shown to exhibit antibiotic properties. For example, iturin A and bacillomycin L and D produced by B. subtilis, and peptidolipin NA produced by Nocardia asteroides all have antifungal activity. These lipopeptides have been shown to dramatically increase the electrical conductance across planar lipid membranes (Maget-Dana *et al.*, 1985a, 1985b). Taking into account the large range of the conductance values, it has been proposed that local modifications in the structure of the bilayer may be induced by interaction with lipopeptide micelles (Maget-Dana *et al.*, 1985b). We have recently observed that the induction of Cal causes depolarization of the inner membrane in *E. coli* cells (manuscript in preparation). Disorganization of the lipid structure of the envelope may be the primary effect of Cal, with the activation of phospholipase A being a consequence of this perturbation. If this is the case it is not surprising that the modification of Cal by lipid, rendering it amphiphilic, is required for its function.

A critical concentration of Cal within the cells appear to be required to promote quasi-lysis. This has been demonstrated by constructing plasmids allowing high level expression of Cal. Two different types of such plasmids have been constructed. In the first one, pAT1, the T1 terminator located downstream of the colicin a gene (caa) and which normally arrests about 80% of the transcripts initiated at the promoter of the *caa-cal* operon (Lloubès *et al.*, 1988) was deleted. In the second one (designated pCK4), the *cal* gene was inserted downstream of the *tac* promoter of an expression vector (Cavard *et al.*, 1989b). These two plasmids were transformed into $PldA^+$ and PldA strains producing an active or inactive phospholipase A. Cal overproducing clones were induced in wild-type and *pld*A mutant cells. The decrease in absorbance after induction was similar with the various cultures, while the timing of this decrease differed. With the overproducing clones pAT1 and pCK4, the decrease in absorbance started after 70 minutes of induction almost one hour earlier than with the normal Cal producer (pColA9 cells). In each case, quasi-lysis was reduced for *pld*A cells, confirming the role of phospholipase activation in the process. We also observed that quali-lysis of lipoprotein-deficient cells was similar to that of lipoprotein-containing cells whatever the plasmid carried by the bacteria (Cavard *et al.*, 1989b). This suggests that the attachment of peptidoglycan to the outer membrane is not important for Cal function.

Cal overproduction not only caused early quasi-lysis but also caused early release of cellular proteins.

The fact that Cal must reach a critical concentration within the cells before occurence of quasi-lysis and colicin release again argues for a direct effect of Cal on the inner membrane. Local modifications in the structure of the bilayer may be induced by the interaction with lipopeptide micelles as previously proposed for Iturin A and bacillomycin L (Maget-Dana *et al.*, 1985b).

Phospholipase A may be located in contact sites between inner and outer membranes as reported by Bayer and Bayer (1985). The activation of this phospholipase following the local modification of the bilayer would have a synergic effect through lysophospholipids produced which have a detergent-like effect. This hypothesis is supported by the fact that the addition of Triton X-100 to *pld*A cells can supplement the lack of formation of lysophospholipids (Pugsley and Schwartz, 1984; Cavard *et al.*, 1989a). The combined effect of Cal and lysophospholipids would result in local membrane permeabilization without transient accumulation of colicin in the periplasmic space as demonstrated by electron microscope studies (Cavard *et al.*, 1984). To conclude, while the mechanism of colicin release to the extracellular medium is mostly elucidated and does not involve any topogenic export signal in the polypeptide chain of colicin A, the mechanism of entry of colicins into sensitive cells is more complex and does involve distinct colicin A domains.

Functional domains of colicin A

Colicins are bactericidal proteins produced by and active against *E. coli* and closely related bacteria. The molecular mechanisms involved in the binding of colicins to their specific receptors and their penetration across the cell envelope to bring them to their biochemical targets are not yet understood. The mode of action of colicins involves several steps of transfer across membranes. They are first produced in the cytoplasm (Varenne *et al.*, 1981) and then subsequently released to the extracellular medium through a mechanism that does not involve any specific region of the polypeptide chain (Baty *et al.*, 1987b). The mode of action of colicins appears to involve 3 steps: i) binding to a specific receptor located in the outer membrane; ii) translocation across the membrane(s); iii) biochemical interaction of the colicin with its target in the cell. Evidence has been put forward that there is a linear organization of three distinct domains along the polypeptide chain of several colicins: A (Martinez *et al.*, 1983; Crozel *et al.*, 1984; Baty *et al.*, 1987a), E1 (Ohno-Iwashita and Imahori, 1982; Cleveland *et al.*, 1983; Brunden *et al.*, 1984; Liu *et al.*, 1986), E2 (Ohno-Iwashita and Imahori, 1980; de Graaf and Oudega, 1986), E3 (Ohno-Iwashita and Imahori, 1980; de Graaf and Oudega, 1986), Ia and Ib (Mankovich *et al.*, 1986), N (Pugsley, 1988b) and cloacin DF13 (De Graaf *et al.*, 1978; Neville and Hudson, 1986). To each of these domains, a specific function has been assigned. The central domain of colicins appears to be involved in receptor binding, the NH2-terminal domain seems to be required for translocation across the outer membrane, and the COOH-terminal domain carries the lethal activity whether it carries an enzymatic (colicins E2, E3, Cloacin DF13) or an ionophoric activity (colicins A, E1, Ia, Ib, N). Most results obtained so far have been obtained by using limited digestion of colicin polypeptide chains. For colicin A for example, we have demonstrated that partial digestion with bromelain or thermolysin allows the isolation of a 20 kDa fragment endowed with ionophoric activity (Martinez *et al.*, 1983; Tucker *et al.*, 1986).

We have constructed many proteins derived from colicin A which lack various regions of the polypeptide chain. These proteins have been purified and their ability to bind to the receptor [constituted by the two proteins OmpF and BtuB (Cavard and Lazdunski, 1981)] and to form ion channels has been tested. All of the constructed proteins were released to the medium and were sufficiently stable to be purified. These proteins were then assayed for their in vivo and in vitro ionophoric activity and for their ability to bind to the colicin A receptor (Fig. 2).

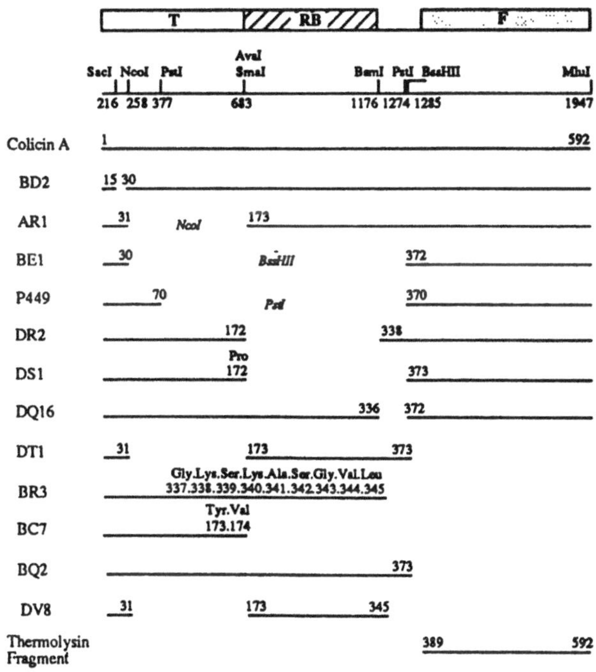

Constructions	Receptor Binding	Activity in vivo	Activity in vitro	Number of A.A.
WT	+	+	+	592
SacI(K)-NcoI(K)	+	--	+	578
NcoI(K)-SmaI	+	--	+	451
NcoI(K)-BssHII(K)	--	--	+	251
PstI-PstI	--	--	+	293
AvaI(K)-BsmI(K)	--	--	+	427
SmaI-BssHII(K)	--	--	+	392
BsmI(K)-BssHII(K)	+	+	+	557
NcoI(K)-SmaI	+	--	--	232
BsmI(K)-SnaBI	+	--	--	345
AvaI(K)-MluI(K)	--	--	--	174
BssHII(K)-SnaBI	+	--	--	373
NcoI(K)-SmaI	+	--	--	204
thermolysin digestion	--	--	+	204

Fig. 2. Functional domains of colicin A.
The functional domains, translocation (T), receptor-binding (RB) and ionophoric activity (F) are indicated at the top of the figure. The numbers of the first nucleotide of the unique enzyme restriction sites are indicated. The details of the constructions are indicated in the first column. WT is the wild-type colicin A, and (K) indicates that the restriction site has been filled in by the Klenow fragment of DNA polymerase I. The amino acid number at the edge of the deletions (represented by broken lines) is indicated. The total number of amino acids for each partially deleted colicin A is indicated in the last column. For BC7 and BR3, the amino acids from the frameshift created by the construction are indicated. + indicates that the protein was active or bound to the receptor, as indicated at the tops of the columns (Reprinted from Baty *et al.*, 1988).

Colicin A derivatives can kill sensitive cells provided that they are able to bind to the receptor, to be translocated across the outer membrane and to depolarize the inner membrane. In contrast, even if they cannot fulfil these requirements, they should be able to form channels in planar lipid bilayers provided that they carry the pore forming region. Only protein DQ16, which lacks the region between residues 336 to 372, was as active in vivo as the wild-type colicin A (Fig.2). This result suggests that this region is not implicated in any of the three steps of the mode of action of colicin A. The other proteins were not active at all in vivo, which suggests that one or several steps were affected. For this reason, it was necessary to study each step of the mode of action individually.

All of the colicin A derivatives carrying at least residues 389 to 592 exhibited ionophoric activity in phospholipid planar bilayers (Fig.2), as did the thermolysin fragment (Tucker *et al.*, 1986). However, most of them (BD2, AR1, BE1, P449, DR2 and DS1), were unable to kill sensitive cells in vivo because one or both of the two first steps of the mode of action were affected. Various regions from residues 1 to 388 have been deleted in these proteins. These deletions did not change the single channel conductance (i. e. the size of the pore was the same as that formed by the wild type colicin A). Proteins lacking the COOH-terminal domain (BC7, BR3, DV8, BQ2 and DT1) were neither active in vivo nor in vitro. These results are in agreement with the fact that the COOH-terminal domain (residues 389 to 592) of colicin A contains the pore forming activity (Martinez *et al.*, 1983; Tucker *et al.*, 1986). The rest of the molecule has an effect on the voltage dependence of the pore (Martinez *et al.*, 1983; Baty *et al.*, 1987a; Collarini *et al.*, 1987).

Proteins which carry the receptor binding domain but which cannot kill sensitive cells should protect these cells from the action of wild-type colicin A by competing for binding to the receptor. Using this test, we observed that proteins BD2, AR1, BR3, DV8, BQ2 and DT1 could provide full protection against colicin A, while BE1, P449, DR2, DS1, BC7 and the thermolysin fragment could not. From the results, it could be deduced that the receptor binding domain must be located between residues 173 and 336 (Baty *et al.*, 1988). The absence of this domain in DS1 and DR1 was verified in bypass experiments in which we found that both were able to kill BtuB mutant cells (H.Benedetti, unpublished result).

The localization of the translocation domain can be inferred from the results presented above. Proteins BD2 and AR1 having both receptor binding and ionophoric activity (in vitro) were unable to kill sensitive cells in vivo. Therefore, this suggested that the region between residues 1 and 172 is responsible for the translocation of colicin A to the inner membrane, although due to the size of the deletions the results must be carefully interpreted.

These studies allowed us to roughly define the domains involved in the various steps of colicin A action on sensitive cells. The NH2-terminal domain, involved in translocation, may comprise residues 1 to 173. The central domain, involved in receptor binding, appears to contain residues 173 to 336, and the ionophoric activity is contained in a region encompassing residues 389 to 592. However, it should be pointed out that these are maximum estimates, and the receptor binding and ionophoric domains may be shorter. It is also difficult to give a definite size for the translocation domain which may be shorter or may contain a sketch downstream from residue 173. The protein BD2, which lacked only 14 residues (residues 16 to 29) in the NH2-terminal domain and yet had lost its activity. This suggests that some "targeting" sequence involved in recognition of envelope proteins in the sensitive cells may be contained in this domain. With regard to this point it is of interest that some point mutations in the NH2-terminal domain of colicin E3 (Escuyer and Mock, 1987) and cloacin DF13 (Verschoor *et al.*, 1988) reduced the translocation of these proteins into susceptible cells.

Finally, it is of interest that none of the colicins isolated so far contains a disulfide bridge. This may allow more flexibility for partial unfolding and refolding that may occur during uptake into sensitive cells.

Roles of BtuB and OmpF proteins in colicin entry into sensitive cells

The receptor of colicin A is composed of two proteins BtuB and OmpF (Cavard and Lazdunski, 1981). BtuB is a receptor shared by many colicins namely, colicins : A, E1, E2, E3, E4, E5, E6, E7 (Mock and Pugsley, 1982). However, differences in the colicin receptor activity of BtuB for various colicins within this group have previously been noted. Outer membranes of sensitive cells inhibited colicin E1 through E7 activities while they have no effect on colicin A. Furthermore this last colicin could only kill cells which produced also the OmpF porin (Pugsley and Schnaitman, 1979). This result suggested that BtuB and OmpF although they are both required as colicin A receptor may play different roles in colicin uptake.

It is possible to distinguish the receptor binding step and the translocation step. The receptor step can be bypassed in sensitive or resistant cells (altered in their receptor) under conditions of osmotic shock (Tilby *et al.*, 1978) or at low ionic strength by a milder treatment named "bypass conditions" (Cavard and Lazdunski, 1981). Using this technique, the respective roles of OmpF and BtuB for binding and translocation of colicins A, E1, E2 and E3 have been investigated. OmpF was found to be required for translocation of colicins A, E2 and E3 but not for colicin E1. For this latter and for colicins E2 and E3, BtuB was required for both binding and translocation (H. Benedetti, unpublished result).

Role of Tol proteins and of the N-terminal region of colicin A

After the initial binding of colicins to the receptors, the occurence of a second step in the entry mechanism is strongly suggested by the existence of mutants which although they have normal receptors are not killed by colicins. These mutants are named "tolerant". A group of proteins, encoded by the *tolQRAB* gene cluster, are necessary for the entry of colicin A and other colicins (the so-called "type A" colicins) from the surface of the cells to their respective targets (Hill and Holland, 1967; Nagel del Zwaig and Luria, 1967; Nomura and Witten, 1967; Reeves, 1966; Sun and Webster, 1986). Some of these proteins are also required for the transport of the genome of filamentous phages into the cytoplasm (Sun and Webster, 1986; Sun and Webster, 1987). However, the role of these proteins in the mechanism of entry of colicins and DNA remains unknown. The Tol proteins probably play an important physiological role in the bacteria because the permeability of the outer membrane is very much altered in *Tol* mutants which are detergent and drug sensitive (Nagel del Zwaig and Luria, 1967); they also release periplasmic proteins to the extracellular medium (Lazzaroni and Portalier, 1981).

As mentioned above, after the initial adsorption, the N-terminal region of colicins is probably involved in translocation. Thus, both Tol proteins and the N-terminal domain of colicins appear to play a role in the second step of the entry mechanism. In contrast to the first step, this second step cannot be bypassed (Bourdineaud *et al.,* 1989). Previous experiments have demonstrated the possibility of killing *omp*F cells with colicin A (Cavard and Lazdunski, 1981). We again used an *omp*F strain as a positive control for the bypass and compared the action of wild-type colicin A, colicin AR1, colicin BD2 and a C-terminal thermolysin fragment (see Fig. 2 for description of these proteins). The results show that when bypassing the outer membrane (i) *omp*F bacteria are not killed by either the deletion derivative AR1 or the colicin A thermolysin fragment; (ii) *omp*F bacteria are killed by wild-type colicin A and to a much lower extent by the BD2 mutant and (iii). Tolerant mutant cells (TPS 13) altered in TolQ are not killed by colicin A. These results thus indicate that parts of colicin A molecule (in the N-terminal region) and the TolQ protein (or an apparatus for which its presence is important) must productively interact to insure translocation to the inner membrane, even if the outer membrane barrier is bypassed.

A model for colicin entry

Since OmpF is necessary for both adsorption and translocation of colicin A as mentioned above but is not required for translocation colicin E1 which is TolB independent, it is tempting to suggest the existence of a complex between receptor proteins and Tol proteins as shown in Fig. 3. Because BtuB protein is required for both adsorption and translocation of colicin E1 which also requires TolC (Davies and Reeves, 1975), we have hypothesized an interaction between BtuB, TolC and TolA. The *Tol* A, B, Q, R genes have now been sequenced and preliminary localization suggests that at least TolA and TolB are partly exposed to the periplasm (Sun and Webster, 1987). More recent experiments from our laboratory indicate that the N-terminal region of TolQ is also exposed to the periplasmic space. In addition, we have obtained preliminary evidence that TolQ is located in adhesion zones between inner and outer membranes (Bourdineaud *et al.*, 1989). We thus hypothesize that TolA, B, Q, R form a complex and that this complex is located in adhesion zones. An analogy may exist between the TonB system, required for entry of "type B" colicins (B, Ia, Ib) and the Tol system necessary for entry of colicins A, E1, E2, E3. A direct interaction between TonB and the BtuB receptor in the outer membrane is strongly suggested by genetic evidence (Heller *et al.*, 1988).

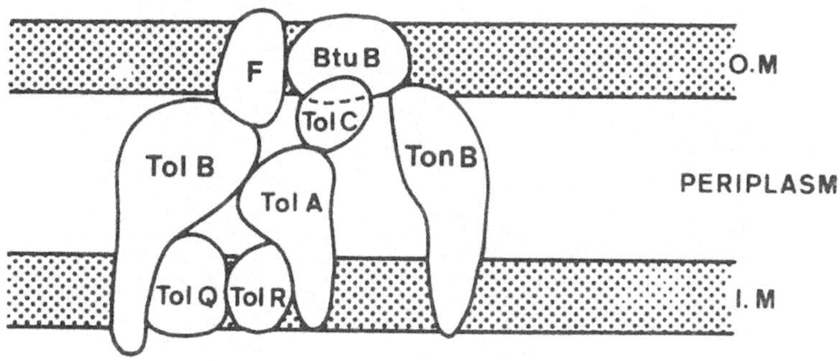

Fig. 3 Hypothetical interactions of proteins involved in colicin entry into sensitive cells.

The three hypothetical steps in colicin entry are summarized in Fig. 4. Colicins would first interact with the receptor (step 1). This adsorption would promote a conformational change in the polypeptide chain making the N-terminal region available for interacting with Tol proteins (step 2). There are evidences for such a productive interaction (Bourdineaud *et al.*, 1989; M. Frenette, H. Benedetti, D. Baty and C. Lazdunski, unpublished results).

In contrast to common belief, we have recently demonstrated that translocation across the outer membrane requires a fluid state but is not energy-dependent. To demonstrate this, kinetics of K^+ efflux caused by colicin A in *E. coli* sensitive cells have been extensively investigated using a K^+ selective electrode (J.P. Bourdineaud P. Boulanger, C. Lazdunski, L. Letellier, submitted for publication). The dependence of K^+ efflux upon multiplicity, pH, temperature and membrane potential ($\Delta\psi$) was determined. The translocation of colicin A from the outer membrane receptor to the inner membrane and insertion into the latter required a fluid membrane but once inserted, the channel properties showed little dependence upon the state of the lipids. At a given multiplicity, the lag time before the onset of K^+ efflux was found to reflect the time required for translocation and/or insertion of colicin into the cytoplasmic membrane. Conditions were designed for closing and reopening of the channel *in vivo*. These conditions allowed us to test separately the $\Delta\psi$ requirement for translocation and channel opening : translocation was shown not to require $\Delta\psi$.

In the last step (step 3), colicin A (the prototype of pore-forming colicins) **would be** inserted into the membrane even though it will spontaneously insert into a bilayer *in vitro* (see below).

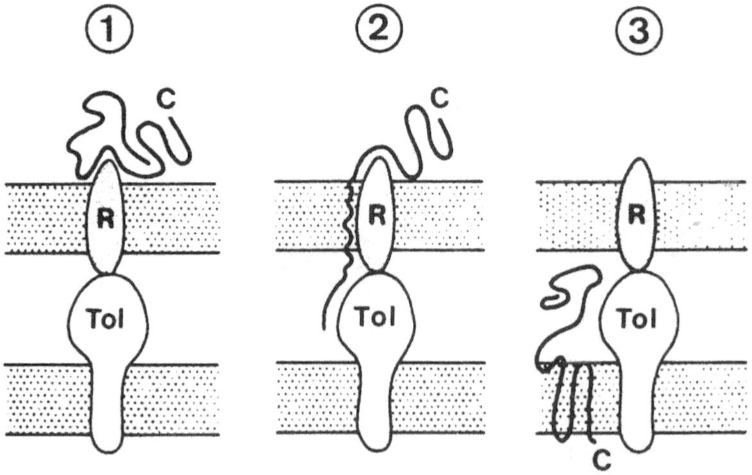

Fig. 4 The three putative stages involved in colicin A entry and insertion into the cytoplasmic membrane.

Insertion of the ionophoric domain of colicin A into the cytoplasmic membrane

The paradox that a colicin can exist as a water-soluble protein and as a membrane protein was resolved by the structure of the pore-forming fragment of colicin A which has been determined by X-ray cristallography and refinement at 2.5 Å resolution (Tucker *et al.*, 1989; Parker *et al.*, 1989). The fragment (204 residues) consists of ten α-helices organized in a three layer structure. Two of the helices are completely buried within the structure and form a hydrophobic hairpin loop similar to that proposed for signal sequences which function in translocation (Fig. 5). Such an hydrophobic strech of residues between 35 and 49 amino acids long exist in all pore-forming colicins related to colicin A (Pattus *et al.*, 1985; Lazdunski *et al.*, 1988). In the three-dimensional structure, the hydrophobic helical hairpin (helices 8 and 9 in Fig. 5) is completely buried within the protein. Thus residues that interact with lipid in the membrane-bound form of the protein are buried within the protein conformation of the water-soluble form.

The mechanism of insertion of colicin A into the lipid bilayer of the cytoplasmic membrane can probably be divided into two steps : binding to the membrane surface and insertion of the protein into the hydrophobic core of the bilayer. The first step is probably electrostatic in nature, whereas the second may be driven by the partioning of hydrophobic residues into the lipid environment of the bilayer. Colicin A and its thermolytic fragment do not bind to neutral bilayers but display a great affinity for negatively charged lipids (Pattus *et al.*, 1983a; Frenette *et al.*, 1989). As mentioned above, the colicin A polypeptide chain (592 aminoacid residues) contains 3 domains which are linearly organized and participate in the sequential steps involved in colicin entry and action. We have compared the penetrating ability in phospholipid monolayers of colicin A and of protein derivatives containing various combinations of its domains. The NH_2-terminal domain (171 amino acid residues), required for translocation across the outer membrane has little affinity for dilauroyl phosphatidyl glycerol (DLPG) monolayers at all pHs tested. The central domain has a pH-dependent affinity, although lower than that of the entire colicin A. The COOH-terminal domain contains a high affinity lipid binding site, but in addition an electrostatic interaction is required as a first step in the process of penetration into negatively charged DLPG films.

Analysis of the distribution of the surface side chains of the pore-forming peptide structure shows that they are randomly distributed with respect to various potential interaction parameters such as charge, hydrophobicity and aromaticity. The exception to this finding occurs at the surface defined by the loop that connects helices 8 and 9 in the structure (Fig. 5). This loop is surrounded by a ring of eight positively charged side chains (indicated by small arrowheads in Fig. 5). The interface defined by this loop, featuring a mean positive charge, this explains the affinity of colicin A and its thermolytic fragment for negatively charged lipids.

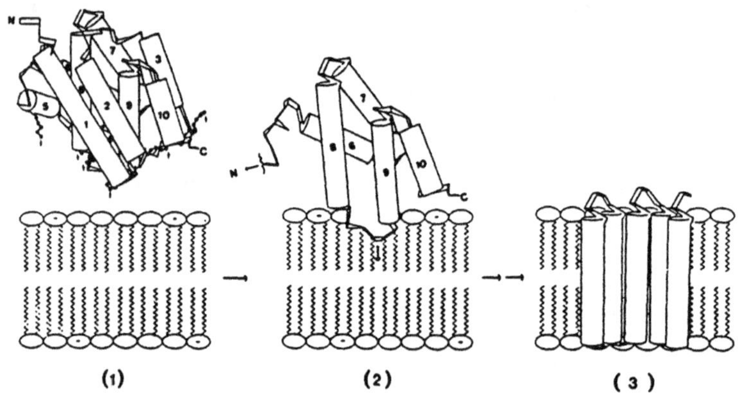

Fig. 5 A model for protein insertion into membranes suggested by the structure of the pore-forming fragment of colicin A. (1) Initial interaction with the membrane occurs as a result of the long range interaction of an electrostatic potential field and shorter-range interactions such as salt bridges. The first step orientates the hydrophobic helical hairpin (helices 8 and 9) so that its axis is perpendicular to the membrane surface. (2) The hairpin spontaneously penetrates into the membrane, and in doing so initiates a conformational change in the protein. (3) It is envisaged that further steps involve insertion of other hairpins which may include oligomerization of the protein molecule. These final steps would lead to the formation of a voltage-gated channel (Reprinted from Parker *et al.*, 1989).

The proposed orientation of binding of the pore-forming domain positions the hydrophobic helical hairpin directly above the membrane surface with helices 8 and 9 sitting perpendicular to the plane of the bilayer. It is likely that at least part of the role of the Tol proteins consists in imposing such an orientation (see Fig. 4). The positioning of the helical hairpin and the hydrophobic nature of the exposed loop connecting the two helices together with the loss of solvatation at the peptide-membrane interface on binding promotes the insertion of the hairpin into the hydrophobic core of the bilayer. This should constitute a first energy independent stage in the insertion of the thermolytic fragment. This stage may constitute the closed form of the channel. Studies on channel functioning in lipid planar bilayers indicated that a membrane potential above 60 mV is required to obtain the open form of the channel (Pattus *et al.*, 1983; Collarini *et al.*, 1987). Using a K^+ selectrode, it has recently been demonstrated that a 85 mV membrane potential across the cytoplasmic membrane of sensitive cells is required for channel opening (Bourdineaud, Boulanger, Lazdunski and Letellier, submitted for publication). Upon switching from the closed form to the open form, the structure of the ionophoric domain must undergo significant conformational changes which is probably triggered by the membrane potential. In this process other helices in addition to the hydrophobic helices 8 and 9 get inserted probably in a pairwise fashion and without large change in secondary structure conformation since the α-helical content of the water soluble form and lipid-bound form of the peptide fragment is identical as estimated by circular dichroism (Pattus *et al.*, 1985). An oligomerization process could have an important role in this insertion of other helical hairpins by shielding their polar faces from contact with the hydrophobic core of the membrane and forming the water-filled pore of the voltage-gated channel in doing so (Pattus *et al.*, 1985; Baty *et al.*, 1987a; Lazdunski *et al.*, 1988). Current work is now directed towards determining the structure of the membrane-bound form of colicin A to understand further the mechanism of insertion and to explain the pore properties of the voltage-gated channel formed.

Acknowledgements

I am greatly indebted to all coworkers cited in the references, who have contributed in Marseille, at the E.M.B.L. and at the University Paris-Sud, over the past 10 years to studies described in this review article. This work was supported by the C.N.R.S., the I.N.S.E.R.M., the M.R.T. and the Fondation pour la Recherche Médicale. I am grateful to M. Payan for preparing the manuscript.

REFERENCES

Baty D, Knibiehler M, Verheij H, Pattus F, Shire D, Bernadac A, Lazdunski C (1987a) Site-directed mutagenesis of the COOH-terminal region of colicin A : effect on secretion and voltage-dependent channel activity. Proc Natl Acad Sci USA 84:1152-1152

Baty D, Lloubès R, Geli V, Lazdunski C, Howard SP (1987b) Extracellular release of colicin A is non-specific. EMBO J 6:2463-2468

Baty D, Frenette M, Lloubes R, Geli V, Howard SP, Pattus F, Lazdunski C (1988) Functional domains of colicin A. Mol Microbiol 2:807-811

Bayer MH, Bayer ME (1985) Phosphoglycerides and phospholipase C in membrane fractions of *Escherichia coli*. J Bacteriol 162:50-54

Bernabeu C, Lake JA (1982) Nascent polypeptide chains emerge from the exit domain of the large ribosomal subunit. Proc Natl Acad Sci USA 79:3111-3115

Bourdineaud JP, Howard SP, Lazdunski C (1989) Localization and assembly into the *Escherichia coli* envelope of a protein required for entry of colicin A. J Bacteriol 171: in press

Braun V, Hantke K (1977) In "Microbial Interactions", JL Reissing (ed), Chapman and Hal : London, p 101-137

Brunden KR, Cramer WA, Cohen FS (1984) Purification of a small receptor binding peptide from the central region of the colicin E1 molecule. J Biol Chem 259:190-196

Cavard D, Lazdunski C (1981) Involvment of BtuB and OmpF proteins in binding and uptake of colicin A. FEMS Microbiol Lett 12:311-316

Cavard D, Bernadac A, Pagès JM, Lazdunski C (1984) Colicins are not transiently accumulated in the periplasmic space before release from colicinogenic cells. Biol Cell 51:79-86

Cavard D, Lloubès R, Morlon J, Chartier M, Lazdunski C (1985) Lysis protein encoded by plasmid ColA-CA31. Gene sequence and export. Mol Gen Genet 199:95-100

Cavard D, Baty D, Howard SP, Verheij HM, Lazdunski C (1987) Lipoprotein nature of the colicin A lysis protein : effect of amino acid substitution at the site of modification and processing. J Bacteriol 169:2187-2194

Cavard D, Howard SP, Lazdunski A (1989a) Functioning of the colicin A lysis protein is affected by Triton X-100 divalent cations and EDTA. J Gen Microbiol 135: in press

Cavard D, Howard SP, Lloubès R, Lazdunski C (1989b) High-level expression of the colicin A lysis protein. Mol Gen Genet : in press

Cleveland M, Slatin S, Finkelstein A, Levinthal S (1983) Structure-function relationships for a voltage-dependent ion channel : properties of COOH-terminal fragments of colicin E1. Proc Natl Acad Sci USA 80:3706-3710

Cole ST, Saint-Joanes B, Pugsley AP (1985) Molecular characterization of the colicin E2 operon and identification of its products. Mol Gen Genet 198:465-472

Collarini M, Amblard G, Lazdunski C, Pattus F (1987) Gating processes and channels induced by colicin A, its C-terminal fragment and colicin E1 in planar bilayers. Eur Biophys J 14:147-153

Crozel V, Lazdunski C, Lloubès R, Cavard D (1984) A colicin A fragment containing the receptor binding domain can be directed to the periplasmic space in *E. coli* through gene fusion. FEBS Lett 172:183-188

Datta DB, Arden B, Henning U (1977) Major proteins of the *Escherichia coli* outer cell envelope membrane as bacteriophage receptors. J Bacteriol 131:821-829

Davies JK, Reeves P (1975) Genetics of resistance to colicins in *Escherichia coli* K12 : cross-resistance among colicins of group B. J Bacteriol 123:96-101

De Graaf FK, Stukart MJ, Boogerd FC, Metselaar K (1978) Limited proteolysis of cloacin DF13 and characterization of the cleavage products. Biochemistry 17:1137-1142

De Graaf FK, Oudega B (1986) Production and release of cloacin DF13 and related colicins. Curr Top Microbiol Immunol 125:183-205

Ebina Y, Nakazawa A (1983) Cyclic AMP-dependent initiation and ρ-dependent termination of colicin E1 gene transcription. J Biol Chem 258:7072-7078

Escuyer B, Mock M (1987) DNA sequence analysis of three muissense mutations affecting colicin E3 bactericidal activity. Mol Microbiol 1:82-85

Frenette M, Knibiehler M, Baty D, Geli V, Pattus F, Verger R, Lazdunski C (1989) Interactions of colicin A domains with phospholipid monolayers and liposomes: relevance to the mechanism of action. Biochemistry 28:2509-2514

Hakkart MJ, Veltkamp E, Nijkamp HJ (1981) Protein H encoded by plasmid Clo DF13 involved in lysis of the bacterial host. I. Localisation of the gene and identification and subcellular localisation of the gene H product. Mol Gen Genet 183:318-325

Hantke K (1976) Phage T6-colicin K receptor and nucleotide transport in *Escherichia coli.* FEBS Lett 70:109-112

Hantke K, Braun V (1975) A function common to iron-enterochelin transport and action of colicins B, I, V in *Escherichia coli.* FEBS Lett 59:277-281

Hantke K, Braun V (1975) Membrane receptor dependent iron transport in *Escherichia coli.* FEBS Lett 49:301-305

Heller KJ, Kadner RJ, Günther K (1988) Suppression of the *btuB*451 : mutations in the *ton*B gene suggests a direct interaction between TonB and TonB-dependent receptor proteins in the outer membrane of *Escherichia coli.* Gene 64:147-153

Hill C, Holland IB (1967) Isolation and properties of colicin refractory mutants and the preliminary mapping of their mutations. J. Bacteriol 94:677-686

Hollifield WC, Neilands JB (1978) Ferric enterobactin transport system in *Escherichia coli* K12. Extraction, assay and specificity of outer membrane receptor. Biochemistry 17:1922-192

Howard SP, Cavard D, Lazdunski C (1989) Amino acid sequence and length requirements for assembly and function of the colicin A lysis protein. J Bacteriol 171:410-418

Kadner RJ, Bassford PJ, Pugsley AP (1979) Colicin receptors and mechanism of colicin uptake. Zentralbl Bakteriol Parasitenkd Infektionskr Hy Abt 244:90-104

Klein PR, Somorjai RI, Lau P (1988) Distinctive properties of signal sequences from bacterial lipoproteins. Protein Eng 2:15-20

Konisky J (1982) Colicins and other bacteriocins with established modes of action. Ann Rev Microbiol 36:125-144

Lau PC, Hefford MA, Klein P (1987) Structural relatedness of lysis proteins from colicinogenic plasmids and icosahedral coliphages. Mol Biol Evol 4:544-556

Lazdunski C, Baty D, Geli V, Cavard D, Morlon J, Lloubès R, Howard SP, Knibiehler M, Chartier M, Varenne S, Frenette M, Dasseux JL, Pattus F (1988) The membrane channel-forming colicin A : synthesis, secretion, structure, action and immunity. Biochim Biophys Acta 947:445-464

Lazzaroni JC, Portalier RC (1981) Genetic and biochemical characterization of periplasmic-leaky mutants of *Escherichia coli* K12. J Bacteriol 145:1351-1358

Liu QR, Crozel V, Levinthal F, Slatin S, Finkelstein A, Levinthal C (1986) A very short peptide makes a voltage-dependent ion channel : the critical length of the channel domain of colicin E1. Proteins 1:218-229

Lloubès R, Baty D, Lazdunski C (1988) Transcriptional terminators in the *caa-cal* operon and *cai* gene. Nucleic Acids Res 16:3739-3749

Lugtenberg B, van Alphen L (1983) Molecular architecture and functioning of the outer membrane of *Escherichia coli* and other gram-negative bacteria. Biochim Biophys Acta 737:51-115

Luirink J, van der Sande C, Tommassen J, Veltkamp E, de Graaf FK, Oudega B (1986) Effects of divalent cations and of phospholipase A activity on excretion of cloacin DF13 and lysis of host cells. J Gen Microbiol 132:825-834

Maget-Dana R, Heitz F, Ptak M, Peypoux F, Guinand M (1985a) Bacteria lipopeptides induce ion-conducting pores in planar bilayers. Biochem Biophys Res Commun 129:965-971

Maget-Dana R, Ptak M, Peypoux F, Michel G (1985b) Pore-forming properties of iturin A, a lipopeptide antibiotic. Biochim Biophys Acta 815:405-409

Mankovich JA, Hsu CH, Konisky J (1986) DNA and aminoacid sequence analysis of structural and immunity genes of colicins Ia and Ib. J Bacteriol 168:228-236

Martinez MC, Lazdunski C, Pattus F (1983) Isolation of molecular and functional properties of the C-terminal domain of colicin A. EMBO J 2:1501-1507

Mock M, Pugsley A (1982) The BtuB group of Col plasmids and homology between colicins they encode. J Bacteriol 150:1069-1076

Nagel del Zwaig R, Luria JE (1967) Genetics and physiology of colicin-tolerant mutants of *Escherichia coli*. J Bacteriol 94:1112-1123

Neville DM, Hudson TH (1986) Transmembrane transport of diphteria toxin, related toxins and colicins. Ann Rev Biochem 55:195-224

Nikaido H, Vaara M (1985) Molecular basis of bacterial outer membrane permeability. Microbiol Rev 49:1-32

Nomura M, Witten C (1967) Interactions of colicins with bacterial cells III. Colicin-tolerant mutations in *Escherichia coli*. J Bacteriol 94:1093-1111

Ohno-Iwashita Y, Imahori K (1980) Assignment of the functional loci in colicin E1 and E3 molecules by the characterization of their proteolytic fragments. Biochemistry 19: 652-659

Ohono-Iwashita Y, Imahori K (1982) Assignment of the functional loci of the colicin E1 molecule by characterization of their proteolytic fragments. J Biol Chem 257:6446-6451

Oudega B, Ykema A, Stegehnis F, de Graaf FK (1984) Detection and subcellular localization of mature protein H involved in excretion of cloacin DF13. FEMS Microbiol Lett 22:101-109

Parker MW, Pattus F, Tucker AD, Tsernoglou D (1989) Structure of the membrane-pore-forming fragment of colicin A. Nature 337:93-96

Pattus F, Martinez MC, Dargent B, Cavard D, Verger R, Lazdunski C (1983a) Interaction of colicin A with phospholipid monolayers and liposomes. Biochemistry 22:5698-5703

Pattus F, Cavard D, Verger R, Lazdunski C, Rosenbuch J, Schindler H (1983b) Formation of voltage-dependent pores in planar bilayers by colicin A. In : Physical Chemistry of transmembrane ions notions. Spach G (ed) Amsterdam Elsevier Biochemical Press, pp 407-413

Pattus F, Heitz F, Martinez C, Provencher SW, Lazdunski C (1985) Secondary structure of the pore-forming colicin A and its C-terminal fragment. Eur J Biochem 152:681-689

Pugsley AP, Schnaitman CA (1979) Factors affecting the electrophoretic mobility of the major outer membrane proteins of *Escherichia coli* in polyacrylamide gels. Biochim

Biophys Acta 581:163-178

Pugsley AP, Schwartz M (1984) Colicin E2 release : lysis, leakage or secretion ? Possible role of a phospholipase. EMBO J 3:2393-2397

Pugsley AP, Cole ST (1987) An unmodified form of the ColE2 lysis protein, an envelope lipoprotein, retains reduced ability to promote colicin E2 release and lysis of producing cells. J Gen Microbiol 133:2411-2420

Pugsley AP (1988a) The immunity and lysis genes of Col N plasmid pCHAP4. Mol Gen Genet 211:335-341

Pugsley AP (1988b) Nucleotide sequencing of the structural gene for colicin N reveals homology between the catalytic-terminal domains of colicins A and N. Mol Microbiol 1:317-325

Pugsley A, Reeves P (1977) The role of colicin receptors in the uptake of ferrienterochelin by *Escherichia coli* K12. Biochem Biophys Res Commun 74:903-911

Reeves P (1966) Mutants resistant to colicin A42-E2 cross resistance and genetic mapping of a special class of mutants. Aust J Exp Med Sci 44:301-316

Sun TP, Webster RE (1986) *fii* (*tolQ*) a bacterial locus required for filamentous phage infection and its relation to colicin-tolerant *tol*A *tol*B. J Bacteriol 165:107-115

Sun TP, Webster RE (1987) Nucleotide sequence of a gene cluster involved in the entry of the E colicins and the single stranded DNA of infecting filamentous phage into *Escherichia coli.* J Bacteriol 169:2667-2674

Tilby M, Hindennach I, Henning U (1978) Bypass of receptor-mediated resistance to colicin E3 in *Escherichia coli* K12. J Bacteriol 136:1186-1191

Toba M, Masaki H, Ohta T (1986) Primary structures of ColE2-P9 and ColE3-CA38 lysis genes. J Biochem 99:591-596

Tommassen J, Lugtenberg B (1980) Outer membrane protein of *Escherichia coli* is co-regulated with alkaline phosphatase. J Bacteriol 143:151-157

Tucker AD, Pattus F, Tsernoglou D (1986) Cristallization of the C-terminal domain of colicin A carrying the voltage-dependent pore activity of the protein. J Mol Biol 190:133-134

Tucker AD, Baty D, Parker MW, Pattus F, Lazdunski C, Ternoglou D (1989) Cristallographic phases through genetic engineering : experiences with colicin A. Prot Eng 2:399-405

Varenne S, Cavard D, Lazdunski C (1981) Biosynthesis and export of colicin A in *Citrobacter freundii* CA31. Eur J Biochem 116:615-620

Verschoor EJ, Luirink J, de Graaf FK, Oudega B (1988) Characterization of a mutation in the cloacin structural gene causing a reduced uptake of cloacin DF13 by susceptible cells. FEMS Microbiol Lett 49:403-409

Wayne R, Neilands JB (1975) Evidence for common binding sites for ferrichrome compounds and bacteriophage Ø 80 in the cell envelope of *Escherichia coli.* J Bacteriol 121:459-503

PROTEIN TRANSPORT INTO AND INSIDE THE CHLOROPLAST

Peter Weisbeek
Department of Molecular Cell Biology
Institute of Molecular Biology
University of Utrecht
Padualaan 8, 3584 CH Utrecht
The Netherlands.

The chloroplast is a complex cellular compartment. It carries out a wide range of metabolic activities, like light capture, photosynthetic CO2 fixation and fatty acid biosynthesis, and it does so in an organelle with an amazingly complex architecture formed by three separate membrane systems which enclose three distinct soluble phases (see Fig. 1).In addition to that the chloroplast contains so little genetic information that it dependends for most of its functions on a large number (>1000) of proteins (and possibly other macromolecules) synthesized outside the organelle which means that there is import of many different proteins in large numbers. Inside the chloroplast these proteins and the ones that are coded by the chloroplast DNA have to find their functional place within this complex total structure of the chloroplast. This requires for many proteins that again membranes have to be recognized and crossed. This problem of recognizing and discriminating between different membranes (e.g. thylakoid and inner membrane) inside the chloroplast can be compared with the task of the cytoplasmic precursor to find the correct organelle, i.e. discriminating between the ER membranes and mitochondrial and chloroplast envelope mebranes. For certain proteins, e.g. the lumen proteins, the same protein has to interact specifically with more than one membrane.

These aspects make the chloroplast one of the most interesting and promising structures for the analysis of how proteins interact with and are translocated across biological membranes.

There are only three sites of protein synthesis in the plant cell; in the cytoplasm, in chloroplasts and in mitochondria. All proteins synthesized inside the mitochondria or chloroplasts remain in these organelles and therefore all proteins found in the other cellular compartments are derived from the cytosol and are nuclear encoded. Proteins synthesized in the cytosol are the major source of transported and secreted proteins.

Mitochondria and chloroplasts are able to synthesize proteins, but the size of the DNA contained in these organelles and therefore their maximal protein-coding capacity is ten to one hundred times lower than what is needed for their functioning. The proteins needed but not synthesized in these organelles are coded for by the nuclear DNA, synthesized in the cytosol, and transported into the organelles. The population of proteins present in

NATO ASI Series, Vol. H 40
Dynamics and Biogenesis of Membranes
Edited by J.A.F. Op den Kamp

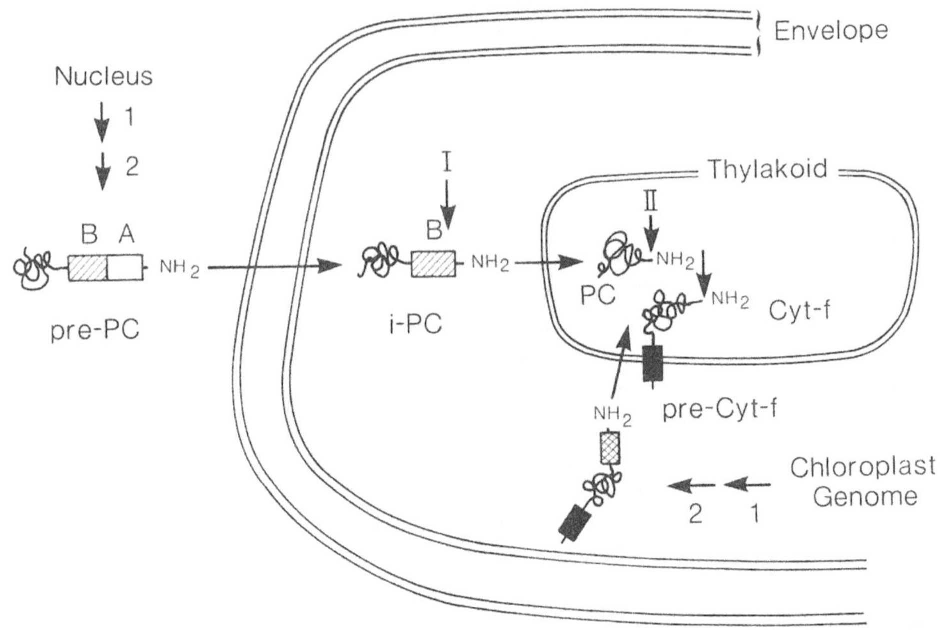

Fig. 1. Routing of the thylakoid lumen specific plastocyanin (PC) and cytochrome f (cyt-f) precursor proteins, encoded by the nuclear- and plastid genomes, respectively. Pre-PC, precursor; i-PC, intermediate; A, chloroplast import domain; B, thylakoid transfer domain. Roman numerals I and II indicate stroma- and thylakoid-associated processing activities, respectively. Transcription (1) and translation (2) are indicated by arrows. The cross hatched and black boxes in the cytochrome f precursor indicate the signal sequence and the stop transfer sequence, respectively.

chloroplasts and mitochondria is therefore of mixed genetic origin; one part is encoded by organellar DNA and the other, much larger part, by nuclear DNA.

Proteins that are functional in other cellular compartments then where they are synthesized must contain information for compartmentalization. In addition, mechanisms are necessary that respond specifically to this information by transporting such proteins toward their proper compartment.

Most proteins found in the chloroplast are synthesized in the cytoplasm as larger precursor molecules.

Proteins targeted towards the chloroplast stroma have to pass the two envelope membranes but proteins active in the chloroplast lumen have to be transferred over the thylakoid membrane also. Recent evidence shows that the routing of proteins towards the lumen involves two post-translational transport processes and is mediated by two different regions of the transit peptide. Concomitant with the transport, the transit peptide is removed in two successive steps. Such transit peptides consist of two functionally different domains. It is postulated that this transport mechanism evolved by the addition of a chloroplast-specific targeting-peptide to the signalpeptide-containing plastocyanin precursor of the endosymbiont after the genetic information for this thylakoid protein had been transferred from the endosymbiont to the nucleus.

Import towards the stroma

Cytosolic proteins that are targeted towards the stroma, have to cross the two membranes that together form the chloroplast envelope. The outer and inner membranes run parralel over the complete length of the envelope with minimal intermembrane space and with multiple sites where they appear to be in close vicinity (contact sites) (Cline et al., 1985). So far all evidence indicate that both membranes are passed in a single translocation step and in fact recent experiment give good evidence that protein import occurs in connection with these sites of contact (Paine & Blobel, 1988). All precursors carry an N-terminal extension, the transit peptide, that is cleaved off during or directly after the passage through the envelope. Transit peptides can vary in lenght between 35-80 amino acid residues , they consist mainly of hydrophobic or hydroxyamino acid residues interspersed with basic amino acids (arginine or lysine) and between different transit peptides there is only very little homology is observed . A recent computer analysis indicates that roughly three domains can be distinguished; an uncharged amino-terminal region of approx. 10 residues, a central region poor in acidic residues but with bacis amino acids and a carboxy-terminal domain with a potential to form an amphiphilic beta-strand (Schmidt & Mishkind, 1986; Karlin--Neumann & Tobin, 1986; von Heijne et al. 1989).

A mutational analysis of the transit peptide of the ferredoxin precursor protein was performed. The ferredoxin precursor DNA in plasmid pFDS22 (Smeekens et al., 1986) was used for the mutagenesis experiments. This yielded the transit peptide mutant series shown Fig. 2. At the fusion point in the mutant DNAs a hybrid codon is created which in some cases codes for an amino acid different from the one in the wildtype precursor. Also three mutants were constructed that have an intact transit peptide fused to mutated mature proteins. Two of these mutants lack 4 and 47 amino acids of the mature protein, respectively (Fig. 2, m4 and m47). In the third mutant (Fig. 2, e13) the transit peptide sequence was fused out of frame to the mature ferredoxin DNA such that a new long open reading frame was formed encoding a very basic protein (calculated precursor MW 12,8 kd, net pos. charge +15). This protein contrasts strongly with the acidic ferredoxin precursor (net neg. charge -10).

These constructs were used as templates for *in vitro* RNA synthesis by the SP6 polymerase and the RNA was translated in a wheat germ system in the presence of [3H]leucine. The wildtype and mutant ferredoxin precursor proteins were tested for binding to and import into isolated pea chloroplasts. For this purpose radiolabeled proteins and intact chloroplasts were mixed and incubated at room temperature for 10 minutes in the light (import experiments) or for 10 minutes on ice in the dark (binding experiments). The chloroplasts that were used in the binding experiments were pretreated with the ionophore nigericin. After incubation, samples of the binding and import reactions were treated with thermolysin to determine whether proteins were exposed on the outside of the chloroplast. Proteins present on the outside are degraded by this protease treatment, whereas internalized proteins are protected.

```
     -48      -40        -30        -20        -10        -1 1   5---98
wt   MASTLSTLSVSASLLPKQQPMVASSLPTNMGQALFGLKAGSRGRVTAM A TYKVTL
t2   MASTLSTLSVSASLLPKQQPMVASSLPTNMGQALFGLKAGSRGRVT   A TYKVTL
t7   MASTLSTLSVSASLLPKQQPMVASSLPTNMGQALFGLKAGS        P TYKVTL
t10  MASTLSTLSVSASLLPKQQPMVASSLPTNMGQALFGLK           A TYKVTL
t13  MASTLSTLSVSASLLPKQQPMVASSLPTNMGQALF              A TYKVTL
t15  MASTLSTLSVSASLLPKQQPMVASSLPTNMGQA                S TYKVTL
t16  MASTLSTLSVSASLLPKQQPMVASSLPTNMGQ                 A TYKVTL
t17  MASTLSTLSVSASLLPKQQPMVASSLPTNMG                  P TYKVTL
t18  MASTLSTLSVSASLLPKQQPMVASSLPTNM                   A TYKVTL
t20  MASTLSTLSVSASLLPKQQPMVASSLPT                     T TYKVTL
t36  MASTLSTLSVSA                                     S TYKVTL

-48- -5  -1 1        10        20        30        40       50-98
wt   GRVTAM ATYKVTLITKESGTVTFDCPDDVTVLDQAEEEGIDLPYSCRAGSCSSCAG
m4   GRVTAM E   VTLITKESGTVTFDCPDDVTVLDQAEEEGIDLPYSCRAGSCSSCAG
m47  GRVTAM A                                               CAG

e13  GRVTAM VPRSQELLHSIARMTCTCLTRLRRRELTYHTRVGRGHALRVLVRWLPVAL
            TSQTRVSLMMTRLKLVGF
```

Figure 2: Ferredoxin transit peptide mutants. Plasmid pFDS22 (Smeekens et al., 1986) was used to generate progressive deletions in the transit peptide. Numbering is relative to the processing site. t: transit peptide mutant, m: mature protein mutant, e13 is the out of frame mutant that codes for the ferredoxin transit peptide fused to a basic protein of 68 amino acids (M.W. 12.8 kd).

The chloroplasts in the binding and import reactions (with or without protease treatment) were centrifuged through a 40 % Percoll cushion to repurify intact organelles only. The chloroplasts were washed once with cold import buffer and solubilized in sample buffer. The radioactive proteins present in each sample were separated by polyacrylamide gel electrophoresis and visualized by autoradiography. The binding and import of the mutant proteins were quantitated and calculated relative to the wild type level (Table I).

The transit peptide mutant that lacks two amino acids still binds to and imports into chloroplasts. This mutant is also processed to a protein similar in size to mature ferredoxin and therefore reacts indistinguishably from the wildtype ferredoxin precursor. In addition, the efficiency of mutant t2 in binding- and import reactions is comparable to the wildtype precursor (Table I).

The other transit peptide mutants from which 7 or more (up to 36) amino acids have been deleted (t7-t36) do not import to a detectable level nor do they bind efficiently (Table I).

	import	binding
wt	100	100
t2	106	64
t7	-	6
t10	-	3
t13	-	-
t15	-	8
t16	-	2
t17	-	-
t18	-	5
t20	-	-
t36	-	-
m4	25	7
m47	-	10
e13	-	169

Table 1: Quantitation of binding and import.
The number of molecules bound or imported was quantitated as described (Cline et al., 1985; Smeekens et al., 1986). Binding and import of mutants is expressed as a percentage of the wildtype level.

Two mutants (m4 and m47) contain an intact transit peptide fused to a truncated mature protein. Mutant m4, which lacks the first four amino acids of the mature protein, is imported and processed by the chloroplasts. This mutant binds and imports at a reduced level when compared to the wildtype ferredoxin (Table I). Mature mutant m47 contains a complete transit peptide but lacks half of the ferredoxin mature protein. This mutant binds at a very reduced level but does not import into chloroplasts. When incubated with a stromal extract, the m47 precursor and its internal translation product are unstable. Thus it is possible that the mutant protein can be imported and subsequently degraded. Therefore, an import timecourse experiment was performed with very short incubation times (0-30-60-90-120 seconds) followed by rapid solubilization in sample buffer. In such an experiment no import of m47 was observed either (results not shown).

Mutant e13 codes for a protein containing an intact ferredoxin transit peptide fused to a positively charged protein, 68 amino acid residues in length (see previous section). This protein binds to chloroplasts with high efficiency but it is not imported (Table I). This protein was completely stable when incubated with a stromal extract for one hour at room temperature (not shown). This result shows that efficient binding does not necessarily result in import. However, at present we do not know whether this strong binding represents a specific interaction of this precursor with chloroplasts or is due to its positive charge. More importantly, in this case the presence of an intact transit peptide does not promote import of a passenger protein.

Our experiments revealed that the amino acid region close to the C-terminus of the transit peptide is important for chloroplast recognition and import. Deletion of the first two amino acids has no effect on binding to and import into the chloroplast but removal of 7 or more amino acids abolishes binding and subsequent import. The finding that the -2/-7 region of the ferredoxin transit peptide is important for the first step of the interaction of a precursor with the chloroplast came as a surprise. In the comparable interaction of nuclear-encoded mitochondrial proteins with mitochondria, this region was found to be completely dipensable. In the mitochondrial system the amino terminal 9-12 amino acids of a presequence can direct a protein into mitochondria (Hurt et al., 1985; Keng et al., 1986).

Recently, transit peptide mutants were constructed and tested for the transit peptide of the small subunit (SS) of ribulosebisphosphate carboxylase fused to either the SS mature protein or to the neomycin-phosphotransferase protein. These mutants were tested in vivo by plant transformation experiments (Kuntz et al., 1986) and in vitro (Reiss et al., 1989). In both investigations the t2-analogs and mutants with small changes around the processing site but not the t15-18-analogs were found inside the chloroplast. These results are in agreement with the results obtained with the ferredoxin transit peptide mutants. The in vitro experiments show that deletion of certain parts of the transit peptide have more severe negative effects than removal of other region but that in all cases the binding and translocation was severly reduced.

The results presented show that sequences in the -2/-7 region of the ferredoxin precursor are essential for interaction with chloroplasts. When compared with many other published transit peptide species, a conserved amino acid sequence is present in this region (see e.g. Lubben et al., 1988; Karlin-Neumann & Tobin, 1986). They contain the sequence 'GRV' or a functionally very similar sequence (see Fig. 3 for examples). In the case of the plastocyanin import domain (Smeekens et al., 1986) the sequence 'PRM' (Fig. 3) constitutes a functional homologous sequence (helix destabilizer/positive charge/hydrophobic amino acid). The conservation of this region among different precursor species suggests an important role in the import process. Secondary structure predictions according to Chou-Fasman for many transit peptide species give high probabilities for turn structures in this conserved region (not shown). The importance of turn structures near processing sites has also been reported for signal sequences (Vlasuk et al., 1984; Inouye et al., 1986).

pSS consensus	S	N	G	G	R	V	*	C ↓	M	
pCAB consensus	*	G	*	G	R	V	T	M ↓	R	
pre-ferredoxin	G	S	R	G	R	V	T	A	M ↓	A
pre-plastocyanin	V	A	T	P	R	M	S	I	K '	A
pSS Chlamydomonas	V	A	A	P	A	Q	A	N	Q ↓	M

Figure 3: Comparison of the sequence in the -2/-7 region of four classes of chloroplast specific transit peptides. The pre-SS and pre-CAB consensus sequences are derived from Karlin-Neumann and Tobin (1986); stars in these sequences denote non conserved residues in the consensus. The sequences shown for preferredoxin and preplastocyanin are from S. pratensis (Smeekens et al., 1985a, 1985b). These are very homologous to sequences from other species for preferredoxin (Arabidopsis thaliana: Vorst et al., in preparation) and preplastocyanin (for a compilation see Vorst et al., 1988). The Chlamydomonas pSS sequence is from

Goldschmidt-Clermont and Rahire (1986). The homologous region is boxed. The arrows indicate the processing sites. For plastocyanin no precise intermediate processing site is available (dashed line).

In *in vitro* processing experiments using a stromal extract we detected abberant processing of the t7 and t10 mutants (not shown). Therefore, we conclude that the -2/-7 region is important for both chloroplast recognition and processing by the stromal peptidase.

The results with mature mutants m4 and m47 suggest that the transit peptide by itself is not the only prerequisite for efficient interaction with the chloroplast. Apparently, sequences present in the mature protein are also important for interaction with chloroplast specific factors. Mutant e13, which contains an intact transit peptide fused to a basic mature protein, binds above wildtype level without being imported. The inability of this mutant to import shows that the presence of an intact transit peptide is not necessarily sufficient for successful import. The high positive charge of the molecule in this case might prevent import even if a specific binding event, which normally leads to import, would occur.

Specific sequences or structures contributed by the mature sequences may play an important role in import. This idea is supported by the results of Wasmann et al. (1986), who showed that in a fusion protein the presence of the first 23 mature amino acids of the mature SS protein greatly enhances the *in vitro* SS transit peptide-dependent import of a passenger protein.

A similar conclusion was also reached by Lubben & Keegstra (1986) who compared import of the SS precursor with that of a fusion protein consisting of the SS transit peptide and a cytosolic heat shock protein. Transit peptides, therefore, might evolve in such a way as to optimally adapt to (or compensate for) the structure of their respective mature proteins. An important implication of this hypothesis is that (foreign) passenger proteins can import with different efficiencies, depending on the transit peptide used.

The structure of the precursor protein during translocation

In recent experiments on protein translocation much attention has been given to the conformation of proteins during membrane passage. Little is known about the structure of the precursor as compared to the mature protein neither do we know if structural changes are required during translocation. Experiments in ER translocation and mitochondrial import (Muller & Zimmermann, 1988; Eilers & Schatz 1986) indicate that the precursor needs a certain degree of flexibility in order to be capable of membrane passage. A model has been put forward in which unfolding is an essential element of the import process (for a review see Eiler & Schatz, 1988). A central observation was that mitochondrial import can be blocked by preventing unfolding of the passenger protein. This has been shown for two artificial precursor proteins containing a mitochondrial presequence attached to dihydrofolate reductase DHFR and copper metallothionein, respectively (Eilers & Schatz,1986; Chen & Douglas,1987). The conformation of these fusion proteins can be stabilized by complexing with specific ligands. Destabilization of the DHFR-MTX complex by denaturation with urea or by mutagenesis of the DHFR protein abolishes this type of translocation arrest (Eilers & Schatz, 1988; Vestweber & Schatz, 1988a). What this exactly means in terms of unfolding is unclear but a too rigid conformation, as obtained after binding of ligands, has been shown to be inhibitory to import although not to binding. However the import into mitochondria of a branched and probably conformationally fixed polypeptide (Vestweber & Schatz, 1988b) was not effected, neither was the MTX complex of a differnt DHFR fusion (Verner & Weber, 1989). The unfolding in the mitochondrial model is thought to be mediated by ATP-dependent unfoldases that act post-translationally in the cytosol or on the target membrane.

For chloroplasts, nothing is known about the folding of proteins during translocation. We have analyzed the structural requirements of chloroplast precursors with fusion proteins in which various parts of the transit peptide of plastocyanin was fused to the complete mouse cytosolic dihydrofolate (DHFR) protein. These fusion proteins bind and are imported with efficiencies similar to the plastocyanin precursor; the precursor is processed and the imported DHFR part is subsequently rapidly cleaved in three discrete fragments (see Fig. 4). When the import reaction is performed in the presence of high concentrations of metothrexate (MTX) than quantitatively translocation occurs with similar efficiency as in the absence of MTX. A number of significant changes are seen. First the fragmentation of the DHFR protein after import is completely abolished; only proteins of the size of DHFR or larger are found. Secondly the processing of the precursor becomes abberant; now three processing products appear instead of the single product observed when import occurs in the absence of MTX.
The binding of MTX to the precursor was determined by proteaseprotection experiments. Liganded DHFR is highly resistant to thermolysin degradation whereas the free DHFR is not. The precursor shows in the presence of MTX the protease resistant characteristics of the complexed molecule.

The conclusion that can be drawn from this experiment is that the presence of MTX is not inhibitory to the import of the precursor. It does not discriminate however between import of the precursor-MTX complex or dissociation of MTX before translocation. Therefore tritium labelled MTX was complexed with DHFR and the import of the labelled MTX was measured, quantitated and compared with the number of protein molecules imported. This gave as result that the intraorganellar amount of 3H-MTX was significantly increased (MTX can not enter the chloroplast by itself) and that the ratio of imported MTX and imported protein was approx. 1.
These results together, import of MTX and the changes in processing and degradation, are strongly indicative for the conclusion that the DHFR-MTX complex as a whole is translocated across the chloroplast envelope membranes.

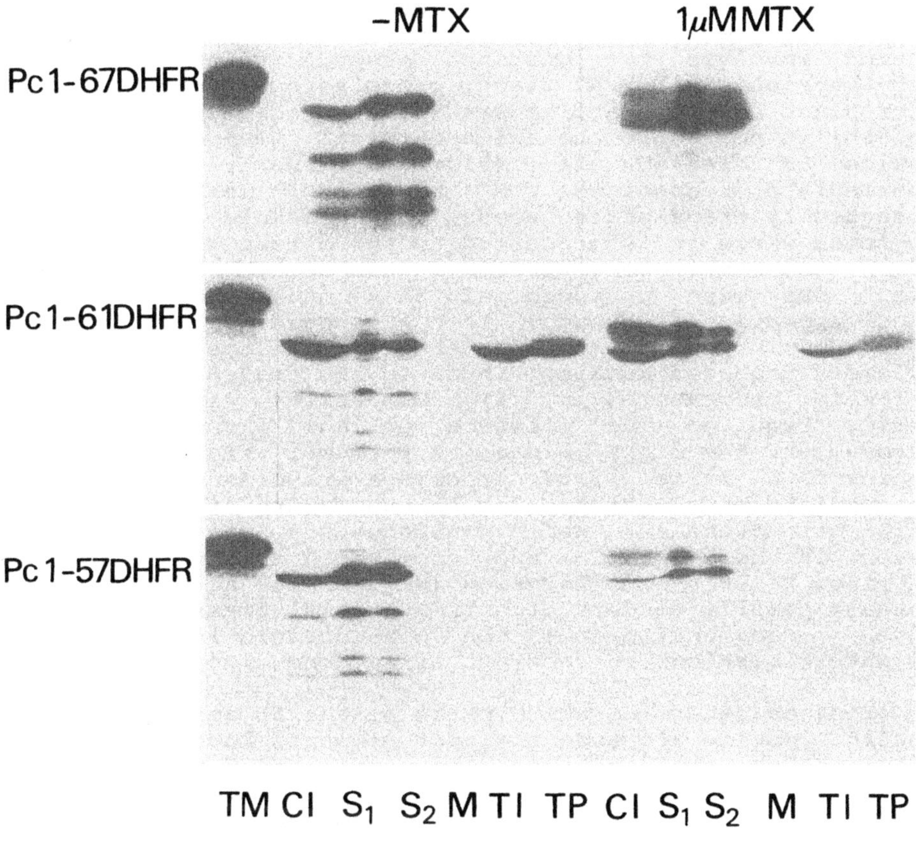

Fig. 4 Import of PC-DHFR fusion proteins into chloroplasts in the absence and presence of metothrexate. TM, translation products; Cl, total chloroplasts after import; S1, stromal fraction; S2, stromal fraction after removal of envelope membranes, M, envelope membranes; Tl, thylakoid fraction; TP, thylakoid fraction protease treated; MTX, metothrexate.

The import characteristics of the precursor of EPSP synthase (pEPSPS) complexed to its competitive inhibitor glyphosate has also been analysed (della-Cioppa & Kishore 1988). The conclusion of this work is that the precursor-inhibitor complex still can be imported but that the rate of entry is significantly reduced. It is unclear whether the imported EPSP synthase still is complexed with glyphosate or that during translocation the ligand has been dissociated.

Thylakoid lumen targeting

The import of pre-plastocyanin and its subsequent routing toward the lumen could be reconstituted using in vitro synthesized radiolabeled precursor proteins and isolated intact chloroplasts (Smeekens et al., 1986). The plastocyanin protein was directed by this system toward the thylakoids, where it is present in the protease-resistant fraction. This routing involves two distinct steps, which both occur post-translationally. At first, pre-plastocyanin binds to the chloroplast envelope and is translocated across the envelope into the stroma where the amino-terminal part of the transit peptide is cleaved off (Fig. 1). Next, this stromal intermediate recognizes the thylakoid membrane and is subsequently translocated across the thylakoid membrane into the lumen where it is processed to the mature size. Based on these, and other, observations a "two domain transit peptide model" has been proposed. In this model the plastocyanin transit peptide is thought to be composed of a chloroplast import domain, which mediates transport to the stroma, and a thylakoid transfer domain, involved in transport across the thylakoid membrane (Fig. 1). Successive removal of these domains requires the presence of the proper processing enzymes at the right place. A stromal transit peptidase (Robinson & Ellis, 1984) processes pre-plastocyanin to the intermediate size. Processing to the mature size requires an additional protease, which subsequently was found to be present in the thylakoids and from which it can be isolated by Triton X-100 solubilization (Hageman et al., 1986). This protease clearly differs from the stromal transit peptidase, since it can only process the plastocyanin intermediate to its mature size but it does not affect the intact precursor.

In order to determine which parts of the 66 amino acids long transit peptide are essential for import into and routing inside the chloroplast, the N-terminal and C-terminal parts of the transit peptide were used to construct fusion proteins. These proteins were subsequently tested in the chloroplast import system.

The first cleavable part is a transit peptide

To test what part of the N-terminus of the plastocyanine transit peptide is sufficient for import of another protein into the chloroplast, a set of N-terminal fragments of the transit peptide was fused to the mature sequence of dihydrofolate reductase (DHFR). The smallest fusion protein of this set contains only the first 39 amino acids of the PC transit peptide and the largest fusion protein contains the first 67 amino acids, the entire PC transit peptide.
Import studies with this set of fusion proteins (Fig. 5) show that the first 53 amino acids of the transit peptide are capable of import that is as efficient as the import of the complete precursor. With 43 amino acids the efficiency goes down considerably, whereas 39 amino acids are not sufficient for significant import into the chloroplast stroma.

The fact that this N-terminal domain can direct the import of a non-plastid protein suggests that its only function in the transport pathway of the plastocyanin precursor is to bring the intermediate protein into the stroma. To test whether this is true, this sequence was replaced by the transit peptide of a stromal protein. Hybrid proteins were

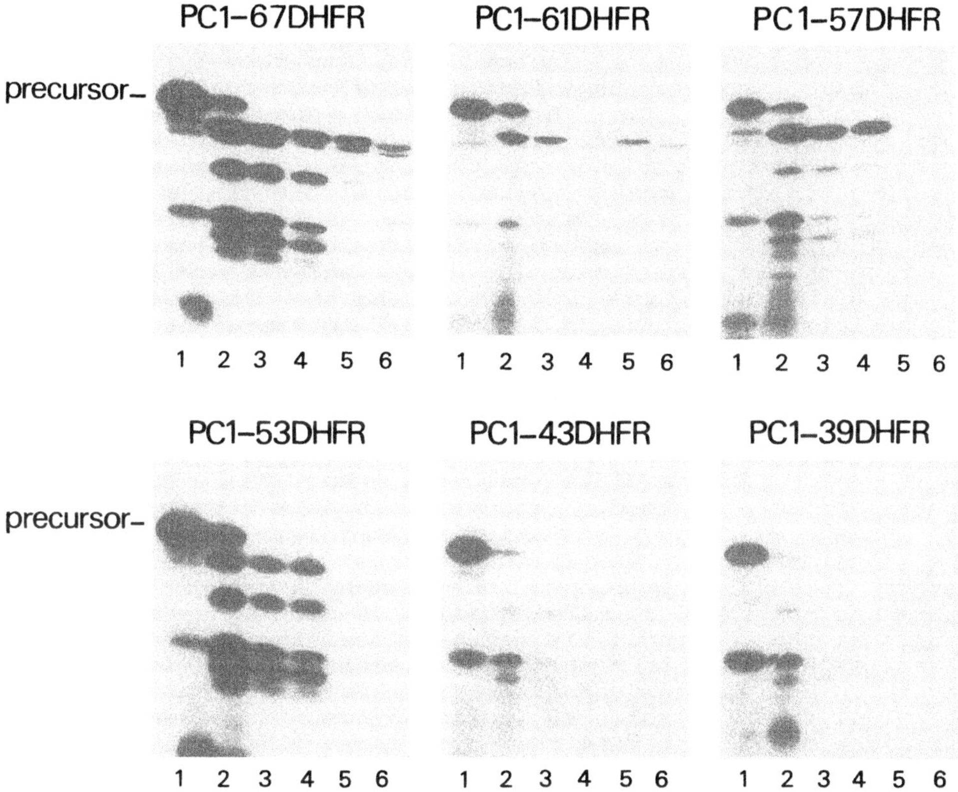

Fig. 5. Import of PC-DHFR fusion proteins into chloroplasts. The radiolabelled proteins were incubated with chloroplasts in the light. After incubatiom chloroplasts were analyzed by polyacrylamide gel electroforeses and fluorography, before and after fractionation. Lanes 1: translation products; lanes 2: chloroplasts before protease treatment; lanes 3: chloroplasts after thermolysin treatment; lanes 4, 5 and 6 are stroma, thylakoid system and thylakoid system treated with thermolysin, respectively. PC1-67, PC1-61 etc. indicates the length of the transit peptide fragment used in that fusion protein, measured from the amino terminus.

constructed in which the N-terminal part of the plastocyanin (PC) transit peptide is replaced by the complete ferredoxin (FD) transit peptide. Two comparable hybrids were constructed, FD-40PC and FD-44PC. FD-40PC is the construct where the FD transit peptide and the first amino acid of the mature ferredoxin is fused to amino acid 40 of the PC precursor. The plastocyanin part in this hybrid protein probably is the complete processing intermediate. In FD-44PC the transit peptide and the first 6 amino acids of mature FD are fused to the PC precursor at position 44. The PC part in this construct is a few amino acids shorter than the PC intermediate and does not contain the intermediate processing site. FD-44PC is one amino acid longer than FD-40PC. Because the FD transit sequence (48 amino acids) is longer than the original chloroplast import domain part (approximately 39 amino acids) FD-40PC and FD-44PC are longer than the wildtype PC precursor (10 and 11 amino acids respectively).

Chloroplast uptake experiments with the fusion proteins FD-40PC and FD-44PC (Fig. 6) show that these proteins are imported and that they are routed correctly. The plastocyanin protein is transported towards the thylakoid system, it is processed to the mature size and it is in a protease protected form in the thylakoid system and therefore in the lumen. The kinetics of import of the fusion proteins, assayed in a time-course experiment (data not shown), are identical to that of the wild type PC, including the occurrence of an intermediate. Import of a functional intermediate into the stroma can therefore be directed by the transit peptide of FD; the subsequent transport step into the lumen can be directed by the intermediate protein.

The intermediate processing in the two hybrids does not occur at the same position as in the wild type. The size of the FD-44PC intermediate suggests that the FD mature processing site is used. Processing at this site yields an intermediate which is a few amino acids larger than the wild type intermediate. The fact that this difference does not prevent correct routing to the lumen suggests that the N-terminus of the C-terminal domain is not as sensitive to changes as the region of the mature processing site.

These experiments show that the N-terminal extension of the plastocyanin precursor consists of two domains. Domain I, the N-terminal first approx. 40 amino acids, is essential and sufficient for import into the chloroplast stroma. Domain II, the C-terminal approx. 27 amino acids, is required only for import into the thylakoid lumen.

Taken together the results presented above warrant a new nomenclature for the routing signals in the plastocyanin presursor. We propose to confine the term <u>transit peptide</u> to the part that directs the proteins across the envelope

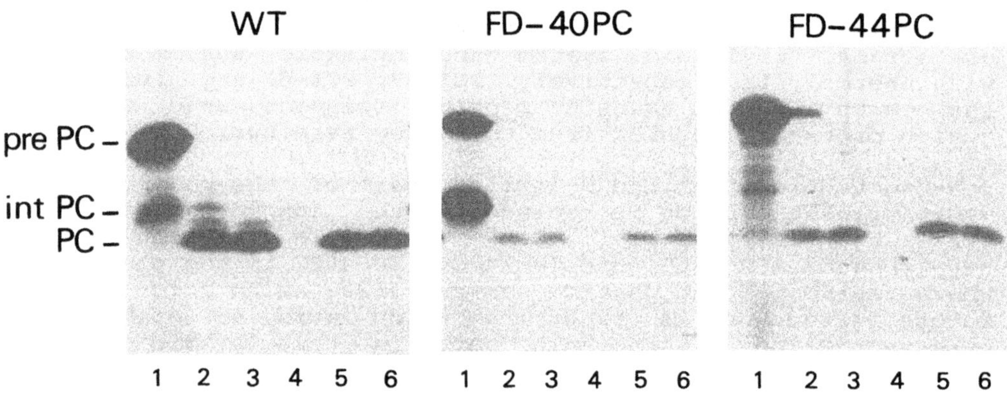

Fig. 6. Import of FD transit PC intermediair fusion proteins into chloroplasts. WT is the complete precursor of plastocyanine; FD-40PC and FD-44PC are fusion proteins with the complete transit peptide of the ferredoxine precursor fused to the precursor of plastocyanine missing the amino terminal 40 and 44 amino acids respectively.

membranes into the stroma only and to use the name thylakoid signal peptide for the part that is needed for transport across the thylakoid membrane.

For the plastocyanin precursor this means that the precursor can be divided into three parts, a transit peptide, a thylakoid signalpeptide and a mature peptide.

The fusion protein PC1-67DHFR contains the complete plastocyanin presequence, including the thylakoid targeting signal. Fig. 5 shows however that no or very little of the DHFR protein reaches the lumen. The protein found in the thylakoid system can be degraded by added protease, indicating association with the thylakoid membrane on the stromal surface. The same holds for PC1-61DHFR. Only the small fraction that is resistant to the protease treatment may have crossed the thylakoid membrane and entered the lumen. DHFR itself is a very protease resistant protein, however, and this may interfere with this conclusion. Therefore, not even the entire extension is enough to direct efficient transport of this non-plastid protein to the lumen. In the case of PC1-67DHFR, the absence of thylakoid translocation is not due to unavailability of the mature processing site. In vitro, PC1-67DHFR is processed at this site by the thylakoidal processing protease (data not shown).

Figure 5 also shows that the length of the extension has a dramatic effect on the processing of the fusion proteins during import. The processing pattern of PC1-67DHFR is completely different from that of PC1-61DHFR and PC1-57DHFR seems to be intermediate. Except PC1-61DHFR, all the PC-DHFR fusion proteins are processed at multiple sites within the mature DHFR part.

Other lumen proteins also have a two domain transit peptide

Since the identification of the two functional domains in the white campion plastocyanin precursor transit peptide, the amino acid sequences for transit peptides of two other plastocyanin precursors (Vorst et al., 1988); (Rother et al., 1986) and of several other lumenal proteins have become available. The latter include the three components of the PS II-associated water splitting complex (Mayfield et al., 1987); (Tyagi et al., 1987); (Jansen et al., 1987) and of the inducible cytochrome c-552 of Chlamydomonas (Merchant & Bogorad, 1987). In all these transit peptides two domains that correspond to the proposed plastocyanin chloroplast import domain and thylakoid transfer domain can clearly be distinguished. This sequence homology indicates a targeting mechanism for these proteins that is very similar to the proposed two step model for plastocyanin. It also makes it likely that imported thylakoid-lumen directed proteins can use a common routing mechanism .

Targeting to other chloroplast compartments

Little is known about the way envelope proteins find their definitive location. No protein active in the innermembrane space has been analyzed so far and only for a single outer membrane protein has it been determined that it is processed during uptake (Flügge and Wessel, 1984). The way the latter protein reaches the outer membrane, either directly from the cytosol or only after it is imported into the stroma, is however unknown. Recently the first cDNA deduced amino acid sequence of the precursor of an innermembrane protein, the phosphate translocator, has been reported (Flügge et al., 1989). It is processed during uptake but it has not been determined yet where this processing occurs. The N-terminal sequence does not resemble a "typical" transit peptide.

The thylakoidal membrane protein LHCP has been analyzed quite extensively. The interesting conclusion from this work is that the transit peptide of preLHCP only contains information for stroma targeting. This part can direct other proteins to the stroma but no further (Van den Broeck et al., 1988, Kohorn & Tobin, 1989). The interaction with the thylakoid membrane is caused by the carboxy terminal 30 amino acids of the mature protein, a membrane spanning segment, and this part is not cleaved off. When it is linked C-terminal to the precursor of the small subunit of RUBISCO (preSSU) this fusion becomes attached to the thylakoid membrane.
Thus the thylakoidal membrane protein LHCP has evolved a sorting signal of a different nature than the lumen proteins.

Evolutionary background of targeting signals

Additional information about the transport routes and signals in the chloroplast can be obtained from comparisons between the genes for chloroplast proteins and cyanobacterial proteins. Cyanobacteria or close relatives thereof are considered to be ancestors of the chloroplasts and they still have much in common with chloroplasts, not in the least their plant type photosynthetic apparatus and a thylakoid membrane system. Many functionally similar proteins have retained a remarkable sequence homology but are now nuclear encoded in the plant. This may give us information how the chloroplast specific protein import and sorting developed. E.g. the plant ferredoxin protein is clearly related to the cyanobacterial protein (Smeekens at al., 1985; Van der Plas et al., 1988). The nuclear plant gene for ferredoxin codes however for a larger protein; the aminoterminal extra 48 amino acids contain the chloroplast import information.

The plastocyanin precursor of plants contains two independent targeting signals (see above), one for chloroplast import and one for thylakoid membrane translocation. In the cyanobacterium plastocyanin was already made as a precursor but with just information for thylakoid transfer (Van der Plas et al., 1989). The same is true for other thylakoid lumen proteins like the 33 kD protein of the oxygen evolving complex and 9 kD protein (Kuwabara et al., 1987; Wallace et al., 1989). So again it appears as if only

information for chloroplast import has been added during the symbiont evolution. One should realise of course that in the chloroplast the thylakoid translocation of the protein is by definition posttranslationally whereas in the cyanobacterium this may be either co- or posttranslationally (compare also with excretion in E.coli). Comparison of both signals for thylakoid transfer however indicates that they interact with a comparable system and that these systems probably are similar to the E.coli inner membrane translocation system.
The tentative conclusion from this is that in evolution only chloroplast import information was developed and linked to already existing mechanisms of intracellulair protein sorting and if this is correct than we can make predictions about the transport signals of proteins that have not yet been analyzed for their transport. When this is applied e.g. to the Rieske Fe-S protein, an integral thylakoid membrane protein, we see that the cyanobacterium gene codes for only the mature protein (Kallas et al., 1988). This means that the thylakoid targeting information probably is internal and not cleaved off.

In cyanobacteria so far the following information about intracellular sorting is available: i) proteins active in the thylakoid lumen are all made as a precursor with a signalpeptide like presequence that is cleaved off (plastocyanin, 33 kD protein, cytochrome c553), ii) thylakoid membrane proteins are found either with a cleavable signalpeptide (cytochrome f) or with an internal signal that is part of the mature protein, iii) envelope inner membrane proteins can have a signalpeptide-like presequence that is cleaved off (one example, the irpA protein, (Reddy et al., 1988)).

Based on the previous arguments this suggests that similar targeting information will be found in chloroplast proteins that are either encoded by nuclear genes or that are derived from genes on the chloroplast DNA.

Discussion

Many aspects of the mechanisms of protein sorting in plant cells still have to be elucidated. It can be expected that the strategies that are employed in other eukaryotic cells can also be used for plants. The routing towards the chloroplast and the sorting within this organelle are however unique mechanisms for the plant cell and the first insights into how cytoplasmically synthesized proteins reach the chloroplast have been obtained.

The results obtained so far with stroma-targeted proteins show that we are able to use transit peptides for import of a variety of foreign proteins into the stroma without specific knowledge about the translocation mechanism by which these peptides function. It enables us to study and manipulate many metabolic processes that occur in the chloroplast.

The results obtained with the plastocyanin precursor and fusion proteins have revealed the basic concept of how proteins reach the thylakoid lumen. Here two different transport steps occur, one for passage of the envelope membranes and the other to pass the thylakoid membrane. They occur independently, as the first part of the plastocyanin import sequence is sufficient for chloroplast import and can be replaced by other transit peptides. The available information with regard to translocation across the thylakoid membrane is very limited. So far it appears to be a very sensitive process. Knowledge of the mechanism is necessary to learn how to overcome the present inability to introduce foreign proteins into this compartment.

The targeting and sorting of proteins of the inner membrane and also of the thylakoid membrane itself, with the exception of the chlorophyll a/b protein, has not been studied yet. The recent experiments (Kohorn & Tobin, 1989) with the chlorophyll a/b protein indicate that it behaves completely differently from plastocyanin. A C-terminal non-cleavable part of the mature protein appears to initiate the interaction with the thylakoids.

The use of such a plastid-specific targeting mechanism for the routing of imported lumen proteins also makes a lot of sense in an evolutionary context; no new mechanism had to be designed for thylakoid transfer when most of the genetic material from the prokaryotic progenitor of the present day chloroplast was relocated to the nucleus. In this view the complete genetic information for such proteins (including their signalpeptide-like thylakoid targeting information) was transferred to the nucleus. Only a chloroplast import domain had to be added to compose a complete precursor that when synthesized in the cytoplasm is targeted for the chloroplast and ends up in the lumen. A similar suggestion for the coupling of an ancient transport system to a newly evolved import mechanism has also been made for transport of nuclear-encoded mitochondrial proteins that are found in the inter-membrane space or on the outer surface of the inner membrane (Hartl et al., 1986). The Fe/S protein of ubiquinol-cytochrome c reductase is transported to the outer surface of the inner membrane in two steps, reminiscent of the plastocyanin transport to the thylakoids.

Translocation of chloroplast imported proteins over the envelope membranes is not blocked by complexing of the proteins with specific ligands. These results completely contrast to what is found for mitochondria, where transfer is made impossible by complex formation. This indicates that the processes at the chloroplast envelope is essentially different from the processes at the outer and inner membrane of mitochondria. It is however not yet clear what the nature of this difference is. For mitochondria it is concluded that the import block is due to the fact that the ligand binding prevents the "unfolding" that is required for transfer. Chloroplasts may posses a more powerful "unfolding" mechanism that is also capable to unfold a protein-ligand complex. It however can not be excluded up till now that such a major conformational change is not required. This means that a protein is able to cross the chloroplast envelope with a ligand as a passenger.

REFERENCES

Baker A and Schatz G (1987) Sequences from a prokaryotic genome or the mouse dihydrofolate reductase gene can restore the import of a truncated precursor protein into yeast mitochondria. Proc Natl Acad Sci USA 84: 3117-3121

Chen W-J and Douglas MG (1987) Phosphodiester bond cleavage outside mitochondria is requierd for the completion of protein import into the mitochondrial matrix. Cell 49: 651-658

Cline K, Werner-Washburne M, Lubben T and Keegstra K (1985) Precursors to two nuclear-encoded chloroplast proteins bind to the outer envelope membrane before being imported into chloroplasts. J Biol Chem 260:3691-3696

Cline K, Keegstra K and Staehelin LA (1985) Freeze fracture electron microscopic analysis of ultrarapidly frozen enevelope membranes on intact chloroplasts and after purification. Protoplasma 125: 111-123

Coruzzi G, Broglie R, Edwards C and Chua N-H (1984) Tissue-specific and light-regulated expression of a pea nuclear gene encoding the small subunit of ribulose-1,5-biphosphate carboxylase. The EMBO J 3: 1671-1679

De Boer D, Cremers F, Teertstra R, Smits L, Hille J, Smeekens S and Weisbeek P (1988) In vivo import of plastocyanin and a fusion protein into developmentally different plastids of transgenic plants. The EMBO J 7: 2631-2635

Della-Cioppa G and Kishore GM (1988) Import of a precursor into chloroplasts is inhibited by the herbicide glyphosate. The EMBO J 7: 1299-1305

Eilers M and Schatz G (1986) Binding of a specific ligand inhibits import of a purified precursor protein into mitochondria. Nature 332: 228-232

Eilers M and Schatz G (1988) Protein unfolding and the energetics of protein translocation across biological membranes. Cell 52: 481-483

Ellis JR (1981) Chloroplast proteins: synthesis, transport and assembly. Ann Rev Plant Physiol 32:111-137.

Flugge UI and Wessel D (1984) Cell-free synthesis of putative precursors for envelope membrane polypeptides of spinach chloroplasts. FEBS 168: 255-259

Flugge UI, Fischer K, Gross A, Sebald W, Lottspech F and Eckerskorn C (1989) The triose phosphate-3-phosphoglycerate-phpsphate translocator from spinach chloroplasts: nucleotide sequence of a full-lenght cDNA clone and import of the in vitro synthsized precursor protein in chloroplasts. The EMBO J 8: 39-46

Goldschmidt-Clermont M, Rahire M (1986) Sequence, evolution and differential expression of the two genes encoding variant small subunits of ribulose bisphosphate carboxylase/oxygenase in Chlamydomonas rheinhardtii. J Mol Biol 191:421-432

Hageman J, Robinson C, Smeekens S and Weisbeek P (1986) A thylakoid located processing protease is required for complete maturation of the lumen protein plastocyanin. Nature 324: 567-569

Hartl F-U, Schmidt B, Wachter E, Weiss H and Neupert W (1987) Transport into mitochondria and intramitochondrial sorting of the Fe/S protein of ubiquinol-cytochrome c reductase. Cell: 939-950

Hurt E, Pesold-Hurt B, Suda K, Opligger W and Schatz G (1985) The first twelve amino-acids (less than half of the pre-sequence) of an imported mitochondrial protein can direct mouse cytosolic dihydropholate reductase into the yeast mitochondrial matrix. EMBO J 4:2061-2068

Inouye S, Duffand G and Inouye M (1986) Structural requirements at the cleavage site for efficient processing of the lipoprotein secretory precursor of Escherichia coli. J Biol Chem 261:10970-10975

Jansen T, Rother C, Steppuhn J, Reinke H, Beyreuther K, Jansson C, Andersson B and Herrmann R (1987) Nucleotide sequence of cDNA clones encoding the complete 23 kda and 16 kda precursor proteins associated with the photosynthetic oxygen-evolving complex from spinach. FEBS Lett 216: 234-240

Kallas T, Spiller S, Malkin R (1988) Primary structure of cotranscribed genes encoding the Rieske Fe-S and cytochrome f proteinsof the cyanobacterium Nostoc PCC 7906. Proc Natl Acad Sci USA 85: 5794-5798

Karlin-Neumann GA and Tobin EM (1986) Transit peptides of nuclear encoded chloroplast proteins share a common amino acid framework. The EMBO J 5: 9-13

Keng T, Alani E and Guarente L (1986) The nine amino-terminal residues of delta-aminolevulinate synthase direct beta-galactosidase into the mitochondrial matrix. Mol Cell Biol 6:355-364

Kohorn BD and Tobin EM (1989) A hydrophobic carboxy-proximal region of a light harvesting chlorophyll a/b protein is necessary for stable integration into thylakoid membranes. The Plant Cell 1: 159-166

Kuntz M, Simons A, Schell J and Schreier PH (1986) Targeting of protein to chloroplasts in transgenic tobacco by fusion to mutated transit peptide. Mol Gen Genet 205: 454-460

Kuwabara T, Reddy and Sherman LA (1987) Nucleotide sequence of the gene from the cyanobacterioum Anacystis nidulans R2 encoding the Mn-stabilizing protein involved in photosystem II water oxidation. Proc Natl Acad Sci USA 84: 8230-8234

Lubben T and Keegstra K (1986) Efficient in vitro import of a cytosolic heat shock protein into pea chloroplasts. Proc Natl Acad Sci USA 83:5502-5506

Lubben T, Theg S and Keegstra K (1988) Transport of proteins into chloroplasts. Photosynthesis Research, in press

Mayfield S, Rahire M, Frank G, Zuber H and Rochaix J-D (1987) Cloning of the OEE2 gene from Chlamydomonas. Proc Natl Acad Sci USA 84: 749-753

Merchant S and Bogorad L (1987) The Cu(III)-repressible plastidic cytochrome c. Cloning and sequence of a cDNA for Pre-apoprotein. J Biol Chem 262: 9062-9067

Muller G and Zimmermann R (1988) Import of honeybee prepromelittin into the endoplasmic reticulum: enegy requirements for membrane insertion. The EMBO J 7: 639-648

Pain D, Kanwar Y and Blobel G (1988) Identification of a receptor for protein import into chloroplasts and its localization to envelope contact sites. Nature 331: 232-237

Reddy KJ, Bullerjahn GS, Sherman DM and Sherman LA (1988) Cloning, nucleotide sequence and mutagenesis of a gene (irpA) involved in iron-deficient growth of the cyanobacterium sp. strain PCC7942. J Bacteriol 170: 4466-4476

Reiss B, Wasmann CC, Schell J and Bohnert HJ (1989) Effect of mutations on the binding and translocation functions of a chloroplast transit peptide. PNAS 86: 886-890

Robinson C and Ellis RJ (1984) Transport of proteins into chloroplasts. Partial purification of a chloroplast protease involved in the processing of imported precursor polypeptides. Eur J Biochem 142: 337-342

Rother C, Jansen T, Tyagi A, Tittgen J and Herrmann R (1986) Plastocyanin is encoded by an uninterrupted nuclear gene in spinach. Curr Genet 11:171-176

Schmidt GW and Mishkind M (1986) The transport of proteins into chloroplasts. Ann Rev Biochem 55: 879-912

Smeekens S, Van Binsbergen J and Weisbeek P (1985) The plant ferredoxin precursor: nucleotide sequence of a full length cDNA clone. Nucl Acids Res 13:3179-3194

Smeekens S, Bauerle C, Hageman J, Keegstra K and Weisbeek P (1986) The role of the transit peptide in the routing of precursors towards different chloroplast compartments Cell 46: 365-375

Tyagi A, Hermans J, Steppuhn J, Jansson C, Vater F and Herrmann R (1987) Nucleotide sequence of cDNA clones encoding the complete 33 kda precursor protein associated with the photosynthetic oxygen-evolving complex from spinach. Mol Gen Genet 207: 288-293

Van den Broeck G, Timko MP, Kausch AP, Cashmore AR, Montagu M van and Herrera-Estrella L (1985) Targeting of foreign protein to chloroplasts by fusion to the transit peptide from the small subunit of ribulpse-1,5-biphosphate carboxylase. Nature 313: 358-363

Van der Plas J, De Groot R, Woortman M, Cremers F, Borrias M, Van Arkel G and Weisbeek P (1988) Genes encoding ferredoxins from Anabaena sp. PCC 7937 and Synechococcus sp. PCC 7942: structure and regulation. Photosynthesis Research 18: 179-204

Van der Plas J, Bovy A, Kruyt F, De Vrieze G, Dassen E, Klein B and Weisbeek P (1989) The gene for the precursor of plastocyanin from the cyanobacterium Anabaena sp. PCC 7937 isolation, sequence and regulation. Molecular Microbiology 3: 275-284

Verner K and Weber M (1989) Protein import into mitochondria in a homologous yeast in vitro system. J Biol Chem 264: 3877-3879

Vestweber D and Schatz G (1988a) Point mutations destabilizing a precurosr protein enhance its post-translational import into mitochondria. The EMBO J 7: 1147-1151

Vestweber D and Schatz G (1988b) A chimeric mitochondrial precursor protein with internal disulfide bridges blocks import of authentic precursors into mitochondria and allows quantitation of import sites. J Cell Biology 107: 2037-2043

Vlasuk G, Inouye S and Inouye M (1984) Effects of replacing serine and threonine residues within the signal peptide on the secretion of the major outer membrane lipoprotein of *Escherichia coli*. J Biol Chem 259:6195-6200

Von Heijne G, Stepphuhn J and Herrmann RG (1989) Domain structure of mitochondrial and chloroplast targeting peptides. Eur J Biochem 180: 535-545

Vorst O, Oosterhoff-Teertstra R, Vankan P, Smeekens S and Weisbeek P (1988) Plastocyanin of Arabidopsis thaliana; isolation and characterization of the gene and chloroplast import of the precursor protein. Gene 65: 59-69

Wallace TP, Stewart AC, Pappin D and Howe CJ (1989) Gene sequence for the 9kd component pf photosystem II from the cyanobacterium Phormidium laminosum indicates similarities between cyanobacyterial and other leader sequences. Mol Gen Genet 216: 334-340

Wasmann C, Reiss B, Bartlett S and Bohnert H (1986) The importance of the transit peptide and the transported protein for protein import into chloroplasts. Mol Gen Genet 205:446-453

IMPORT OF PREPROCECROPIN A AND RELATED PRECURSOR PROTEINS INTO THE ENDOPLASMIC RETICULUM

Gabriel Schlenstedt, Elmar Wachter, Maria Sagstetter, Frederic Morel, Richard Zimmermann
Institut für Physiologische Chemie, Physikalische Biochemie und Zellbiologie
Universität München
Goethestr. 33
D-8000 München 2
Federal Republic of Germany

and

Gudmundur H. Gudmundsson, Hans G. Boman
Department of Microbiology
University of Stockholm
Svante Arrheniusväg 16
S-10691 Stockholm
Sweden

The cecropins are a family of potent antibacterial proteins that constitute a main part of the inducible cell-free immunity in insects (Boman and Hultmark, 1981, 1987). The secretory proteins cecropins A, B, and D from cecropia moth (Hyalophora cecropia) vary in size from 35 to 37 amino acid residues and are made as preproproteins with a size of 62 to 64 residues (Lidholm et al., 1987). Besides of the biological function of the mature forms, this protein family is interesting with respect to the processing steps involved in maturation of the prepro-forms to the biologically active proteins and with respect to the mechanism of import of the precursor proteins into the endoplasmic reticulum.

NATO ASI Series, Vol. H 40
Dynamics and Biogenesis of Membranes
Edited by J.A.F. Op den Kamp

On the basis of our previous work on the mechanism of import of small precursor proteins into the mammalian microsomes we concluded that there are at least two different mechanisms for the import of proteins into the endoplasmic reticulum (Müller and Zimmermann, 1987, 1988a, 1988b; Sagstetter and Zimmermann, 1988; Schlenstedt and Zimmermann, 1987; Watts et al., 1983; Wiech et al., 1987; Zimmermann and Mollay, 1986; Zimmermann et al., 1988a, 1988b); a very similar situation seems to exist for the export of proteins in Escherichia coli (Cobet et al., 1989). We proposed that both mechanisms depend on the presence of a -typically transient- signal peptide on the respective precursor protein and some type of signal peptide receptor on the cis-side and a so called signal peptidase on the trans-side of the microsomal membrane. The one mechanism was suggested to be used by precursor proteins larger than about 75 amino acid residues and to involve the two ribonucleoparticles, signal recognition particle (SRP) and ribosome, and their receptors on the microsomal surface, docking protein and ribosome receptor. The other one was thought to be used by small precursor proteins and to depend on the mature part within the precursor molecule and a cytoplasmic ATPase. The idea was that the small precursors have a mature part which allows a competent conformation of the precursor which otherwise was supposed to be preserved by the ribosome/SRP-interaction. On the other hand, an antifolding or unfolding system (ATPase) was suggested to be involved at least in the case of small precursors in order to facilitate membrane insertion which does not allow a high degree of tertiary structure. We used the preprocecropins as tools to reexamine these interpretations (Schlenstedt et al., 1989).

Preprocecropin A is processed by and imported into mammalian microsomes

The 64 amino acid long preprocecropin A is an extremely short precursor protein (Fig. 1) carrying a typical cleavable signal sequence (Lidholm et al., 1987; Boman et al., 1989). A plasmid carrying the cDNA coding for ppcec A under the control of the strong bacteriophage SP6 promoter, was transcribed in vitro. Translations in different cell-free

```
    +                             - + +  ++ -+  +  +     +          - +
MNFSRIFFFVFALVLALSTVSAAPEPKWKVFKKIEKMGRNIRNGIVKAGPAIAVLGEAKALG

    +                             - + +  ++ -+       +-    +                +
MNFSRIFFFVFACLTALAMVNAAPEPKWKLFKKIEKVGQNIRDGIIKAGPAVAVVGQATQIAKG

    +                                    +- -+   + +-                      + +
MNFTKILLFVVACVFAMGTVSAAP    WNPFKELEKVGQRVRDAVISAGPAVATVAQATALAKGK
```

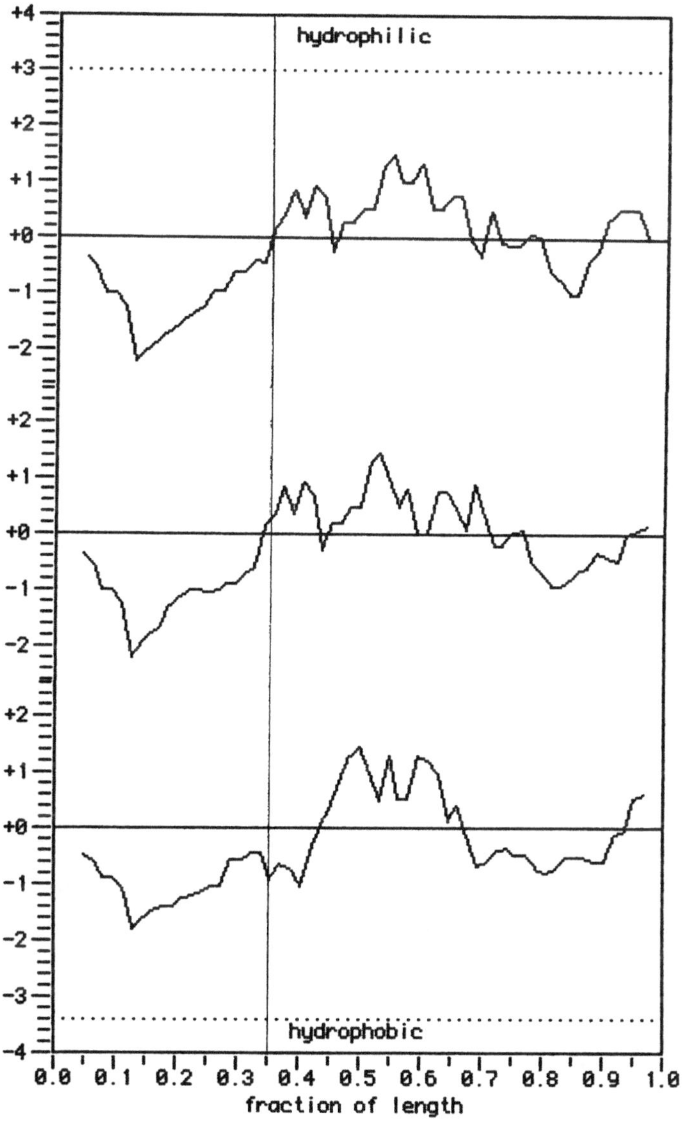

Figure 1. Structure of preprocecropin (ppcec) B, A, and D (preceding page)

Upper panel: Amino acid sequences of ppcec B, A, and D shown in one letter code (Lidholm et al., 1987). Positively (+) and negatively (–) charged amino acid residues as well as the predicted cleavage sites for signal peptidase (underlined) are indicated; ppcec D is given as to show maximal alignment to ppcec B and A.
Lower panel: Hopp-Woods hydrophilicity plots for ppcec B, A, and D; the predicted cleavage site for signal peptidase is indicated at 0.35 fraction of length.

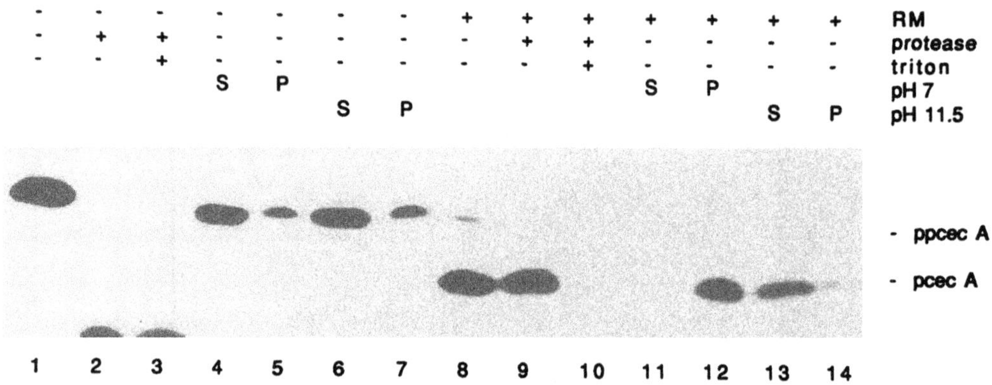

Figure 2. Import of ppcec A into dog pancreas microsomes in a rabbit reticulocyte lysate

Ppcec A was synthesized in rabbit reticulocyte lysates in the presence of [3H]proline either in the absence (lanes 1-7) or presence of dog pancreas microsomes (RM) (lanes 8-14). The concentration of microsomes given as adsorbance at 280 nm in 2 % SDS was 0.6. The reaction mixtures were incubated for 60 min at 37°C and then devided into five aliquots. Three aliquots were further incubated without protease (lanes 1 and 8), with a combination of proteinase K and trypsin (lanes 2 and 9) or with protease in the presence of Triton X-100 (lanes 3 and 10) for 60 min at 0°C as described (Schlenstedt et al., 1989). The other two aliqouts were fractionated into supernatant (S) and pellet (P) fractions by incubation and centrifugation at neutral or alkaline pH-values (lanes 4-7 and 11-14) as described previously (Müller and Zimmermann, 1987). All samples were analyzed by gel electrophoresis and fluorography.

systems programmed with these synthetic transcripts yielded a protein with the expected molecular weight (Fig. 2, lane 1). In the presence of dog pancreas microsomes, preprocecropin A (ppcec A) was processed to procecropin A (pcec A) which was translocated across the microsomal membrane (Fig. 2, lane 8). This was shown i) by protease resistance of the processed form to added protease in the absence of detergent (Fig. 2, lane 9) and its sensitivity in the presence of detergent (Fig. 2, lane 10), and ii) by fractionation in the pellet after centrifugation at neutral pH-values and fractionation in the supernatant at alkaline pH-values (Fig. 2, lanes 11-14). The processing site for signal peptidase was determined by radiosequencing of the processed protein which carried proline residues in positions 2 and 4 (Fig. 3). This is in agreement with the predicted cleavage site of signal peptidase.

Ppcec A and a related hybrid protein, ppcecDHFR, are imported into microsomes as completed polypeptides in an energy-dependent manner

We compared ppcec A with an artificially constructed hybrid protein with respect to the requirements for translocation across dog pancreas microsomal membranes. The hybrid protein ppcecDHFR (252 amino acids) comprises the ppcec A sequence in the N-terminal part and is connected with the mouse dihydrofolate reductase (DHFR) by a threonine residue. DHFR is a cytosolic protein which is known to be compatible with import into microsomes (Müller and Zimmermann, 1987). To test whether import of ppcec A and ppcecDHFR can take place in the absence of protein synthesis, reticulocyte lysate translation reactions performed in the absence of microsomes were terminated by addition of cycloheximide and RNase. Processing and sequestration of both proteins occured following the addition of microsomes (Fig. 4, panel B and D). We conclude that ppcecDHFR represents the first precursor protein which is larger than 75 amino acid residues and is imported into mammalian microsomes in the absence of ongoing protein synthesis. The import efficiency in such a posttranslational assay was compared with a cotranslational assay (Fig. 4, panel A and C). The two precursors showed a difference in the import efficiency in a posttranslational assay, whereas in a cotranslational assay both

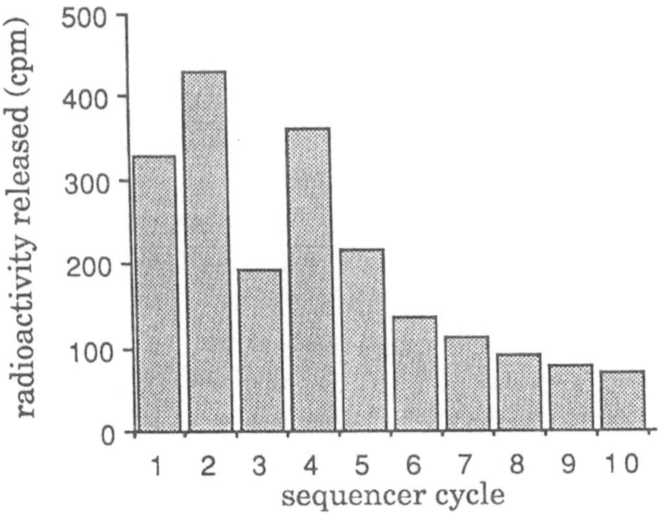

Figure 3. Amino terminus of pcec A generated by dog pancreas microsomes in a rabbit reticulocyte lysate

Ppcec A was synthesized in a rabbit reticulocyte lysate in the presence of [3H]proline and dog pancreas microsomes for 60 min at 37°C. After precipitation with ammonium sulfate as described (Müller and Zimmermann, 1987) the pcec A was subjected to solid phase amino acid sequencing. The samples from the sequencing steps were analyzed for radioactivity.

Figure 4. Processing and sequestration of ppcec A and ppcecDHFR in cotranslational and posttranslational assays in reticulocyte lysate (following page)

A and C: Translation of ppcec A in the presence of [3H]proline (A) and ppcecDHFR in the presence of [35S]methionine (C) was performed in reticulocyte lysate in the absence or presence of dog pancreas microsomes (RM) for 30 min at 37°C. B and D: Translation of ppcec A in the presence of [3H]proline and ppcecDHFR in the presence of [35S]methionine was carried out for 15 min at 37°C; translation was terminated by addition of cycloheximide and RNase A; aliquots were further incubated for 30 min at 37°C in the absence or presence of microsomes. The concentration of microsomes which was used in the respective reaction is given as absorbance at 280 nm. Each aliquot was devided into two halfs and further incubated in the presence or absence of protease. All samples were analyzed by gel electrophoresis and fluorography.

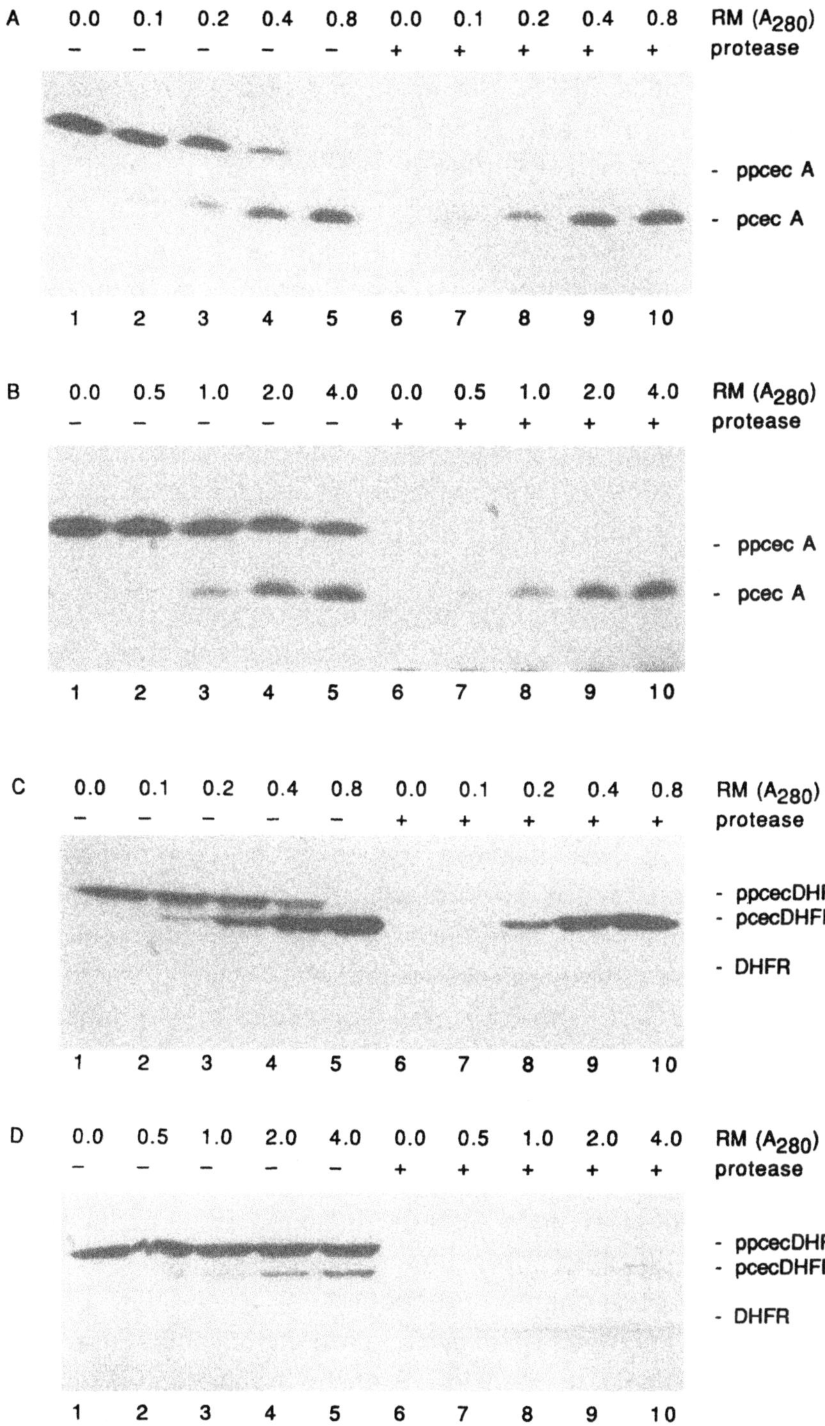
A
0.0 0.1 0.2 0.4 0.8 0.0 0.1 0.2 0.4 0.8 RM (A_{280})
– – – – – + + + + + protease
- ppcec A
- pcec A
1 2 3 4 5 6 7 8 9 10
B
0.0 0.5 1.0 2.0 4.0 0.0 0.5 1.0 2.0 4.0 RM (A_{280})
– – – – – + + + + + protease
- ppcec A
- pcec A
1 2 3 4 5 6 7 8 9 10
C
0.0 0.1 0.2 0.4 0.8 0.0 0.1 0.2 0.4 0.8 RM (A_{280})
– – – – – + + + + + protease
- ppcecDHFR
- pcecDHFR
- DHFR
1 2 3 4 5 6 7 8 9 10
D
0.0 0.5 1.0 2.0 4.0 0.0 0.5 1.0 2.0 4.0 RM (A_{280})
– – – – – + + + + + protease
- ppcecDHFR
- pcecDHFR
- DHFR
1 2 3 4 5 6 7 8 9 10

precursors were translocated almost completely. The processing efficiency in a posttranslational assay reached about 50% for ppcec A and about 25% for ppcecDHFR. However, only about half of the pcecDHFR became protease resistant. Therefore, we tested the effect of methotrexate on posttranslational import of ppcecDHFR into microsomes. We observed a complete loss of the protease resistance of pcecDHFR (however, there was only a slight effect on cotranslational import). Half-maximal inhibition of translocation of pcecDHFR was obtained at a methotrexate concentration of approximately 10 nM (Fig. 5). All other tested natural or artificial substrates of DHFR (dihydrofolate, tetrahydrofolate, trimethoprim and aminopterin) gave similar results, although at different concentrations (not shown). Furthermore, translation in reticulocyte lysate followed by incubation of ppcecDHFR with methotrexate caused almost complete resistance of the DHFR-domain during protease-treatment (Fig. 5). We conclude that folate analogues bind to the DHFR-part within ppcecDHFR and stabilize its native conformation; this apparently allows membrane insertion, as assayed as processing, but blocks completion of translocation. Since methotrexate can only bind to completed and folded polypeptides, one can conclude that the posttranslational import is ribosome-independent.

Is docking protein involved in ribosome-independent import? To answer this question, we included trypsin pretreated microsomes (i.e. microsomes which were depleted of docking protein) in a posttranslational assay. No difference in import efficiencies of ppcec A and ppcecDHFR was obtained as compared to import into untreated microsomes (data not shown). We conclude that posttranslational translocation is independent of docking protein and SRP.

To study the energy-dependence of import in a posttranslational assay, lysates were depleted of ATP and ADP with potato apyrase. Following the addition of microsomes, processing and translocation of ppcec A and ppcecDHFR were strongly inhibited (Fig. 6, lanes 2 and 5 versus 3 and 6). Furthermore, import was reconstituted after ATP-depletion (by the spun-column procedure) by readdition of ATP (not shown). Various ionophores (calimycin, carbonyl cyanide m-chlorophenyl hydrazone, gramicidin D, monensin, nigericin, nonactin, valinomycin, and a combination of nigericin plus valinomycin) as well as the sodium/potassium-ATPase-inhibitor ouabain were tested but did not show any inhibitory effect. We conclude that energy in form of ATP is required and that the ATP is not used to generate a membrane potential.

SRP can interact with the signal sequence of nascent ppcec A and ppcecDHFR polypeptide chains in a productive manner

Addition of SRP causes a signal sequence specific elongation arrest of protein synthesis which is released by docking protein. To investigate the effect of SRP on translation, ppcec A and ppcecDHFR were synthesized together with rabbit globin (a cytosolic protein as a control) in wheat germ lysates (Fig. 7). The amount of transcipts used was limiting, i.e. ribosomes were in excess. Only the amount of added SRP was varied. Translations were synchronized by blocking further initiation after 90 seconds translation with 7-methylguanosine monophosphate. Thus, only a few ribosomes are allowed to translate the mRNAs. Synthesis of globin which served as an internal control was not affected by the addition of purified SRP. The translation of ppcecDHFR was inhibited by SRP in a concentration-dependent manner. SRP showed a similar inhibitory effect on the translation of preprolactin, a precursor which strictly requires SRP for translocation (not shown). We conclude that SRP mediates a typical arrest of ppcecDHFR translation. However, much more SRP is needed to inhibit the translation of ppcec A. The amount which is necessary for half-maximal inhibition is about 3 times higher as compared to ppcecDHFR. This difference must be due to the size of the two precursors since the N-terminal part of ppcecDHFR is identical to the ppcec A sequence. Thus, we can rule out that both precursor proteins differ in their affinity to SRP.

To test whether SRP has an effect on import, wheat germ lysate translation reactions were supplemented with salt-washed (i.e. SRP-depleted) canine microsomes. These microsomes were inactive when tested for import of preprolactin (not shown). Significant processing and sequestration of ppcec A and ppcecDHFR was caused by salt-washed microsomes. Addition of purified canine SRP which reconstitutes the system, however, led to a stimulation of processing and sequestration. Another approach to test the involvement of SRP is to inactivate docking protein by limited protease treatment of microsomes. We

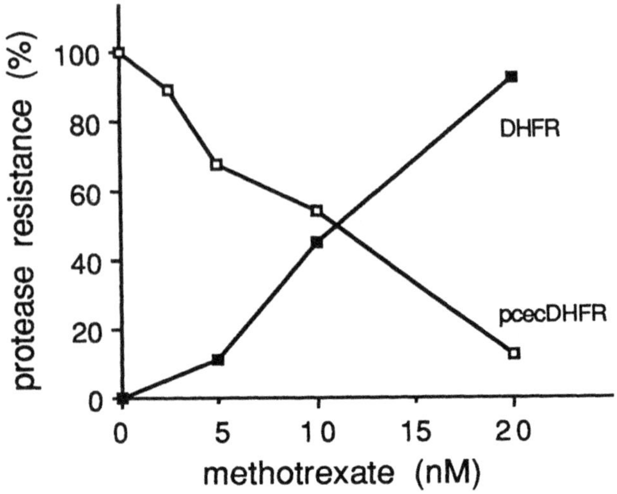

Figure 5. Effect of methotrexate on sequestration of pcecDHFR and on protease sensitivity of the DHFR-domain within ppcecDHFR

Translation of ppcecDHFR was performed in a rabbit reticulocyte lysate in the presence of [35S]methionine for 15 min at 37°C; synthesis was terminated; the reaction mix was devided into five aliquots and these were supplemented with methotrexate at the indicated concentrations. After 5 min microsomes were added; following incubation for 30 min at 37°C the aliquots were devided into two halfs and further incubated in the presence or absence of proteinase K. All samples were subjected to gel electrophoresis, fluography and densitometry. The amount of protease resistant pcecDHFR generated in the absence of methotrexate was set to 100 % and served as control. The protease resistant DHFR-domain was compared with the undigested ppcecDHFR; the proportion of resistant DHFR in the absence of methotrexate was subtracted.

Figure 6. Processing and sequestration of ppcecDHFR after ATP-depletion in posttranslational assays in reticulocyte lysate (following page)

Translation of ppcecDHFR in the presence of [35S]methionine was carried out for 15 min at 37°C; translation was terminated; the reaction mixture was divided into three aliquots, one of these was treated with apyrase (80 units/ml); the different aliquots were further incubated for 30 min at 37°C in the absence or presence of microsomes (RM). Each aliquot was devided into two halfs and further incubated in the presence or absence of protease. All samples were analyzed by gel electrophoresis and fluorography.

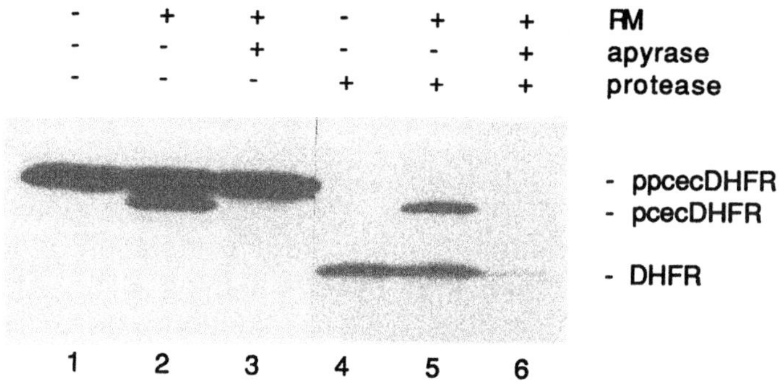
- + + - + + RM
- - + - - + apyrase
- - - + + + protease
- ppcecDHFR
- pcecDHFR
- DHFR
1 2 3 4 5 6

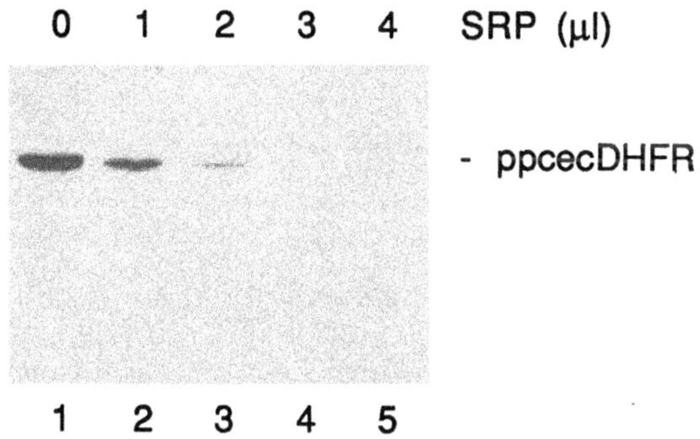
0 1 2 3 4 SRP (μl)
- ppcecDHFR
1 2 3 4 5

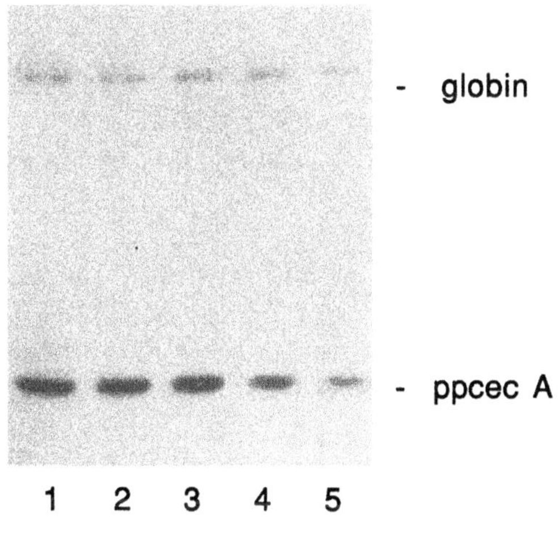
- globin
- ppcec A
1 2 3 4 5

Figure 7. Effect of canine SRP on synthesis of ppcec A, ppcecDHFR and rabbit globin in a wheat germ lysate
(preceding page)

Limiting amounts of synthetic transcripts coding for ppcec A or ppcecDHFR were translated together with rabbit globin poly(A)+ RNA in the presence of [35S]methionine. Each 25µl translation reaction contained the indicated amounts of purified canine SRP. After incubation for 90 seconds at 30°C, the translations were synchronized by addition of 7-methylguanosine-5′-monophosphate (final concentration 6 mM). After further incubation for 30 min at 30°C the samples were analyzed by gel electrophoresis and fluorography.

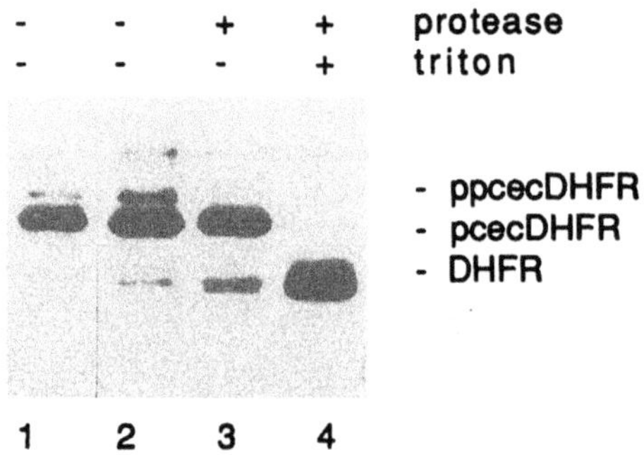

Figure 8. Processing and sequestration of ppcecDHFR in Saccharomyces cerevisiae

A culture of yeast strain DMM1-15A, transformed with a derivative of plasmid YEp51 which contained the DNA coding for ppcecDHFR under the control of the GAL10 promoter, was grown overnight in YP-medium (containing 2 % glucose) at 30°C; after dilution with SD-medium (containing 2 % galactose as well as adenin, uracil, tryptophan and histidin) which resulted in an absorbance at 600 nm of 1.0 the incubation was continued until an absorbance of 1.5 was reached; then [35S]methionine was added (35 µCi/ml); after further incubation for 10 min a 500-fold molar excess of methionine was added and the incubation was continued for 20 min; then the cells were converted to spheroplasts according to standard procedures; the spheroplasts were either immediately subjected to TCA-precipitation (lane 1) or were gently lysed and incubated in the absence or presence of protease minus or plus detergent and then subjected to TCA-precipitation (lanes 2-4); all precipitates were analyzed by immunoprecipitation with antiserum directed against DHFR and Protein A-Sepharose and gel electrophoresis.

tested microsomes which were pretreated with different trypsin concentrations. At a concentration of 3μg trypsin/ml during the pretreatment essentially no intact docking protein could be detected. This is in agreement with previously published results. The import of preprolactin tested in reticulocyte lysate was inhibited to the same extent as docking protein was inactivated. Both precursor proteins ppcec A and ppcecDHFR were imported into docking protein-depleted microsomes during translation in reticulocyte lysate. However, the efficiency was reduced as compared to import into untreated microsomes. When these microsomes were tested in wheat germ lysate, we obtained the same result as with salt-washed microsomes. This rules out the possibility that residual SRP is responsible for the import of ppcec A and ppcecDHFR into these microsomes. Addition of SRP did not stimulate import into trypsin pretreated microsomes indicating that the effect of SRP is due to an interaction with docking protein. We conclude that SRP and docking protein are not necessarily required for the import of ppcec A and ppcecDHFR but can act stimulatory. However, both precursors differed in the import efficiency in the absence of SRP and docking protein. Again, this must be due to the size of the precursors.

Open questions

Because of their inefficiency to use the SRP/ribosome-system the small precursors have apparently evolved with a special role of their mature part (Müller and Zimmermann, 1987, 1988a, 1988b). In prepromelittin and related precursor proteins a single (or cluster of) negatively charged amino acid residue(s) near the amino terminus of the mature part must be balanced by a single (or cluster of) positively charged amino acid residue near the carboxy terminus in order to create a competent precursor protein. While ppcec B and A could easily be fitted into a similar scheme, this is apparently not the case for ppcec D (Fig. 1). Therefore, it should be interesting to compare ppcec A and D with respect to their molecular requirements for import into microsomes. This should shed light onto the structural requirements in the case of the preprocecropins. Along these lines, the cDNA coding for ppcec D was transferred into the plasmid containing the SP6 promoter; future experiments will be directed towards the import requirements of this precursor protein.

Some type of signal peptide receptor appears to be involved in the import of small precursor proteins into mammalian microsomes (Müller and Zimmermann, 1988a, 1988b). In order to characterize and, eventually, purify the receptor from microsomal membranes chemical amounts of precursor proteins are required. Therefore, we are employing chemical synthesis of small precursor proteins as a means of obtaining large amounts of precursor (Boman et al., 1989). These precursor proteins will be characterized as such in competition experiments with in vitro synthesized precursors and after reductive methylation. Furthermore, they will be used to titrate receptor sites and as an affinity probe in a purification.

The observation that methotrexate allowed membrane insertion of ppcecDHFR but caused a block in completion of translocation of pcecDHFR even under cotranslational import conditions confirmed our earlier conclusions on the negative role of a high degree of tertiary structure in a protein which is supposed to become imported into microsomes (for review see: Zimmermann and Meyer, 1986). We are currently trying to characterize the processed but protease-sensitive protein as an intermediate in the import pathway. If this is the case future work will employ the putative intermediate as a tool to get hold of the membrane components which are associated with it. It, furthermore, enabled us to ask if this effect can be reproduced in vivo. The cDNA, coding for ppcecDHFR was excised from the plasmid which contains the SP6 promoter and was transfered to a plasmid which contains the GAL10 promoter and is suitable for inducible expression in Saccharomyces cerevisiae. In preliminary experiments we observed that the hybrid protein was expressed and efficiently imported into the endoplasmic reticulum, as assayed as signal peptide processing and sequestration (Fig. 8). Future experiments will employ aminopterin, a drug which is related to methotrexate in its action on DHFR and is known to effect intact yeast cells.

ACKNOWLEDGEMENTS

This work was supported by the "Sonderforschungsbereich 184: Molekulare Grundlagen der Biogenese von Zellorganellen" (grant B10), by the "Fonds der Chemischen Industrie" and by the Swedish Natural Science Research Council (grant BU 2453).

REFERENCES

Boman HG, Hultmark D (1981) Cell-free immunity in insects. Trends Biochem Sci 6:306-309

Boman HG, Hultmark D (1987) Cell-free immunity in insects. Ann Rev Microbiol 41: 103-126

Boman HG, Boman IA, Andreu D, Li Z-q, Merrifield RB, Schlenstedt G, Zimmermann R (1989) Chemical synthesis and enzymic processing of precursor forms of cecropin A and B. J Biol Chem 264: 5852-5860

Cobet WWE, Mollay C, Müller G, Zimmermann R (1989) Export of honeybee prepromelittin in E. coli depends on the membrane potential but does not depend on proteins secA and secY. J Biol Chem 264: in press

Lidholm D-A, Gudmundsson GH, Xanthopoulos KG, Boman HG (1987) Insect immunity: cDNA clones coding for the precursor forms of cecropins A and D, antibacterial proteins from Hyalophora cecropia. FEBS Lett 226:8-12

Müller G, Zimmermann R (1987) Import of honeybee prepromelittin into the endoplasmic reticulum: Structural basis for independence of SRP and docking protein. EMBO J 6:2099-2107

Müller G, Zimmermann R (1988a) Import of honeybee prepromelittin into the endoplasmic reticulum: Energy requirements for membrane insertion. EMBO J 7:639-648

Müller G, Zimmermann R (1988b) Import of honeybee prepromelittin into the endoplasmic reticulum. In: Bissel M, Deho G, Sironi G, Torriani A (eds) Gene expression and regulation. Elsevier North Holland Amsterdam, pp 199-208

Sagstetter M, Zimmermann R (1988) Assembly of M 13 and M 13 am8H1R1 procoat protein into microsomes is stimulated by rabbit reticulocyte lysate and ATP. Biochem Biophys Res Comm 153:498-501

Schlenstedt G, Zimmermann R (1987) Import of frog prepropeptide GLa into microsomes requires ATP but does not involve docking protein or ribosomes. EMBO J 6:699-703

Schlenstedt G, Gudmundsson GH, Boman HG, Zimmermann R (1989) Binding of folate analogues inhibits import of a preprocecropin

A/dihydofolate reductase hybrid protein into mammalian microsomes. Submitted for publication

Watts C, Wickner W, Zimmermann R (1983) M 13 procoat protein and preimmunoglobulin share processing specificity but use different membrane receptor mechanisms. Proc Natl Acad Sci USA 80:2809 2813

Wiech H, Sagstetter M, Müller G, Zimmermann R (1987) The ATP requiring step in the assembly of M 13 procoat protein into microsomes is related to preservation of transport competence of the precursor protein. EMBO J 6:1011-1016

Zimmermann R, Meyer DI (1986) 1986: A year of new insights into how proteins cross membranes. Trends Biochem Sci 11:512-515

Zimmermann R, Mollay C (1986) Import of honeybee prepromelittin into the endoplasmic reticulum: Requirements for membrane insertion, processing and sequestration. J Biol Chem 261:12889-12895

Zimmermann R, Sagstetter M, Lewis MJ, Pelham HRB (1988a) Seventy kilodalton heat shock proteins and an additional component from reticulocyte lysate stimulate import of M 13 procoat protein into microsomes. EMBO J 7:2875-2880

Zimmermann R, Sagstetter M, Schlenstedt G, Wiech H, Kaßeckert B, Müller G (1988b) Import of small secretory and plasma membrane proteins into the endoplasmic reticulum. In: Op den Kamp JAF (ed.) Membrane Biogenesis. Springer Berlin Heidelberg New York, pp 337-350

SOLUBLE AND MEMBRANE-ASSOCIATED FACTORS REQUIRED FOR PROTEIN TRANSLOCATION INTO THE YEAST ENDOPLASMIC RETICULUM

R.J. Deshaies, A. Eun, B.D. Koch, J.A. Rothblatt, S. Sanders, C. Stirling, and R. Schekman
Department of Biochemistry
University of California
Berkeley, CA 94720
USA

SUMMARY

A genetic approach aimed at defining molecules that participate in protein translocation across the membrane of the endoplasmic reticulum (ER) has led to the identification of three new genes. Mutant cells that fail to translocate secretory proteins into the lumen of the ER have been selected by a procedure that demands retention of a signal peptide-containing hybrid protein in the cytosol. Temperature sensitive mutations were obtained in which precursors of secreted and vacuolar proteins accumulate in forms that are devoid of any post-translational modifications normally coupled to translocation into the ER. Pre-proteins that accumulate at the mutant block sediment with membranes, but are exposed to the cytosol as judged by sensitivity to exogenous protease. The precursor of α-factor is arrested in a form that can complete translocation and secretion during an extended period of chase, confirming *in vitro* results that show protein synthesis and membrane assembly are not obligately coupled.

The effects of two of the mutations, *sec62* and *sec63*, have been reproduced *in vitro*. α-factor precursor, synthesized in an S-100 fraction of either wild-type or *sec* mutant cells, is assembled normally into wild-type ER membranes. Mutant membranes, however, display low and thermolabile translocation activity. Molecular cloning, DNA sequencing and molecular genetic analyses indicate that *SEC61*, *SEC62*, and *SEC63* are essential genes that probably encode integral membrane proteins. Hence, the Sec61, Sec62, and Sec63 proteins may be components of the secretory protein permease.

NATO ASI Series, Vol. H 40
Dynamics and Biogenesis of Membranes
Edited by J.A.F. Op den Kamp

Several lines of evidence suggest that secretory and mitochondrial precursor proteins require cytosolic factor(s) that stabilize a conformation that is compatible with membrane penetration. Among candidates for the cytosolic factors we have considered the 70 kD heat shock (stress) proteins which are thought to influence protein folding. Four genes (*SSA1* to *4*) in *Saccharomyces cerevisiae* encode interchangeable homologues of the 70 kD stress protein family. Deletion of the *SSA1*, *2*, and *4* genes leads to cell inviability which can be overcome by introduction of *SSA1* on a plasmid. Repression of the plasmid-encoded copy causes depletion of the remaining 70 kD stress protein. During such depletion, cells accumulate untranslocated forms of α-factor precursor, and the precursor of the $F_1\beta$ subunit of mitochondrial ATPase. From this we suggest that the 70 kD stress proteins serve to unfold secretory and mitochondrial precursors so that a compact globular structure, incompatible with membrane penetration, is prevented from forming.

INTRODUCTION

Recently, several groups have reconstituted protein translocation into the yeast ER *in vitro* (Rothblatt and Meyer, 1986a; Hansen et al., 1986; Waters and Blobel, 1986). A yeast translation extract programmed with prepro-α-factor mRNA directs the synthesis of an intact precursor which can insert co- or post-translationally into yeast microsomes in an ATP-dependent manner and become core-glycosylated. Import into the yeast ER requires both cytosolic (Waters et al., 1986) and membrane (Sanz and Meyer, 1988) proteins. The reconstitution of yeast protein translocation *in vitro* presents an opportunity to combine a biochemical analysis of protein translocation with a genetic approach aimed at identifying genes whose products participate in the reconstituted reaction.

RESULTS

A Selection Scheme for Yeast Translocation Mutants.

Among a large collection of temperature-sensitive, secretion-defective mutants of *S. cerevisiae* isolated in this laboratory (Novick and Schekman, 1979; Novick et al., 1980; Ferro-Novick et al., 1984), members of only two complementation groups (*sec53* and *sec59*) affect early events in protein secretion, though neither mutation blocks protein translocation into the ER (Feldman et al., 1987; Bernstein et al., 1989). In an attempt to identify genes required for the translocation event, we have developed a direct selection for temperature-sensitive (Ts) mutants of *S. cerevisiae* that are defective in the import of secretory precursor proteins into the lumen of the ER (Deshaies and Schekman, 1987).

Numerous experiments have demonstrated that a signal sequence can direct cytoplasmic proteins to the secretory pathway *in vivo* (Emr et al., 1984; Sharma et al., 1985; Simon et al., 1987). Based on these data, we reasoned that if a signal sequence were attached to a cytoplasmic enzyme required for the formation of an essential nutrient, the hybrid protein would be directed to the ER. If the substrate of the enzyme were limited to the cytoplasm, the cells would not grow unless they were supplied with the nutrient. Mutations that block hybrid protein import into the ER would allow cells to grow on the substrate.

We chose the *HIS4* gene because it encodes a trifunctional cytoplasmic polypeptide that can sustain amino-terminal protein fusion events and retain histidinol dehydrogenase activity associated with the carboxy-terminal domain (Keesey et al., 1979; Parker and Guthrie, 1985). Histidinol dehydrogenase catalyzes the last step in histidine biosynthesis, the conversion of histidinol to histidine. Cells that express wild-type His4 protein therefore are able to convert exogenously supplied histidinol to histidine.

We predicted that if a signal sequence were fused to the amino terminus of the His4 protein, it would be translocated into the lumen of the ER and glycosylated, as shown schematically in Fig 1A. Wild-type cells (bearing a chromosomal *his4* deletion) expressing this plasmid-encoded fusion protein would not convert histidinol to histidine in that the histidinol dehydrogenase would be sequestered from the cytoplasm, and charged, polar molecules such as histidinol or histidine are unlikely to penetrate across the ER membrane. Therefore these cells would not grow on medium containing histidinol in place of histidine. If cells containing this gene fusion were selected for growth on histidinol, mutants that

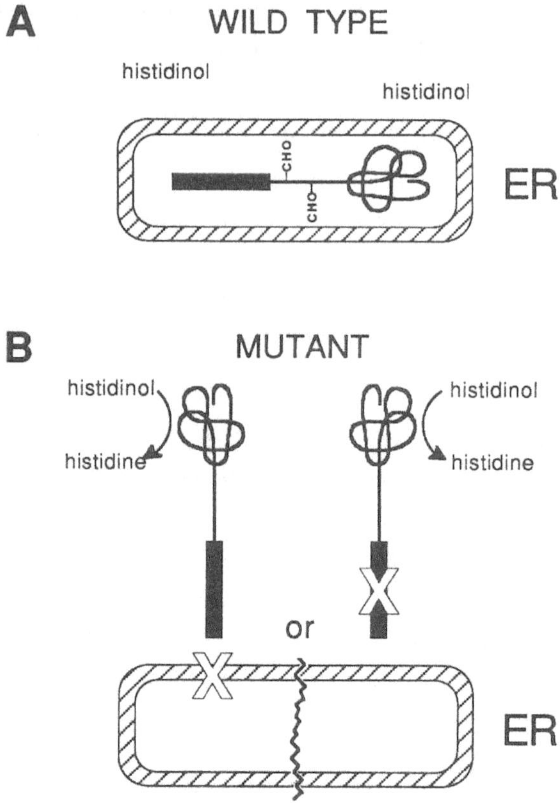

Figure 1. Predicted topology of a signal peptide-containing His4 hybrid protein. The predicted subcellular location and enzymatic properties of a histidinol dehydrogenase fusion protein are shown for both (A) wild-type and (B) mutant cells. (CHO) Asparagine-linked carbohydrate; (▬) a signal sequence; and (folded black trace) the catalytic domain of the fusion protein. (X) A mutation that inactivates either the signal sequence or some cellular component (membrane-bound or cytosolic) required for protein translocation.

mislocalized the fusion protein to the cytoplasm would convert the histidinol to histidine (Fig. 1B) and grow.

Potentially this procedure would select for mutations that disrupted signal sequence function or mutations that disabled the cellular machinery responsible for targeting secretory proteins to the ER. These alternative possibilities are indicated by X marks in Fig. 1B. Genetic tests (ie, plasmid linkage) may be used to distinguish between these two classes of mutations. We focused on mutations that caused pleiotropic defects in the ER targeting apparatus.

In order to recover viable, translocation-defective cells, it was necessary to strike a balance between the lethal effect of secretory mutations and the requirement for sufficient cytosolic histidinol dehydrogenase activity. For this purpose, the growth temperature (24°C-30°C) represented a semipermissive condition where the mutant phenotype was only partially expressed. To identify mutants with more severe mislocalization defects, clones selected for growth on histidinol at 24°C-30°C were screened on rich medium for Ts growth at 37°C, presumably owing to an exaggeration of the partial defect expressed at lower temperatures.

Among a large collection of Ts growth mutants selected by this procedure (~800 independent isolates), about 15% were defective in the translocation of precursor proteins into the secretory pathway. Genetic analysis of these selected mutants established three complementation groups each defined by fifteen or more alleles. A representative of each group was further evaluated for linkage of the Ts growth and protein secretion defects. These phenotypes coincided for each group, and the mutant genes were named *sec61*, *sec62* and *sec63*.

Additional Ts alleles of *sec61* and Ts mutations in a gene named *sec65* have recently been isolated using a modified form of the histidinol selection procedure. This new scheme employed a fusion protein containing the first six transmembrane domains of HMG-CoA reductase (encoded by *HMG1*, see Basson et al., 1988) fused amino-terminal to histidinol dehydrogenase. The *HMG1*-based gene fusion directed the synthesis of a fusion protein that was targeted to the ER membrane. The topology of the chimeric protein was such that the histidinol dehydrogenase domain resided in the lumen of the ER.

Defective Protein Translocation *in vivo*

The biogenesis of the secreted mating pheromone α-factor was compared in wild-type and *sec* mutant strains. At the semi-permissive and restrictive temperatures, unglycosylated α-factor precursor accumulated in a form lacking any of the covalent modifications that occur on the lumenal surface of the ER. More direct analysis indicated that the accumulated precursor was exposed to the cytosol in homogenates of mutant cells. A similar pattern of accumulation was observed for precursors of the vacuolar protein, carboxypeptidase Y, and the secreted proteins acid phosphatase and invertase. Mutant *sec62* cells also displayed a considerable defect in the proper integration of the Golgi body membrane protein Kex2. These findings suggest that the *sec* gene products are required for efficient translocation of soluble and membrane-spanning precursor proteins into the ER.

Whereas the translocation of some precursor proteins (prepro-carboxypeptidase Y, pre-acid phosphatase and prepro-α-factor) was strongly blocked in *sec* mutant cells, other proteins (pre-invertase and dipeptidyl aminopeptidase B) were only weakly blocked or not affected at all. A recently isolated *sec61* mutant strain displayed a severe defect in translocation into the ER of all precursor proteins examined, including dipeptidyl aminopeptidase B. The differential effects on protein translocation observed with the original *sec* mutants are not understood, though the results obtained with the new allele of *sec61* suggest that at least the Sec61 protein may be a general translocation factor required for the import of most, if not all, membrane-spanning and soluble proteins into the ER.

Translation and membrane translocation of α-factor precursor are not obligately coupled (Hansen et al., 1986; Waters and Blobel, 1986; Rothblatt and Meyer, 1986b). Precursor synthesized *in vitro* may be assembled across membranes added either during or after completion of protein synthesis. *In vivo*, the precursor accumulated in *sec61* mutant cells was slowly, but efficiently (~80 complete) translocated into and transported through the secretory pathway, emerging as mature pheromone secreted into the growth medium. Hence, the *sec61* defect represents a kinetic block that does not permanently disable either the translocation apparatus or the α-factor precursor.

Membrane-localized Defects in Translocation

A more precise definition of the lesions in *sec62* and *sec63* mutant cells comes from a comparison of α-factor precursor translocation into mutant and wild type membranes *in vitro*. Precursor translated in a soluble lysate from mutant or wild type cells was efficiently translocated into wild type membranes. In contrast, precursor made in either lysate was inefficiently imported by membranes isolated from *sec62* or *sec63* mutant cells grown at a semi-permissive temperature (24°C). The low level of translocation into *sec63* membranes was abolished by preincubation at 30°C. Mutant *sec62* membranes displayed the same pattern of selectivity in translocation of different precursor proteins *in vitro* as was observed with intact cells. Prepro-α-factor and prepro-carboxypeptidase Y were poorly assembled into *sec62* microsomes, whereas pre-invertase was imported at nearly wild-type levels. The thermolability observed with *sec63* microsomes and precursor selectivity observed with *sec62* microsomes suggest that the defects obtained in reconstituted reactions containing mutant components were directly a consequence of the *sec* mutations.

Molecular Analysis of *SEC* Genes

The results described above imply that some of the *SEC* gene products are required for the proper elaboration of membrane-associated functions involved in translocation of prepro-α-factor into the ER lumen. To test whether *SEC62* encodes a potential ER membrane spanning protein, we isolated the *SEC62* gene,

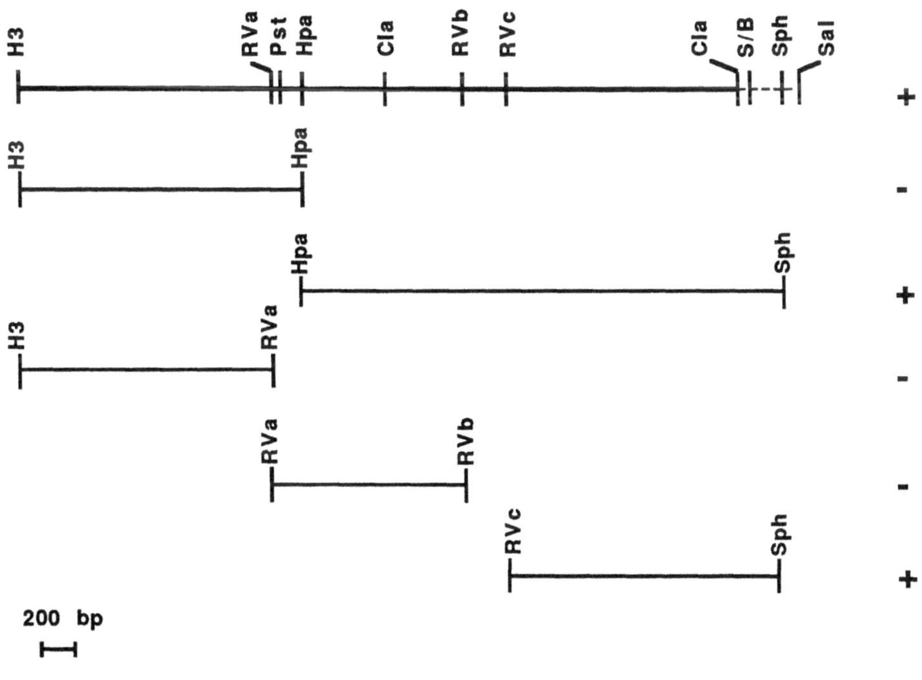

Figure 2. Complementation of the *sec62* Ts growth defect by different subclones of pSEC6207. A restriction map of the genomic yeast DNA insert of the *sec62*-complementing plasmid pSEC6207 is shown on the top line. Different regions of this insert were subcloned into the centromere-containing plasmid pSEYc68 and tested for their ability to complement the Ts growth defect of *sec62* strains. Results of the complementation experiments are indicated in the right hand column. (+), complements the Ts growth defect of *sec62* strains; (-), fails to complement the Ts growth defect of *sec62* strains; H3, HindIII; RVabc, distinct Eco RV sites; Pst, PstI; Hpa, HpaI; Cla, ClaI; S/B, the SauIIIA/BamHI boundary of genomic and vector sequences; Sph, SphI; Sal, SalI. The dotted line indicates sequences derived from the Tet region of YCp50.

determined its nucleotide sequence and examined the structure of the predicted *SEC62* polypeptide (Sec62p). We have also cloned the *SEC61* and *SEC63* genes and determined the DNA sequence of *SEC61*.

DNA fragments that complemented the Ts growth phenotypes of *sec61*, *sec62* and *sec63* strains were isolated by transformation with a yeast genomic bank in a single copy centromeric plasmid. In each case, subcloning identified a small region (<3 kb) of genomic DNA capable of conferring Ts+ growth (see Fig 2 for a restriction map and subclones of *SEC62*). Besides curing the Ts growth defects of the *sec* strains, these plasmids restored proper translocation of secretory precursors into the ER. Restriction mapping of the complementing plasmids indicated that they were distinct from each other. Homologous integration of the cloned genomic sequences at their corresponding chromosomal loci confirmed that we had isolated the authentic *SEC61*, *SEC62*, and *SEC63* (Sadler and Silver, personal communication) genes.

The nucleotide sequences of *SEC61* and *SEC62* were determined, and each gene was found to contain a long open reading frame. *SEC62* potentially encoded a 283 amino acid, 32 kD polypeptide with an unusually basic isoelectric point of 10.7. Hydropathy analysis revealed that the predicted Sec62p contained two stretches of amino acids that were sufficiently long and hydrophobic to span a lipid bilayer (amino acids 159-178, 187-218). Notably, Sec62p lacked an amino-terminal hydrophobic stretch characteristic of signal peptides. Based on the hydropathy analysis, we predict that Sec62p spans the ER membrane twice, leaving its amino and carboxy terminal domains cytoplasmically-oriented. A model of the predicted topology of Sec62p is shown in Fig 3.

SEC61 potentially encoded a 480 amino acid, 53 kD polypeptide. Hydropathy analysis indicated that Sec61p contained numerous hydrophobic stretches of amino acids capable of spanning a lipid bilayer. Presumably, Sec61p is a multi-spanning ER membrane protein. The overall predicted topology of Sec61p is similar to that of SecY (Akiyama and Ito, 1987), which is a multi-spanning cytoplasmic membrane protein of *Escherichia coli* required for export and assembly of periplasmic, inner membrane and outer membrane proteins. Comparison of the Sec61p and SecY protein sequences identified a region of 23.8% identity over 122 residues near the amino termini of both proteins. Perhaps Sec61

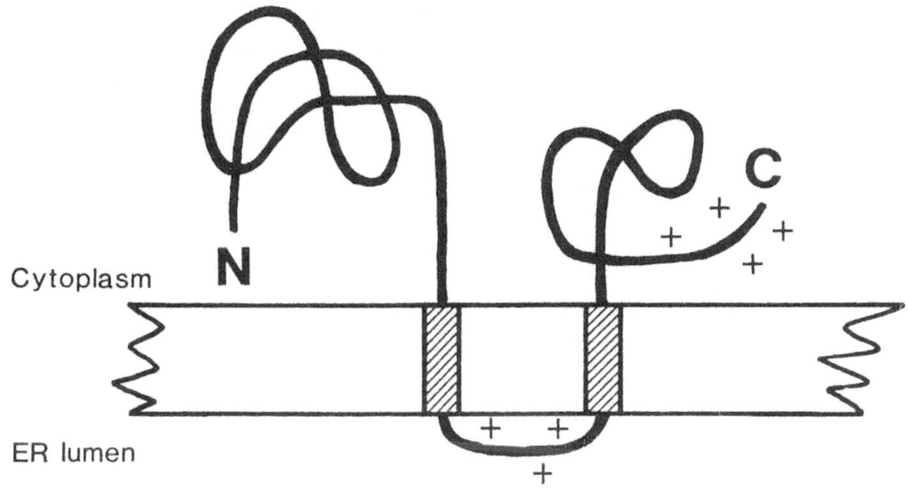

Figure 3. Proposed orientation of Sec62p in the endoplasmic reticulum membrane. N and C refer to the amino and carboxy termini of Sec62p, respectively. The predicted transmembrane domains are designated as hatched boxes. (+) refers to clusters of positive charges at the C-terminus of Sec62p, and to three arginine residues situated between the two hydrophobic domains.

and SecY perform similar functions in protein translocation. The polytopic nature of these proteins makes it tempting to speculate that they are components of a proteinaceous pore.

SEC63 has also been cloned independently in another laboratory. Preliminary sequence data indicated that the predicted Sec63p contained hydrophobic segments capable of functioning as transmembrane domains (P. Silver, personal communication)

We originally adopted a genetic approach to identify and characterize novel components of the protein translocation machinery of the yeast ER membrane. Specifically, we hoped to discover membrane-associated translocator proteins, since these have proved to be especially intractable to conventional biochemical analysis. The biochemical and nucleotide sequence data described above suggest that our approach has indeed led to the identification of membrane-localized translocation factors. The availability of cloned DNA encoding Sec61p, Sec62p and

Sec63p should allow for production of specific antisera, determination of transmembrane topologies and mapping of functional domains of these polypeptides. This information should lead to a more refined picture of the structure and function of the ER membrane secretory protein translocator.

Soluble Proteins Required for Translocation

Soluble proteins, aside from SRP, have been reported to stimulate post-translational translocation of precursor proteins across various target membranes. One of these, trigger factor, forms a complex with renaturing pro-OmpA to facilitate translocation into E. coli cytoplasmic membrane vesicles (Crooke et al., 1988). Import of the mitochondrial F_1 ATPase ß subunit is stimulated by a soluble factor found in yeast or reticulocyte lysates (Ohta and Schatz, 1984; Chen and Douglas, 1987b). α-factor precursor translocation seems also to depend on a soluble factor that is present in yeast but is much less abundant in a wheat germ lysate (Waters et al., 1986).

What purpose might these soluble factors serve? Post-translational import of precursor proteins into mitochondria requires both ATP hydrolysis outside of the mitochondrial inner membrane (Pfanner et al., 1987; Eilers et al., 1987; Chen and Douglas, 1987a) and unfolding of the transported polypeptide (Eilers and Schatz, 1986). Rothman and Kornberg (1986) suggested that the energy of phosphodiester bond cleavage may be harnessed by a factor to unravel mitochondrial precursors ensuring a translocation competent conformation.

70 kD Heat Shock Proteins

We have considered the possibility that members of the 70 kD heat shock, or stress protein (hsp70) family play a role in stabilizing secretory precursor polypeptides. Genes encoding these proteins were first isolated from *Drosophila*, and have since been found in every organism tested, including *E. coli*, yeast, and humans. Eukaryotic organisms typically contain several hsp70 related sequences. Regulation of these genes is complex; some hsp70 genes are expressed only after cells are stressed, whereas others are expressed constitutively. A role for hsp70 homologues in normal processes is indicated by the fact that two subfamilies of hsp70 genes are essential for yeast cell growth (Werner-Washburne et al., 1987; Craig et al., 1987).

A variety of functions have been postulated for 70 kD stress proteins based on their biochemical characteristics. A common property of hsp70 proteins is their ability to bind ATP with high affinity. A member of the 70 kD stress protein family catalyzes an ATP-dependent depolymerization of clathrin from coated vesicles *in vitro* (Chappell et al., 1986; Schlossman et al., 1984; Ungewickell, 1985). ATP hydrolysis has also been implicated in the functions of the major heat-inducible hsp70, which is released from tight binding sites in the nucleolus by micromolar concentrations of hydrolyzable ATP (Lewis and Pelham, 1985). Lastly, ATP disrupts the association between the hsp70-like immunoglobulin heavy chain binding protein and immunoglobulin heavy chains in IgG immunoprecipitates (Munro and Pelham, 1986). These properties of the 70 kD stress proteins led Pelham to suggest that hsp70s couple ATP hydrolysis to the disruption of quaternary structure in multisubunit complexes or denatured aggregates. A simple extension of this model could accomodate a function for hsp70 in the relaxation of tertiary structure within a single polypeptide chain. We speculated that an hsp70-like protein might function as an 'unfoldase' and that hsp70s might also act in the post-translational import of certain secretory proteins into the lumen of the ER. To test this hypothesis, we examined ER and mitochondrial biogenesis in *S. cerevisiae* strains bearing mutated hsp70 genes.

hsp70 Depletion *in vivo*

A major subfamily of hsp70 genes in yeast is defined by four loci referred to as *SSA1*-*SSA4* (*SSA*, stress subfamily A). The interactions between these genes suggest that the *SSA* gene products perform interchangeable functions *in vivo* (Werner-Washburne et al., 1987). *SSA1* and *SSA2* are transcribed constitutively, though the level of *SSA1* mRNA increases several fold after a heat shock. *SSA3* and *SSA4* are only expressed in cells exposed to high temperatures (Werner-Washburne et al., 1987). Whereas yeast strains bearing single disruptions in any one of the four genes are phenotypically indistinguishable from wild type, *ssa1 ssa2* strains grow slowly at 24°C and do not grow at 37°C, and *ssa1 ssa2 ssa4* cells are inviable. Thus, the *SSA* genes define a subgroup of hsp70-related loci in *S. cerevisiae* that perform an essential function (Werner-Washburne et al., 1987).

To examine the consequences of an *SSA* deficiency in relation to protein translocation, a strain that conditionally expresses a single *SSA* gene was generated. This strain harbors chromosomal disruptions of the *SSA1*, *SSA2*, and *SSA4* genes, but is rescued by a single copy of *SSA1* present on a plasmid (Werner-

Washburne et al., 1987). The plasmid contains the coding region of *SSA1* fused to the yeast *GAL1* promoter. This allows expression of the *SSA1* gene product (Ssa1p) on galactose-containing medium, and repression of Ssa1p synthesis on glucose-containing medium. Cells grown on galactose overproduce Ssa1p, but lose the protein after 8 hours of incubation in glucose.

Extracts prepared from cells at various stages of Ssa1p depletion contained immunoreactive forms of α-factor precursor identical to those accumulated in *sec61*, *sec62*, and *sec63* cells. This accumulation depended on depletion of all members of the Ssa protein family. No block in α-factor secretion was seen in strains containing a deletion of *SSA1*, *SSA2*, or *SSA4*, or in a strain containing a deletion of *SSA1* and *SSA2*. Kinetic radiolabeling experiments demonstrated that Ssa1p depletion led to a progressive block in glycosylation and secretion of α-factor which was virtually complete after 8 hours of growth on glucose. A defect in translocation was established with the demonstration that α-factor precursor was degraded by trypsin added to gently prepared lysates of Ssa1p-depleted cells.

A requirement for hsp70 homologues in the assembly of other membranes was indicated by the accumulation of a mitochondrial precursor polypeptide during Ssa1p depletion. Mitochondrial F_1 ATPase ß subunit precursor (F_1ß) accumulated roughly in parallel with the block in α-factor maturation as mutant cells were incubated in glucose. A block in F_1ß assembly was not seen in *sec61*, *sec62*, or *sec63* cells, suggesting that the secretory mutant defects were restricted to the ER, while hsp70 homologues participated in diverse translocation events.

<u>Ssa1p and Translocation *in vitro*</u>

A requirement for Ssa proteins in translocation of prepro-α-factor into yeast microsomal vesicles *in vitro* has also been shown (Chirico et al., 1988; Deshaies et al., 1988). α-factor precursor synthesized in a wheat germ lysate was poorly assembled into yeast ER vesicles, but uptake was stimulated approximately five-fold by a soluble extract from yeast (Waters et al., 1986). Purification of this cytosolic translocation-promoting activity yielded two related 70 kD polypeptides (Chirico et al., 1988). These proteins cross-reacted with antibodies raised against the *Drosophila* heat shock protein hsp70 and bound ATP, suggesting that they were members of the yeast hsp70 family. Comparison (by two dimensional gel electrophoresis) of the purified 70 kD proteins with extracts of yeast cells lacking individual members of the hsp70 family established that the translocation-promoting polypeptides corresponded to the constitutively expressed *SSA1* and

SSA2 gene products (Chirico et al., 1988). In addition, ATP-binding proteins isolated from wild type cells stimulated import of wheat germ derived prepro-α-factor into yeast microsomes whereas ATP-binding proteins isolated from Ssa1p depleted cells failed to support translocation. Ssa proteins also stimulated import of delta^1-pyrolline-5-carboxylate dehydrogenase (Put2p) into mitochondria *in vitro*. Besides Ssa proteins, maximal import of Put2p requires an additional N-ethyl maleimide (NEM) -sensitive factor (Murakami et al., 1989).

Besides the Ssa proteins, one or more NEM-sensitive soluble factors (factor X) are required for maximal import of wheat germ synthesized prepro-α-factor into yeast microsomes (Chirico et al., 1988; Deshaies et al., 1988). Preliminary characterization indicated that factor X is likely to be a protein, since it is trypsin- and heat-sensitive. Gel filtration chromatography revealed that factor X had an apparent molecular weight of 500 kD, suggesting that factor X activity may reside in a multimeric complex. It is not known whether factor X is related to the NEM-inactivated component required for import of Put2p into mitochondria.

A Model of hsp70 Function in Protein Translocation

Members of the hsp70 family may be responsible for a variety of cellular processes including the recovery of nucleolar morphology following heat stress (Pelham, 1984) and the removal of clathrin triskelions from endocytic coated vesicles (Chappell et al., 1986; Ungewickell, 1985). Given the large number of hsp-related sequences in the yeast, *Drosophila*, human, and mouse genomes, it seems reasonable to propose an even wider and more ubiquitous role for these proteins. The high degree of sequence homology detected among yeast or *Drosophila* hsp70 genes implies that some hsp70 polypeptides share overlapping activities or perform similar activities in different compartments. *SSA* gene products, for example, appear to function both in mitochondrial and ER import pathways, though each may have distinct primary responsibilities while retaining the capacity to act interchangeably.

The simultaneous requirement for hsp70 function in the proper localization of proteins to distinct subcellular compartments suggests that hsp70s fulfill a general mechanistic requirement for the transfer of precursor proteins across target membranes. A model that extends upon previous work and incorporates these new proposals is presented in Fig 4. This hypothesis can easily be accomodated within a model for hsp70 function suggested by Pelham (1986). The ATP-driven cycle of hsp70 binding to and release from substrates targeted to the

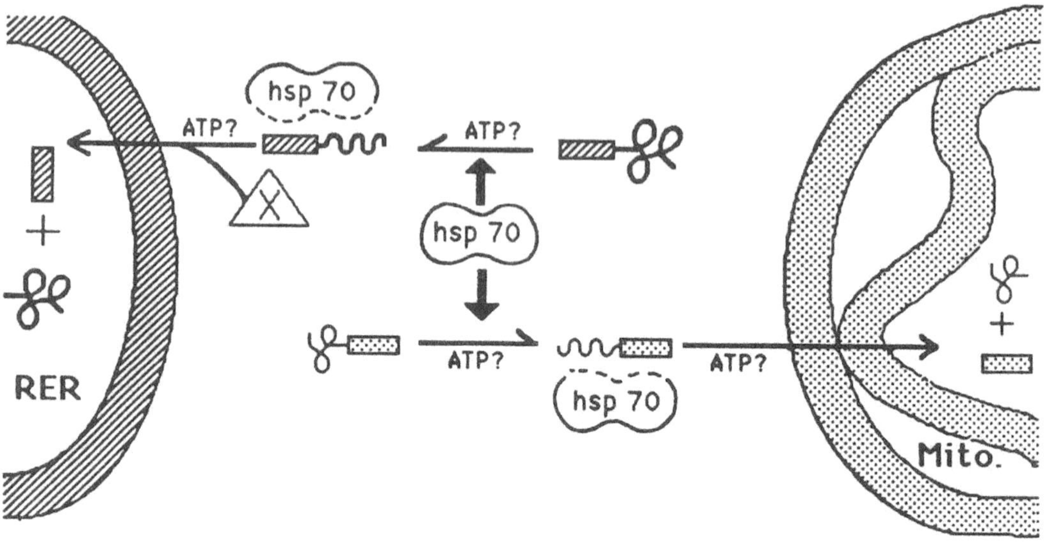

Figure 4. Proposed role of hsp70s in protein translocation. One or more hsp70 binds to a secretory or mitochondrial precursor polypeptide. The position of this interaction is unknown, as indicated by the dotted lines at the hsp70-precursor interface. Binding may occur as the precursor is being made or after completion of the polypeptide. The initial interaction may consume ATP, or ATP may be bound but not hydrolyzed by the complex. For import into the ER, at least one additional yeast lysate component is required (factor X). We propose that this factor(s) targets secretory proteins to the ER membrane. A translocation competent complex then associates via a signal or transit peptide with the proper target membrane and as penetration begins the complex dissociates allowing the hsp70 to be recycled. ATP hydrolysis may facilitate this dissociation step.

mitochondria or ER may act to dissolve aggregates/multimers formed by untranslocated precursors (Chen and Douglas, 1987b), or to relax the tertiary structure of polypeptides prior to membrane penetration. Alternatively, hsp70 may contact unfolded nascent chains and prevent them from adopting conformations incompatible with translocation. Lastly, hsp70 action may result in a specific configuration required for membrane assembly. The interaction of hsp70s with precursor proteins and the conformation that precursors adopt under the influence of hsp70s can now be examined with purified components.

REFERENCES:

Akiyama Y and Ito K (1987) Topology analysis of the SecY protein, an integral membrane protein involved in protein export in Escherichia coli. EMBO J. 6: 3465-3470

Basson ME, Thorsness M, Finer-Moore J, Stroud RM and Rine J (1988) Structural and functional conservation between yeast and human 3-hydroxy-3-methylglutaryl coenzyme A reductases. Mol. Cell. Biol. 8: 3797-3808

Bernstein M, Kepes F and Schekman R (1989) SEC59 encodes a membrane protein required for core glycosylation in Saccharomyces cerevisiae. Mol. Cell. Biol. 9: 1191-1199

Chappell TG, Welch WJ, Schlossman DM, Palter KB, Schlesinger MJ and Rothman JE (1986) Uncoating ATPase is a member of the 70 kilodalton family of stress proteins. Cell 45: 3-13

Chen W-J and Douglas MG (1987a) Phosphodiester bond cleavage outside mitochondria is required for the completion of protein import into the mitochondrial matrix. Cell 49: 651-658

Chen W-J and Douglas MG (1987b) The role of protein structure in the mitochondrial import pathway. J. Biol. Chem. 262: 15598-15604

Chirico WJ, Waters MG and Blobel G (1988) 70 K heat shock related proteins stimulate protein translocation into microsomes. Nature 332: 805-810

Craig EA, Kramer J, and Kosic-Smithers J (1987) SSC1, a member of the 70-kDa heat shock protein multigene family of Saccharomyces cerevisiae, is essential for growth. Proc. Natl. Acad. Sci. USA 84:4156-4160

Crooke E, Guthrie B, Lecker S, Lill R and Wickner W (1988) proOmpA is stabilized for membrane translocation by either purified E. coli trigger factor or canine SRP. Cell 54: 1003-1011

Deshaies RJ and Schekman R (1987) A yeast mutant defective at an early stage in import of secretory protein precursors into the endoplasmic reticulum. J. Cell Biol. 105: 633-645

Deshaies RJ, Koch BD, Werner-Washburne M, Craig EA, and Schekman R (1988) 70 kD stress protein homologues facilitate translocation of secretory and mitochondrial precursor polypeptides. Nature 332: 800-805

Eilers M and Schatz G (1986) Binding of a specific ligand inhibits import of a purified precursor protein into mitochondria. Nature 332: 228-232

Eilers M, Ipplinger W and Schatz G (1987) Both ATP and an energized inner membrane are required to import a purified precursor protein into mitochondria. EMBO J 6: 1073-1077

Emr SD, Schauer I, Hansen W, Esmon P and Schekman R (1984) Invertase-β-galactosidase hybrid proteins fail to be transported from the endoplasmic reticulum in Saccharomyces *cerevisiae*. Mol. Cell. Biol. 4: 2347-2355

Feldman RI, Bernstein M and Schekman R (1987) Product of SEC53 is required for folding and glycosylation of secretory proteins in the lumen of the yeast endoplasmic reticulum. J. Biol. Chem. 262: 9332-9339

Ferro-Novick S, Novick P, Field C and Schekman R (1984a) Yeast secretory mutants that block the formation of active cell surface enzymes. J Cell Biol. 98: 35-43

Hansen W, Garcia PD and Walter P (1986) In vitro translocation across the yeast endoplasmic reticulum: ATP-dependent post-translational translocation of the prepro-α-factor. Cell 45: 397-406

Keesey JK, Bigelis R, and Fink GR (1979) The product of the his4 gene cluster in Saccharomyces cerevisiae. J. Biol. Chem. 254: 7427-7433

Lewis MJ and Pelham HRB (1985) Involvement of ATP in the nuclear and nucleolar functions of the 70 kd heat shock protein. EMBO J. 4: 3137-3143

Munro S and Pelham HRB (1986) An hsp70-like protein in the ER: identity with the 78 kd glucose-regulated protein and immunoglobulin heavy chain binding protein. Cell 46: 291-300

Murakami H, Pain D and Blobel G (1988) 70-kD heat shock-related protein is one of at least two distinct cytosolic factors stimulating protein import into mitochondria. J. Cell Biol. 107: 2051-2057

Novick P and Schekman R (1979) Secretion and cell-surface growth are blocked in a temperature-sensitive mutant of Saccharomyces cerevisiae. Proc. Natl. Acad. Sci. USA 76: 1858-1862

Novick P, Field C and Schekman R (1980) Identification of 23 complementation groups required for post-translational events in the yeast secretory pathway. Cell 21: 202-215

Ohta S and Schatz G (1984) A purified precursor polypeptide requires a cytosolic protein fraction for import into mitochondria. EMBO J 3: 651-657

Parker R and Guthrie C (1985) A point mutation in the conserved hexanucleotide at a yeast 5' splice junction uncouples recognition, cleavage, and ligation. Cell 41: 107-118

Pelham HRB (1986) Speculations on the functions of the major heat shock and glucose-regulated proteins. Cell 46: 959-961

Pfanner N, Tropschug M and Neupert W (1987) Mitochondrial protein import: nucleoside triphosphates are involved in conferring import-competence to precursors. Cell 49: 815-823.

Rothblatt JA and Meyer DI (1986a) Secretion in yeast: reconstitution of the translocation and glycosylation of α-factor and invertase in a homologous cell-free system. Cell 44: 619-628

Rothblatt J and Meyer DI (1986b) Secretion in yeast: translocation and glycosylation of prepro-α-factor in vitro can occur via an ATP-dependent post-translational mechanism. EMBO J. 5: 1031-1036

Rothman JE and Kornberg RD (1986) An unfoloding story of protein translocation. Nature 322: 209-210

Sanz P and Meyer DI (1988) Signal recognition particle (SRP) stabilizes the translocation-competent conformation of pre-secretory proteins. EMBO J 7: 3553-3557

Schlossman DM, Schmid SL, Braell WA and Rothman JE (1984) An enzyme that removes clathrin coats: purification of an uncoating ATPase. J Cell Biol. 99: 723-733

Sharma S, Rodgers L, Brandsma J, Gething M-J and Sambrook J (1985) SV40 T antigen and the exocytotic pathway. EMBO J. 4: 1479-1489

Simon K, Perara E and Lingappa, VR (1987) Translocation of globin fusion proteins across the endoplasmic reticulum membrane in Xenopus laevis oocytes. J. Cell Biol. 104: 1165-1172

Ungewickell E (1985) The 70-kd mammalian heat shock proteins are structurally and functionally related to the uncoating protein that releases clathrin triskelia from coated vesicles. EMBO J 4: 3385-3391

Waters MG and Blobel G (1986) Secretory protein translocation in a yeast cell-free system can occur post-translationally and requires ATP hydrolysis. J. Cell Biol. 102: 1543-1550

Waters MG, Chirico WJ and Blobel G (1986) Protein translocation across the yeast microsomal membrane is stimulated by a soluble factor. J. Cell Biol. 103: 2629-2636

Werner-Washburne M, Stone DE and Craig EA (1987) Complex interactions among members of an essential subfamily of hsp70 genes in Saccharomyces cerevisiae. Mol. Cell Biol. 7: 2568-2577

DYNAMIC INTERACTIONS BETWEEN MEMBRANE CONSTITUENTS STUDIED BY BIOPHYSICAL TECHNIQUES AND FUNCTIONAL RECONSTITUTION OF MEMBRANE PROTEINS INTO LIPID BILAYERS

Norbert A. Dencher
Biophysics Group, Department of Physics
Freie Universität Berlin
Arnimallee 14
D-1000 Berlin 33
FRG

Biological membranes are the largest and most active organ of any organism. Their building blocks are proteins and lipids, the former are responsible for most of the vital processes accomplished. Photosynthesis and oxidative phosphorylation, the two most important energy conversion mechanisms in biological systems, as well as sensory transduction are carried out by membranes, i.e., by the proteins residing in the lipid phase. Furthermore, membrane proteins, which are in part or completely embedded in the lipid matrix mainly via helical amino acid segments (Fig. 1), act as pores, pumps, and transport systems for ions and nutriments. They are also very selective receptors for hormones, neurotransmitter, drugs, and they mediate specific cell-cell recognition. Biological membranes are highly dynamic systems and both the lipids and the proteins can undergo fast rotational and lateral diffusion so long their mobility is not restricted by the cytoskeleton (Fig. 1).

The properties of membrane proteins are more or less influenced by the microenvironment. Not only the aqueous solvent containing various ions interacts with the amino acids protruding from the membrane, but especially the lipid phase, in which the protein is embedded, will modulate its function. In addition, protein-protein interactions can be of relevance. If and to what extent this can occur will be exemplified for various functional properties of the light-driven proton pump bacteriorhodopsin. On the other hand, the membrane protein itself can influence the properties of the lipid phase, e. g., the order and dynamics of the lipid molecules and the permeability of the bilayer. Spectroscopic techniques such as steady-state and time-resolved fluorescence polarization as well as circular dichroism and flash-induced transient linear dichroism allow the investigation of protein-protein and lipid-protein interactions. These biophysical methods will be introduced by demonstrating their successful application in the investigation of the effect of an intrinsic membrane protein on the order and dynamics of the lipids and on the cooperativity of the gel to liquid-crystalline phase transition.

Membrane biology has gained enormous progress by the application of reconstituted protein-lipid vesicles. Reconstituted vesicles enable the study of structural and functional properties of isolated membrane proteins, the investigation of mechanisms

NATO ASI Series, Vol. H 40
Dynamics and Biogenesis of Membranes
Edited by J.A.F. Op den Kamp

involved in energy coupling, and the analysis of lipid-protein and protein-protein interactions. Furthermore, reconstituted vesicles can be used for the formation of planar lipid bilayers with inserted transmembrane proteins (Schindler, 1980; Bamberg et al., 1981). Various methods for functional reconstitution of membrane proteins into lipid vesicles will be discussed. Recently, new reconstitution techniques have been developed which omit detergents or other harsh treatments (Dencher, 1986; Dencher, 1989; Scotto and Zakim, 1986). These new methods are gentle and fast, furthermore they might also be of relevance for our understanding of membrane biogenesis.

Bacteriorhodopsin, a prototype of an integral membrane protein

In various membrane studies, bacteriorhodopsin is considered as a prototypic membrane protein. This integral membrane protein is the only protein species of the purple membrane and functions as a light-energized proton pump (Oesterhelt and Stoeckenius, 1973). The protein consists of a single polypeptide chain of 248 amino acids (M_r 26486 Da) traversing the lipid bilayer (about 8 lipids per bacteriorhodopsin) in seven α-helical segments (Fig. 1). Photons are absorbed by the chromophore retinal (Fig. 1), which is covalently linked to lysine 216 of the protein moiety via a protonated Schiff's base. The antenna retinal is responsible for the characteristic purple colour of bacteriorhodopsin with its absorption maximum at 568 nm. In the purple membrane, bacteriorhodopsin is arranged as clusters of three molecules in a two-dimensional hexagonal lattice. Upon photon absorption, bacteriorhodopsin undergoes a cyclic photoreaction with an overall half-time of about 10 ms via a series of at least five intermediates of different colour (Dencher and Wilms, 1975) having rise times of femto- to milliseconds. During the photochemical cycle, the protein changes its conformation (Dencher et al., 1989) and one proton is vectorially translocated across bacteriorhodopsin (Grzesiek and Dencher, 1986), causing the formation of an electrochemical proton gradient. This light-generated proton electrochemical potential across the energy-transducing membranes is utilized by the halobacteria as driving force for ATP synthesis, active transport processes, and rotation of flagella. (Further information is reviewed by Dencher, 1983).

The lipid bilayer can be influenced by integral membrane proteins

To study the influence of membrane proteins on the order and dynamics of the lipids and on the cooperativity and transition temperature (T_C) of the gel to liquid-crystalline phase transition, reconstituted DMPC-bacteriorhodopsin vesicles can be used. The rotational motion of the rodlike fluorophore 1,6-diphenyl-1,3,5-hexatriene (DPH) (Fig. 1), the most frequently employed fluorescence polarization probe, allows determination of both the order parameter and viscosity of the lipid phase. This probe incorporates very efficiently into the hydrophobic hydrocarbon core of the lipid bilayer which is displayed by a sharp increase (approx. 1000-fold) in the fluorescence signal

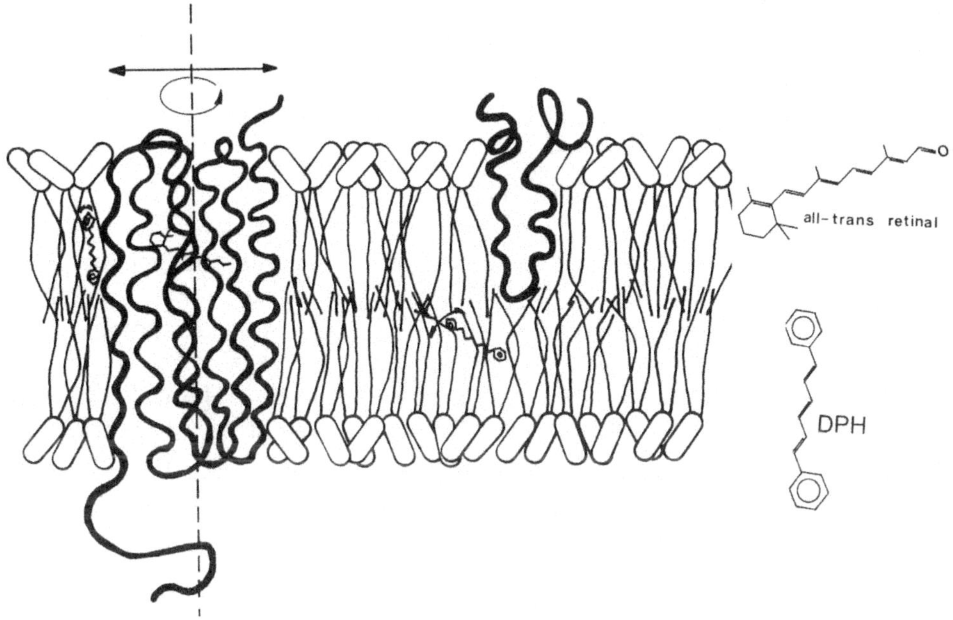

Fig. 1: Schematic drawing of a biological membrane with the native chromophore retinal and the fluorophore DPH.

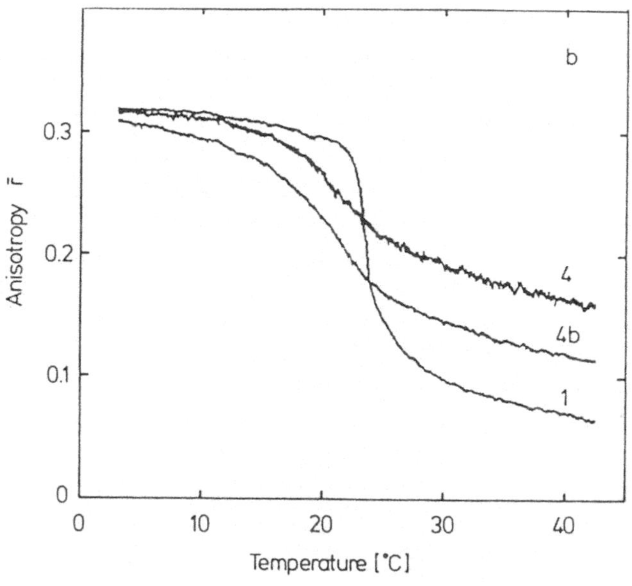

Fig. 2: Comparison of the effect of bacteriorhodopsin and bacteriorhodopsin without chromophore on the phase-transition curve of DMPC. Curve 1, DMPC vesicles; curve 4, DMPC/BR = 102; curve 4b, DMPC/BR without retinal = 102. (From Rehorek et al., 1983).

(Shinitzky and Barenholz, 1978). Details of the experimental methods are given elsewhere (Heyn et al., 1981; Rehorek et al., 1983; Rehorek et al., 1985). At a first step, the steady-state fluorescence anisotropy **r** of DPH embedded in DMPC-bacteriorhodopsin vesicles of various lipid-to-protein ratios as a function of temperature is measured. **r** is defined by

$$r = (F_{\parallel} - F_{\perp})(F_{\parallel} + 2F_{\perp})$$

where $F_{\parallel}$ and $F_{\perp}$ are the two fluorescence intensities with the analyzer parallel and perpendicular, respectively, to the vertical excitation polarizer. For the protein-free vesicles, the lipid phase transition as monitored by the fluorescence anisotropy of DPH occurs at 23.5°C (Fig. 2). With decreasing lipid to protein ratio, the transition broadens, but its midpoint appears to remain close to 23°C. The most interesting aspect is the fact that the r values far above T_c (when the proteins are monomeric) strongly increase when the bacteriorhodopsin content is increased. The interpretation of this observation is not unambiguous, since according to the expression for the steady-state fluorescence anisotropy

$$r = r_{\infty} + (r_o - r_{\infty})(\tau_c/(\tau_c + \tau_F)),$$

the effect (i.e., the marked increase in **r** with decreasing lipid/bacteriorhodopsin) can be due to an increase in r_{∞} (r_{∞} describes the fact that the rotational diffusion of the label is restricted and that its equilibrium orientational distribution is anisotropic. r_{∞} is a measure of the order parameter of the lipids in its neighbourhood.), to an increase in τ_c (τ_c is the correlation time for the approach to the anisotropic equilibrium distribution; it characterizes the dynamics of the restricted rotational motion and is proportional to the viscosity experienced by DPH), or to a reduction in the fluorescence lifetime τ_F. (r_o = 0.4 is the initial value of the fluorescence anisotropy). **It is apparent from this equation that the information on the order and on the dynamics is not separable in steady-state experiments.**

Figure 3 illustrates fluorescence lifetime measurements of DPH in DMPC vesicles of varying bacteriorhodopsin/DMPC ratios at 35°C. At this temperature bacteriorhodopsin is monomeric. For vesicles without protein the decay is single exponential with a lifetime of 8.24 ns. As the bacteriorhodopsin/DMPC ratio increases, the decay becomes non-single exponential and the mean lifetime gradually decreases to 2.0 ns for the vesicles with a molar bacteriorhodopsin to DMPC ratio of 1/102. However, when the chromophore retinal of bacteriorhodopsin is removed, the lifetime for each of the samples returns to the value for the bacteriorhodopsin-free vesicles. These measurements clearly indicate that energy transfer from DPH to the chromophore retinal of bacteriorhodopsin occurs and is the reason for the decrease in lifetime with increasing surface density of the acceptor bacteriorhodopsin (Rehorek et al., 1983). Energy transfer through its reduction in the lifetime (a shorter lifetime of the donor DPH reduces the time over which rotational diffusion can lead to depolarization) leads to an increase in the steady-state anisotropy, r, of DPH. At all temperatures the

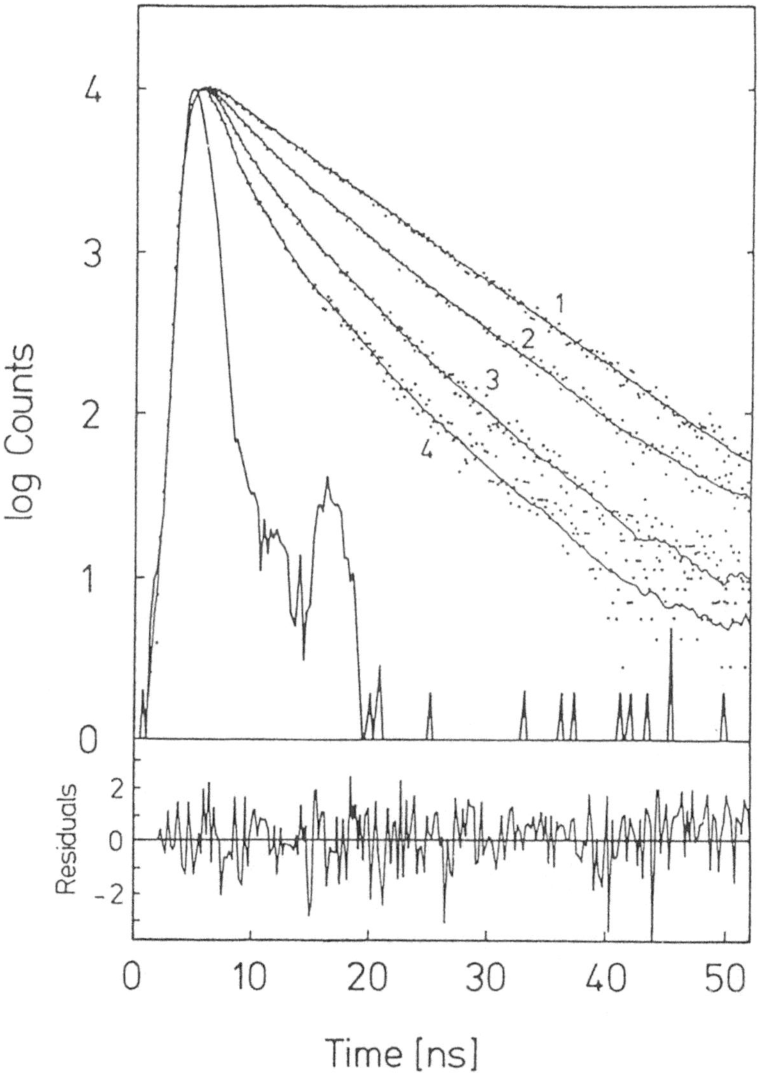

Fig. 3: Lamp pulse and fluorescence decays (···) of DPH in DMPC vesicles as a function of the molar BR/DMPC ratio at 35°C. The molar ratios are (top to bottom) 0, 1:331, 1:171, and 1:102. (From Rehorek et al., 1983).

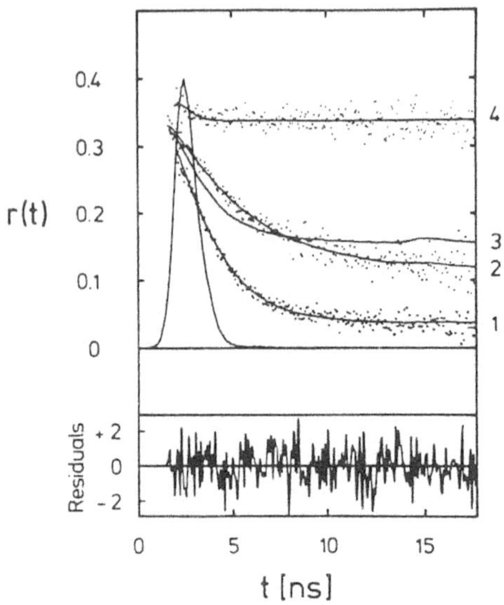

Fig. 4: Experimental (···) and calculated (—) anisotropy decay curves for pure DMPC vesicles at 35°C (curve 1) and 0.5°C (curve 4), and for vesicles with a L/BR of 52 (curve 3) and the same vesicles without chromophore (curve 2) at 35°C. A normalized lamp pulse is shown on a linear scale. (From Rehorek et al., 1985).

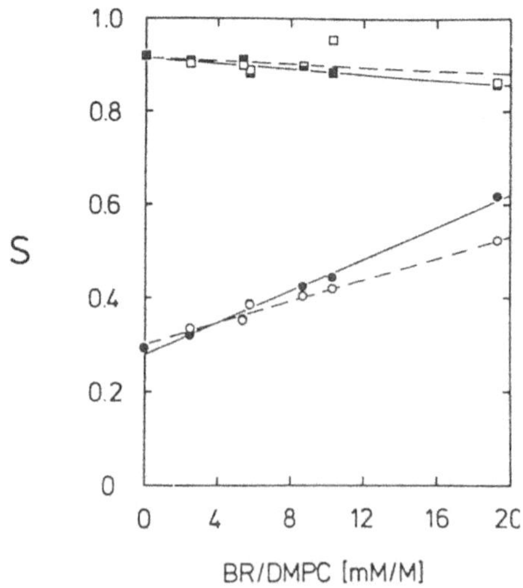

Fig. 5: Order parameters of DPH incorporated in DMPC vesicles containing bacteriorhodopsin at various protein/lipid ratios at 0.5 (■,□) and 35°C (o,●). Open symbols: bleached vesicles. The straight lines correspond to least-squares fits to the data. Dashed lines: bleached vesicles. (Rehorek et al., 1985).

anisotropy for vesicles with acceptor is considerably higher than for the same vesicles after removal of the acceptor retinal. The effect is largest at high temperatures where energy transfer is most effective (Fig. 2). Therefore, if the steady-state anisotropy is used to calculate the lipid order parameter and the correlation time, the data have to be corrected for the effect of energy transfer. The same is true for other membrane systems with natural acceptors, e.g., cytochromes, rhodopsins.

Time-dependent fluorescence anisotropy measurements with DPH provide both dynamical information on the rate of rotation of the label ("membrane viscosity") and structural information on the extent to which this rotational motion is restricted by the anisotropic environment (order parameter, $S = (r_\infty/r_0)^{1/2}$). Curves 1-3 of figure 4 show the effect of bacteriorhodopsin, in the presence and absence of the acceptor retinal, on the anisotropy decay of DPH at 35°C. The incorporation of proteins into DMPC vesicles leads to a marked increase in the end-value of the decay, r_∞, which represents a direct measure of the order parameter. From pure lipid vesicles (curve 1) to vesicles containing a large amount of bacteriorhodopsin (lipid/bacteriorhodopsin = 52, curve 3), an increase in r_∞ of 450% can be observed. Removing the chromophore (curve 2) leads only to a small decrease in r_∞ of 30%. For comparison, the data at 0.5°C are shown for DPH in pure DMPC vesicles (curve 4), which are identical to those of vesicles containing bacteriorhodopsin. The order parameter shows a large linear increase from 0.29 for pure DMPC to 0.62 for a molar bacteriorhodopsin to DMPC ratio of 1/52 at 35°C (Fig. 5). Far below the phase transition temperature of the lipids, the order parameter is very high ($\approx$ 0.9) and almost independently of the bacteriorhodopsin to DMPC ratio. The mean decay constant, τ_C, which provides information on the mobility or viscosity of the lipid phase, is about the same in the absence or presence of proteins (Fig. 4), indicating that the membrane viscosity at 35°C is almost independent on the bacteriorhodopsin/DMPC ratio. The viscosity calculated from the rotational diffusion constant of DPH is approx. 0.1 Poise at 35°C. It can be concluded that the observed increase in the steady-state fluorescence anisotropy is thus due to a protein-induced increase in lipid order (Rehorek et al., 1985). Furthermore, energy transfer experiments show that the range of this perturbation is larger than 45 Å and that bacteriorhodopsin affects the order of all the lipids in roughly the same way (Rehorek et al., 1985).

Determination of the aggregation state of membrane proteins

In the native biological membrane, proteins are either aggregated or are present as monomers, which can undergo fast translational and rotational diffusion in the membrane plane if not immobilized by the cytoskeleton. Bacteriorhodopsin, in contrast to the mobile monomeric visual pigment rhodopsin of vertebrates, is arranged in a two-dimensional hexagonal lattice of protein trimers which are separated from neighbouring trimer clusters by one shell of lipids. However, the aggregation state of bacteriorhodopsin reconstituted into vesicular membranes can be monomeric or aggregated, depending on the condition selected. It is possible to assay the actual

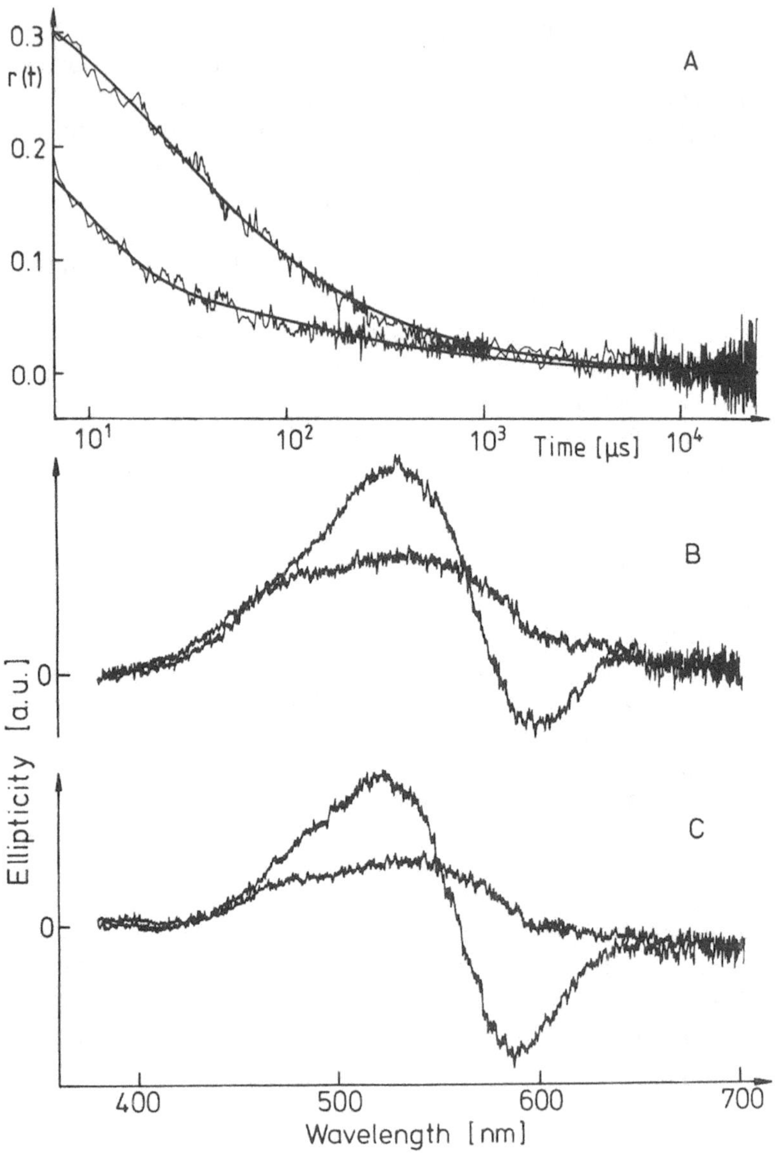

Fig. 6: (A) Anisotropy decay [r(t)] curves for bacteriorhodopsin in soybean PL vesicles at 25°C. Upper trace, molar L/BR = 35; lower trace, molar L/BR = 195; monitoring wavelength of 540 nm. Smooth curves are fits. (B) Circular dichroism spectra of bacteriorhodopsin in soybean PL, vesicles at 22°C. Bilobic trace, molar L/BR = 35; monolobic trace, molar L/BR = 175; a.u., arbitrary units. (C) CD spectra of bacteriorhodopsin in DMPC/PS vesicles, molar L/BR = 120. Bilobic trace at 0.3°C, monolobic trace at 32°C. (From Grzesiek and Dencher, 1988).

aggregation state by two different methods (Fig. 6), i.e., by the rotational diffusion of the protein within the lipid bilayer and by the changes within the visible circular dichroism spectrum of bacteriorhodopsin which are induced by the excitonic coupling of the chromophores within the aggregates (Heyn et al., 1975).

Flash-induced transient linear dichroism can be applied for the study of the rotation of bacteriorhodopsin around the normal to the membrane (Cherry, 1979). Spectroscopic methods of measuring rotation depend on photoselection whereby an oriented population of excited molecules is optically selected from an initial random distribution. This is achieved by excitation of a chromophore with plane polarized light. Bacteriorhodopsin has the intrinsic chromophore retinal and excitation yields a photoproduct of sufficiently long lifetime. Most membrane proteins contain no intrinsic chromophore, however, they can be labeled, e.g., with derivatives of the triplet probe eosin (Cherry, 1979). For monitoring the rotational diffusion of bacteriorhodopsin in DMPC vesicles, the sample is excited by a linearly polarized laser light pulse of 5 ns duration and 580 nm wavelength and the transient absorbance changes due to ground-state depletion are detected at 540 nm. The signals are analyzed by calculating the absorption anisotropy

$$r(t) = [A_{\parallel}(t) - A_{\perp}(t)]/[A_{\parallel}(t) + 2A_{\perp}(t)]$$

where $A_{\parallel}(t)$ and $A_{\perp}(t)$ are the absorbance changes at time t after the flash for light polarized parallel and perpendicular with respect to the polarization of the exciting laser pulse. If the time dependence of the absorption anisotropy r(t) of bacteriorhodopsin's 540 nm depletion signal is measured in reconstituted unilamellar dimyristoylphosphatidylcholine (DMPC) vesicles, with increasing temperature the initial decay of r(t) becomes faster, corresponding to an increase in the rotational diffusion constant $D_{\parallel}$ for rotation of bacteriorhodopsin around an axis perpendicular to the bilayer (Fig. 7). As the temperature is lowered from 25 to 12°C, the diffusion constant decreases by more than 3 orders of magnitude, in agreement with the expected immobilization of bacteriorhodopsin in the protein lattice (Fig. 8). The rotational relaxation time $1/D_{\parallel}$ changes from about 15 μs well above T_C to about 30 ms well below T_C (Heyn et al., 1981). These relaxation times correspond to the rotation of monomeric bacteriorhodopsin molecules in the lipid phase and to the rotation of the vesicles in the aqueous solution, respectively. The protein aggregation transition as monitored by the immobilization of the rotational motion of bacteriorhodopsin occurs below the lipid phase transition temperature, T_C, of 23.5 °C, i.e., at 14-16 °C (Fig. 8). The diffusion coefficient $D_{\parallel}$ is related to the size of the protein (idealized by a cylinder of radius **a** embedded in the bilayer to the depth **h**) and the membrane viscosity η:

$$D_{\parallel} = (kT)/4\pi a^2 h\eta \, .$$

Due to the dependency of $D_{\parallel}$ on the square of the protein radius, this technique is not only able to distinguish between protein monomers, dimers, trimers and larger

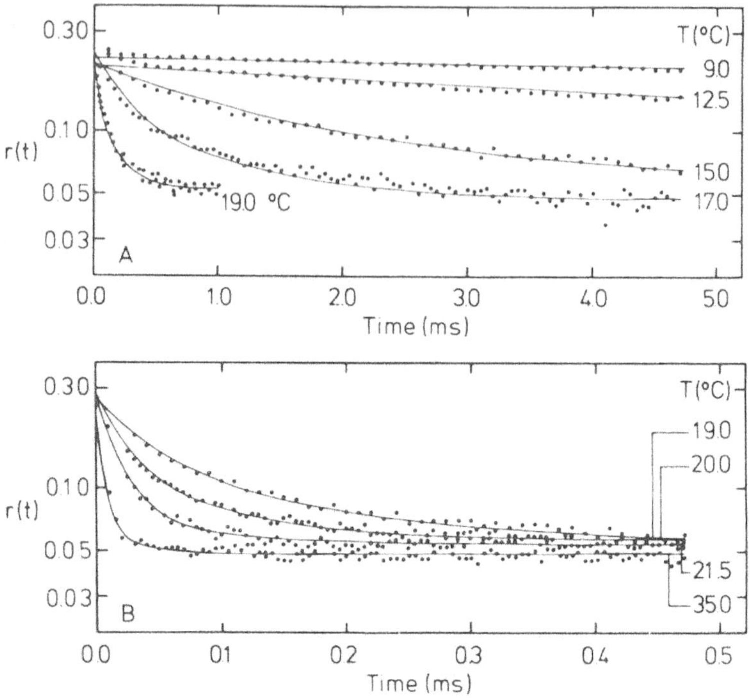

Fig. 7: Time dependence of the absorption anisotropy r(t) of bacteriorhodopsin-DMPC vesicles at various temperatures. Molar L/BR = 111. Time scale: 0 - 5.0 ms (A) and 0 - 0.5 ms (B). (From Heyn et al., 1981).

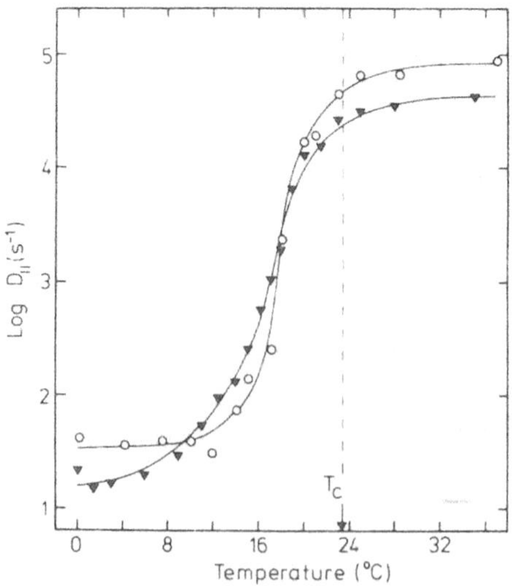

Fig. 8: Temperature dependence of the diffusion constant $D_{\parallel}$ for rotation of bacteriorhodopsin around an axis normal to the vesicle bilayer. The arrow labelled with T_c at 23.5°C indicates the position of the midpoint of the lipid-phase transition. (▾) L/BR = 111. (o) L/BR = 168. (From Heyn et al., 1981).

aggregates, but might be sufficient sensitive to detect even conformational changes of the protein. By this equation, it is possible to interpret the measured values for $D_{||}$ in terms of a membrane viscosity η. The viscosity of about 3.5-8.5 Poise above T_C (Heyn et al., 1981), calculated from the rotational diffusion constant of the large protein bacteriorhodopsin with its highly irregular surface structure is about an order of magnitude larger than the one probed by the small rodlike molecule DPH (see above).
If the appropriate conditions are chosen, reconstituted samples can be prepared containing either monomeric or aggregated bacteriorhodopsin. At molar lipid to protein ratios exceeding 60-70, the aggregation state is reversibly altered from an aggregated lattice to mobile monomers by merely changing the temperature from about 15°C below the lipid phase transition to above (Fig. 6C, Fig. 8) (Dencher and Heyn, 1979). In preparations of lipids in the liquid-crystalline phase, bacteriorhodopsin's state of aggregation is determined only by the lipid to protein ratio. Below a molar ratio of about 50, the protein molecules are aggregated whereas for higher ratios bacteriorhodopsin is monomeric (Dencher et al., 1983). This is illustrated in figure 6A,B. Fig. 6A depicts the decay of flash-induced transient linear dichroism within two samples of reconstituted bacteriorhodopsin in soybean PL vesicles at high (lower trace) and low (upper trace) molar lipid to protein ratio. The anisotropy decay revealed in the lower trace (L/P = 195) has an exponential decay time, τ, of 8.3 μs and can be ascribed to the rotation of bacteriorhodopsin monomers. Anisotropy decay of the low L/P = 35 sample (upper trace) could be well fitted by a model distribution of exponential functions with a mean τ of 62 μs, corresponding to rotating clusters of 8 aggregated bacteriorhodopsin molecules (Grzesiek and Dencher, 1988).
The CD spectra in Fig. 6B,C illustrate an assay of the state of bacteriorhodopsin aggregation by excitonic coupling of the chromophores. These spectra show the monolobic shape for the monomeric bacteriorhodopsin in soybean PL at a molar L/P of 175 and the bilobic shape at the lower molar L/P of 35, indicating bacteriorhodopsin aggregation (Fig. 6B). In Fig. 6C, the aggregation state is altered by merely changing the temperature from below to above T_C of DMPC.

Membrane organization and the function of bacteriorhodopsin: Application of optical spectroscopy with pH-indicators

Although in the native purple membrane the bacteriorhodopsin molecules are present as hexagonally aggregated trimers, the mobile monomeric bacteriorhodopsin is able to pump protons (Dencher and Heyn, 1979) and to generate a transmembrane electrochemical potential difference with an efficiency very similar to that of the aggregates (Grzesiek and Dencher, 1988). This could be demonstrated by the application of time-resolved laser spectroscopy in combination with optical pH-indicators that allows determination of the kinetics and stoichiometry of light-induced proton release from purple membrane sheets, vesicles, and cells in single-turnover experiments (Grzesiek and Dencher, 1986, 1988). The photocycle of

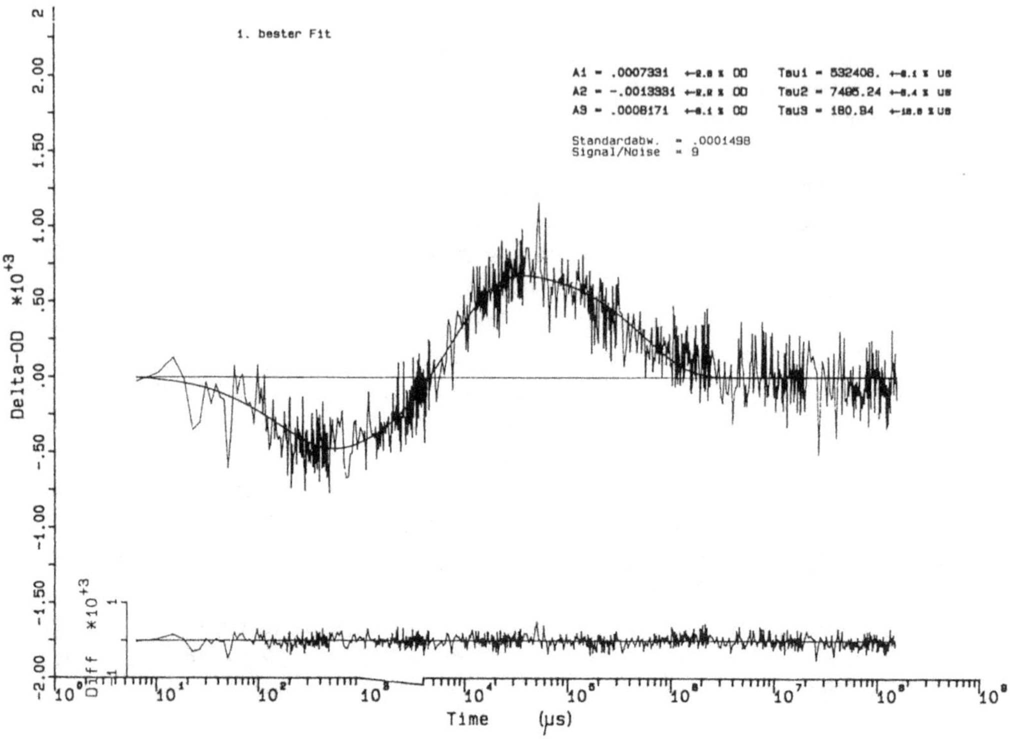

Fig. 9: Absorbance changes at 450 nm of pyranine (25°C). Negative signals correspond to an acidification of the extravesicular space of bacteriorhodopsin-soybean phospholipid vesicles, which have been reconstituted by an octylglucoside dilution procedure (molar lipid/BR = 94, monomeric BR). Abscissa: logarithmic time scale from 10^0 to 10^9 μs. (From Dencher and Grzesiek, 1988).

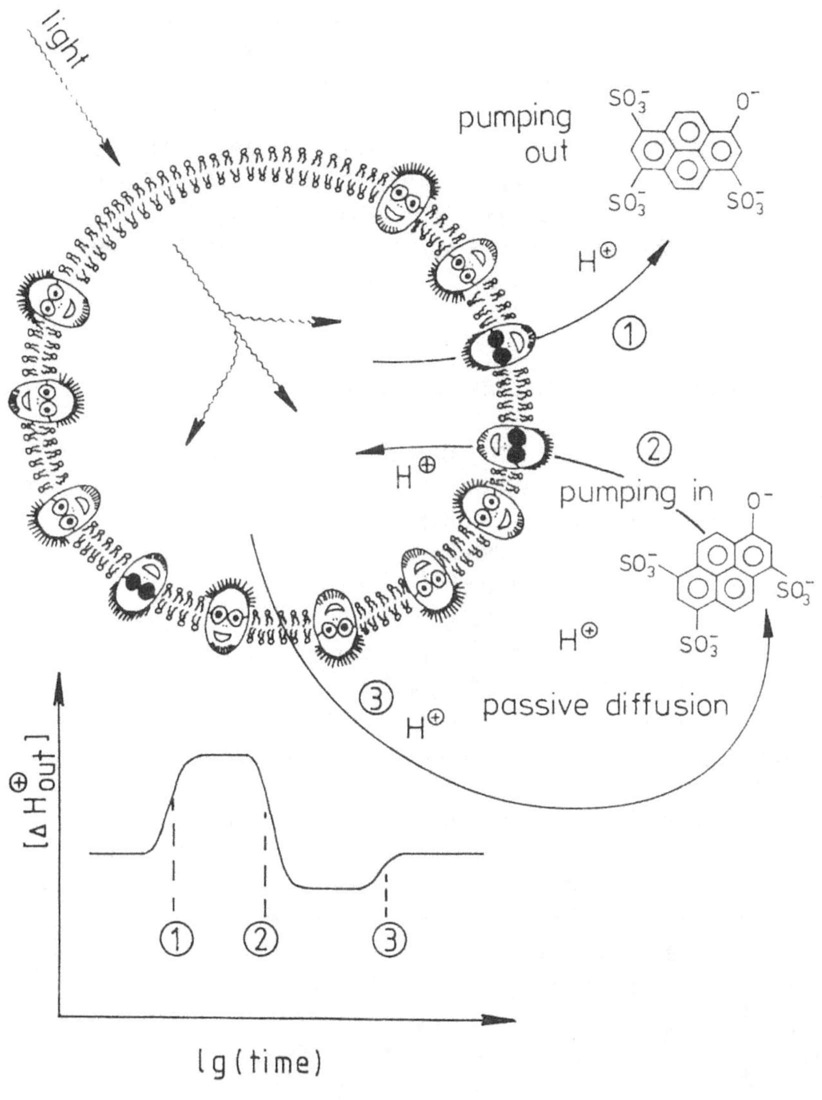

Fig. 10: Scheme of transient changes in extravesicular pH, caused by bacteriorhodopsin's transmembrane proton pumping. Bacteriorhodopsin molecules may be oriented right-side out or inside-out in the vesicle membrane (net 71% inside out). Photocycling bacteriorhodopsin molecules are marked by dark glasses. (1) pumping-out by right-side out oriented bacteriorhodopsin molecules, time range 100 to 300 μs (not time-resolved due to diffusion between membrane surface and pyranine in the bulk phase. (2) pumping-in by inside out oriented bacteriorhodopsin molecules, time range 5 to 10 ms (end of photocycle). (3) passive diffusion of protons across the membrane, dissipating the pH-gradient generated by the difference in the number of pumped protons by process 1 and 2. (From Dencher and Grzesiek, 1988).

bacteriorhodopsin molecules is excited by a laser flash and subsequent changes of pH occurring in the 10 μs to 10 s range are monitored in the aqueous medium as absorbance changes of the pH-indicator pyranine at 450 nm (Fig. 9 and 10). (Further information concerning the application of optical pH-indicators is presented in ref. Dencher et al, 1986). Other functional properties of bacteriorhodopsin, however, are influenced by protein-protein interactions (Dencher et al. 1983). Furthermore, whereas the physical state of the lipid phase has no effect on bacteriorhodopsin's function, the chemical nature of the lipids used for reconstitution strongly influences various kinetic parameters of the pump (Dencher et al., 1983).

Functional reconstitution of membrane proteins

Various methods can be applied to reconstitute membrane proteins into lipid vesicles as will be exemplified for bacteriorhodopsin (from Dencher, 1987 and 1988).

Sonication. This is the most often used (Hwang and Stoeckenius, 1977; Racker, 1979, Dencher, 1986), but unfortunately also most unsuitable procedure. Purple membrane patches are incorporated into preformed small sonicated lipid vesicles by prolonged sonication with a powerful tip-sonicator. The reconstituted vesicles obtained are quite small and therefore less suitable for transport studies. Furthermore, they are very heterogeneous with respect to the protein-to-lipid ratio; a high percentage of the vesicles (up to 50%) does not contain bacteriorhodopsin at all (van Dijk and van Dam, 1982). During prolonged sonication, part of the bacteriorhodopsin is destroyed.

Freeze-thaw sonication. Purple membrane patches are inserted into preformed small sonicated lipid vesicles by slowly cooling the sample to about -70°C in a refrigerator or by rapid freezing in liquid nitrogen (- 196°C). Thereafter the sample is warmed up to room temperature and then shortly sonicated to convert multibilayer vesicles into unilamellar vesicles (Racker, 1979). To increase the amount of incorporated bacteriorhodopsin, the freeze-thawing step should be repeated 2-3 times.

Reverse phase evaporation. Phospholipid in diethylether is mixed (3:1,v/v) with an aqueous purple membrane suspension. The resulting two-phase system is sonicated for 3 minutes and thereafter the organic solvent is removed under reduced pressure (Rigaud et al., 1983). Since many membrane proteins will not tolerate the organic solvent, application of this procedure is limited.

Reconstitution by aid of detergents. By using various detergents (e.g., octyl-ß-glucopyranoside (octylglucoside), Triton X-100, deoxycholate/cholate), intact purple membrane patches as well as monomeric bacteriorhodopsin can be reconstituted into lipid vesicles upon detergent removal by different methods.

Detergent dilution: Bacteriorhodopsin in purple membrane sheets can be incorporated into phospholipid vesicles by a detergent dilution method (Racker et al., 1979; Dencher et al., 1986). Lipids (e.g., 27 mM soybean phospholipids) are sonicated to clarity in a bath-type sonicator. 1.1 mg of the vesicles obtained are mixed with 30-300 μg of bacteriorhodopsin (purple membrane) in a final volume of 400 μl containing buffer (e.g., 1 mM HEPES, pH 7.4, and 75 mM KCl), and 1.25% octylglucoside. Two minutes prior to the addition of the lipid vesicles, purple membrane is mixed with the detergent solution and sonicated for 20 sec. The bacteriorhodopsin-phospholipid-octylglucoside mixture is incubated in the dark at 20°. After 10-180 min, 40 μl samples are diluted with 1 ml buffer, and the vesicles are ready for use.

Detergent dialysis: Monomeric bacteriorhodopsin can be reconstituted into phospholipid vesicles by mixing lipid and detergent-solubilized protein together and removing the detergent by dialysis(Dencher et al., 1986; Heyn and Dencher, 1982). Bacteriorhodopsin is solubilized to the state of monomers in the nonionic detergents Triton X-100 or octylglucoside (Dencher and Heyn, 1982). For DMPC and DPPC the procedure works best with Triton X-100. Soybean phospholipid, diphytanoyl PC, and phosphatidylserine vesicles can be prepared only with bacteriorhodopsin solubilized in octylglucoside. Monomeric bacteriorhodopsin (at a concentration required to achieve the desired lipid-to-protein ratio after mixing) is added to the lipids and incubated with occasional shaking at a temperature above the lipid-phase transition for 20 min. The mixture is poured into a dialysis bag and dialyzed for at least 2 days or 8 days in the case of octylglucoside and Triton X-100, respectively, as detergent. In order to remove residual traces of detergent, Bio-Beads SM-2 are added to the dialysis buffer. The vesicles are fractionated by centrifugation on a 5-40% (w/w) sucrose gradient (250000 g for 5 hr). They are unilamellar, with radii between 35 and 300 nm.

Gel filtration: Reconstituted vesicles can be obtained in 2-5 minutes by passing a mixture of bacteriorhodopsin/lipid/detergent through a short Sephadex G-25 or G-50 column (Krupinski and Hammes, 1986). The detergent, e.g., octylglucoside, is nearly completely removed and the vesicles formed are eluted with the void volume. If dye molecules such as the hydrophilic, pH-sensitive probe pyranine are present in the initial mixture, the dye is entrapped in the internal aqueous compartment of the vesicles and the external free dye molecules are separated.

Adsorption: Rapid removal of detergent from mixtures of bacteriorhodopsin /lipid/detergent and vesicle formation in the time-course of minutes can be achieved by adsorption to nonpolar polymeric adsorbents, e.g., Bio-Beads SM-2 (Bio-Rad) or Amberlite XAD-2 (Sigma). Either a short column made of this resin is used or the adsorbents are added directly to the sample.

Reconstitution by aid of short-chain phospholipids. Recently, a new technique has been developed for gentle and rapid functional reconstitution of integral membrane proteins, such as bacteriorhodopsin and the H^+-ATPase of yeast plasma membrane, into lipid vesicles (Dencher, 1986 and 1989). Spontaneous insertion of membrane proteins into lipid vesicles is facilitated by the short-chain phospholipid diheptanoylphosphatidylcholine (DHPC). The long-chain and the short-chain phospholipids are dissolved in a small volume of chloroform and separately spread as a thin film on the wall of two glass test tubes by evaporation of the solvent under a stream of nitrogen. The dried lipid films are hydrated in an appropriate aqueous solvent. The aqueous suspension of the respective long-chain phospholipid (DMPC,

DPPC, or soybean phospholipid SBPL), final concentration of 20 mM after addition of all constituents, is mixed with a concentrated aqueous suspension of purple membranes. To this mixture of long-chain phospholipid and purple membranes, the aqueous micellar 50 mM DHPC solution is added, yielding a final DMPC concentration of 5 mM (20 mol% of total lipid), and the sample is immediately vortexed. All manipulations should be performed at temperatures above the lipid phase transition of the respective long-chain phospholipids, e.g., room temperature for SBPL, 30°C for DMPC, and 44°C for DPPC. Transmembrane insertion of bacteriorhodopsin into the lipid vesicles occurs immediately upon addition of DHPC.
Another gentle reconstitution procedure utilized myristate or cholesterol as fusogen in combination with cooling to temperatures below the gel to liquid-crystalline phase transition (Scotto and Zakim, 1986).

ACKNOWLEDGEMENTS:

This work was supported by the Deutsche Forschungsgemeinschaft (SFB 312 B4, Heisenberg Grant De 300). All figures are reproduced by copyright permission of the American Chemical Society (Biochemistry), the Biophysical Society (Biophysical Journal), the National Academy of Sciences (PNAS), and studia biophysica.

REFERENCES:

Bamberg E, Dencher NA, Fahr A and Heyn MP (1981) Transmembraneous incorporation of photoelectrically active bacteriorhodopsin in planar lipid bilayers. Proc Natl Acad Sci USA 78: 7502-7506

Cherry RJ (1979) Rotational and lateral diffusion of membrane proteins. Biochim Biophys Acta 559: 289-327

Dencher NA (1983) The five retinal-protein pigments of halobacteria: Bacteriorho dopsin, Halorhodopsin, P 565, P 370, and slow-cycling rhodopsin. Photochem Photobiol 38: 753-767

Dencher NA (1986) Spontaneous transmembrane insertion of membrane proteins into lipid vesicles facilitated by short-chain lecithins. Biochemistry 25: 1195-1200

Dencher NA (1987) The light-energized H^+-pump bacteriorhodopsin: A model system for functional transmembrane reconstitution of ion channels, receptors, and pumps, in: Receptors and Ion Channels, Y.A. Ovchinnikov and F. Hucho, ed. Walter de Gruyter, Berlin/New York, pp 265-273

Dencher NA (1988) Biotechnological application of membrane proteins reconstituted into vesicular and planar lipid bilayers, in: Biotechnological Application of Lipid Microstructures. Plenum Publishing Corporation, pp 269-278

Dencher NA (1989) Gentle and fast transmembrane reconstitution of membrane proteins. Methods Enzymol 171: 265-274

Dencher NA, Burghaus PA and Grzesiek S (1986) Determination of the net proton -hydroxide ion permeability across vesicular lipid bilayers and membrane proteins by optical probes. Methods Enzymol 127: 746-760

Dencher NA, Dresselhaus D, Zaccai G and Büldt G (1989) Structural changes in bacteriorhodopsin during proton translocation revealed by neutron diffraction. Proc Natl Acad Sci USA 86: 7876-7879

Dencher NA and Grzesiek S (1988) Proton/hydroxide ion translocation across membranes and at the membrane surface studied by optical probes. studia biophysica 127: 19-28

Dencher NA and Heyn MP (1979) Bacteriorhodopsin monomers pump protons. FEBS Lett 108: 307-310

Dencher NA and Heyn MP (1982) Preparation and properties of monomeric bacteriorhodopsin. Methods Enzymol 88: 5-10

Dencher NA, Kohl KD and Heyn MP (1983) Photochemical cycle and light-dark adaptation of monomeric and aggregated bacteriorhodopsin in various lipid environments. Biochemistry 22: 1323-1334

Dencher NA and Wilms M (1975) Flash photometric experiments on the photochem ical cycle of bacteriorhodopsin. Biophys Struct Mechanism 1: 259-271

Grzesiek S and Dencher NA (1986) Time-course and stoichiometry of H^+-release and uptake during the photochemical cycle of bacteriorhodopsin. FEBS Lett 208: 337 -342

Grzesiek S and Dencher NA (1988) Monomeric and aggregated bacteriorhodopsin: single-turnover proton transport stoichiometry and photochemistry. Proc Natl Acad Sci USA 85: 9509-9513

Heyn MP, Bauer PJ and Dencher NA (1975) A natural CD label to probe the structure of the purple membrane from Halobacterium halobium by means of exciton coupling effects. Biochem Biophys Res Commun 67: 897-903

Heyn MP, Cherry RJ and Dencher NA (1981) Lipid-protein interactions in bacteriorhodopsin-dimyristoylphosphatidylcholine vesicles. Biochemistry 20: 840 -849

Heyn MP and Dencher NA (1982) Reconstitution of monomeric bacteriorhodopsin into phospholipid vesicles. Methods Enzymol 88: 31-35

Hwang SB and Stoeckenius W (1977) Purple membrane vesicles: Morphology and proton translocation. J Membrane Biol 33: 325-350

Krupinski J and Hammes GG (1986) Steady-state ATP synthesis by bacteriorhodop sin and chloroplast coupling factor co-reconstituted into asolectin vesicles. Proc Natl Acad Sci USA 83: 4233-4237

Oesterhelt D and Stoeckenius W (1973) Functions of a new photoreceptor membrane. Proc Natl Acad Sci USA 70: 2853-2857

Racker E (1979) Reconstitution of membrane processes. Methods Enzymol 55: 699 -711

Racker E, Violand B, O'Neal S, Alfonzo M and Telford J (1979) Reconstitution, a way of biochemical research; some new approaches to membrane-bound enzymes. Arch Biochem Biophys 198: 470-477

Rehorek M, Dencher NA and Heyn MP (1983) Fluorescence energy transfer from diphenylhexatriene to bacteriorhodopsin in lipid vesicles. Biophys J 43: 39-45

Rehorek M, Dencher NA and Heyn MP (1985) Long-range lipid-protein interactions. Evidence from time-resolved fluorescence depolarization and energy transfer experiments with bacteriorhodopsin-dimyristoylphosphatidylcholine vesicles. Biochemistry 24: 5980-5988

Rigaud JL, Bluzat A and Bushlen S (1983) Incorporation of bacteriorhodopsin into large unilamellar liposomes by reverse phase evaporation. Biochem Biophys Res Commun 111: 373-382

Schindler H (1980) Formation of planar bilayers from artificial or native membrane vesicles. FEBS Lett 122: 77-79

Scotto AW and Zakim D (1986) Reconstitution of membrane proteins: catalysis by cholesterol of insertion of integral membrane proteins into preformed lipid bilayers. Biochemistry 25: 1555-1561

Shinitzky M and Barenholz Y (1978) Fluidity parameters of lipid regions determined by fluorescence polarization. Biochim Biophys Acta 515: 367-394

van Dijk PWM and van Dam K (1982) Bacteriorhodopsin in phospholipid vesicles. Methods Enzymol 88: 17-25

NATO ASI Series H

Vol. 1: **Biology and Molecular Biology of Plant-Pathogen Interactions.** Edited by J.A. Bailey. 415 pages. 1986.

Vol. 2: **Glial-Neuronal Communication in Development and Regeneration.** Edited by H.H. Althaus and W. Seifert. 865 pages. 1987.

Vol. 3: **Nicotinic Acetylcholine Receptor: Structure and Function.** Edited by A. Maelicke. 489 pages. 1986.

Vol. 4: **Recognition in Microbe-Plant Symbiotic and Pathogenic Interactions.** Edited by B. Lugtenberg. 449 pages. 1986.

Vol. 5: **Mesenchymal-Epithelial Interactions in Neural Development.** Edited by J.R. Wolff, J. Sievers, and M. Berry. 428 pages. 1987.

Vol. 6: **Molecular Mechanisms of Desensitization to Signal Molecules.** Edited by T.M. Konijn, P.J.M. Van Haastert, H. Van der Starre, H. Van der Wel, and M.D. Houslay. 336 pages. 1987.

Vol. 7: **Gangliosides and Modulation of Neuronal Functions.** Edited by H. Rahmann. 647 pages. 1987.

Vol. 8: **Molecular and Cellular Aspects of Erythropoietin and Erythropoiesis.** Edited by I.N. Rich. 460 pages. 1987.

Vol. 9: **Modification of Cell to Cell Signals During Normal and Pathological Aging.** Edited by S. Govoni and F. Battaini. 297 pages. 1987.

Vol. 10: **Plant Hormone Receptors.** Edited by D. Klämbt. 319 pages. 1987.

Vol. 11: **Host-Parasite Cellular and Molecular Interactions in Protozoal Infections.** Edited by K.-P. Chang and D. Snary. 425 pages. 1987.

Vol. 12: **The Cell Surface in Signal Transduction.** Edited by E. Wagner, H. Greppin, and B. Millet. 243 pages. 1987.

Vol. 13: **Toxicology of Pesticides: Experimental, Clinical and Regulatory Perspectives.** Edited by L.G. Costa, C.L. Galli, and S.D. Murphy. 320 pages. 1987.

Vol. 14: **Genetics of Translation. New Approaches.** Edited by M.F. Tuite, M. Picard, and M. Bolotin-Fukuhara. 524 pages. 1988.

Vol. 15: **Photosensitisation. Molecular, Cellular and Medical Aspects.** Edited by G. Moreno, R.H. Pottier, and T.G. Truscott. 521 pages. 1988.

Vol. 16: **Membrane Biogenesis.** Edited by J.A.F. Op den Kamp. 477 pages. 1988.

Vol. 17: **Cell to Cell Signals in Plant, Animal and Microbial Symbiosis.** Edited by S. Scannerini, D. Smith, P. Bonfante-Fasolo, and V. Gianinazzi-Pearson. 414 pages. 1988.

Vol. 18: **Plant Cell Biotechnology.** Edited by M.S.S. Pais, F. Mavituna, and J.M. Novais. 500 pages. 1988.

Vol. 19: **Modulation of Synaptic Transmission and Plasticity in Nervous Systems.** Edited by G. Hertting and H.-C. Spatz. 457 pages. 1988.

Vol. 20: **Amino Acid Availability and Brain Function in Health and Disease.** Edited by G. Huether. 487 pages. 1988.

NATO ASI Series H

Vol. 21: **Cellular and Molecular Basis of Synaptic Transmission.** Edited by H. Zimmermann. 547 pages. 1988.

Vol. 22: **Neural Development and Regeneration. Cellular and Molecular Aspects.** Edited by A. Gorio, J. R. Perez-Polo, J. de Vellis, and B. Haber. 711 pages. 1988.

Vol. 23: **The Semiotics of Cellular Communication in the Immune System.** Edited by E. E. Sercarz, F. Celada, N. A. Mitchison, and T. Tada. 326 pages. 1988.

Vol. 24: **Bacteria, Complement and the Phagocytic Cell.** Edited by F. C. Cabello und C. Pruzzo. 372 pages. 1988.

Vol. 25: **Nicotinic Acetylcholine Receptors in the Nervous System.** Edited by F. Clementi, C. Gotti, and E. Sher. 424 pages. 1988.

Vol. 26: **Cell to Cell Signals in Mammalian Development.** Edited by S. W. de Laat, J. G. Bluemink, and C. L. Mummery. 322 pages. 1989.

Vol. 27: **Phytotoxins and Plant Pathogenesis.** Edited by A. Graniti, R. D. Durbin, and A. Ballio. 508 pages. 1989.

Vol. 28: **Vascular Wilt Diseases of Plants. Basic Studies and Control.** Edited by E. C. Tjamos and C. H. Beckman. 590 pages. 1989.

Vol. 29: **Receptors, Membrane Transport and Signal Transduction.** Edited by A. E. Evangelopoulos, J. P. Changeux, L. Packer, T. G. Sotiroudis, and K. W. A. Wirtz. 387 pages. 1989.

Vol. 30: **Effects of Mineral Dusts on Cells.** Edited by B. T. Mossman and R. O. Bégin. 470 pages. 1989.

Vol. 31: **Neurobiology of the Inner Retina.** Edited by R. Weiler and N. N. Osborne. 529 pages. 1989.

Vol. 32: **Molecular Biology of Neuroreceptors and Ion Channels.** Edited by A. Maelicke. 675 pages. 1989.

Vol. 33: **Regulatory Mechanisms of Neuron to Vessel Communication in Brain.** Edited by F. Battaini, S. Govoni, M. S. Magnoni, and M. Trabucchi. 416 pages. 1989.

Vol. 34: **Vectors as Tools for the Study of Normal and Abnormal Growth and Differentiation.** Edited by H. Lother, R. Dernick, and W. Ostertag. 477 pages. 1989.

Vol. 35: **Cell Separation in Plants: Physiology, Biochemistry and Molecular Biology.** Edited by D. J. Osborne and M. B. Jackson. 449 pages. 1989.

Vol. 36: **Signal Molecules in Plants and Plant-Microbe Interactions.** Edited by B. J. J. Lugtenberg. 425 pages. 1989.

Vol. 37: **Tin-Based Antitumour Drugs.** Edited by M. Gielen. 226 pages. 1990.

Vol. 38: **The Molecular Biology of Autoimmune Disease.** Edited by A. G. Demaine, J-P. Banga, and A. M. McGregor. 404 pages. 1990.

Vol. 39: **Chemosensory Information Processing.** Edited by D. Schild. 403 pages. 1990.

Vol. 40: **Dynamics and Biogenesis of Membranes.** Edited by J. A. F. Op den Kamp. 367 pages. 1990.

www.ingramcontent.com/pod-product-compliance
Lightning Source LLC
Chambersburg PA
CBHW061155240926
51846CB00006B/1024
* 9 7 8 3 6 4 2 7 4 1 9 6 8 *